KINETIC ANATOMY

FIFTH EDITION

Rachel M. Koldenhoven, PhD, ATC
Texas State University
Robert S. Behnke, HSD
Indiana State University, Professor Emeritus

Library of Congress Cataloging-in-Publication Data

Library of Congress Cataloging-in-Publication information is available. LCCN 2026003248 (print).

ISBN: 978-1-7182-4123-7 (print)

The web addresses cited in this text were current as of September 2025, unless otherwise noted.

Acquisitions Editor: Brittany Phillips
Publication Services Manager: Anna Lan Seaman
Permissions Manager: Laurel Mitchell
Cover Designer: Keri Evans
Cover Design Specialist: Susan Rothermel Allen
Photograph (cover): Morsa Images/DigitalVision/Getty Images
Photographs (interior): © Human Kinetics, unless otherwise noted
Photo Asset Manager: Laura Fitch
Photo Production Manager: Jason Allen
Senior Art Manager: Kelly Hendren
Illustrations: © Human Kinetics, unless otherwise noted
Editorial and Composition Services: S4Carlisle Publishing Services
Printer: Versa Press

Printed in the United States of America 10 9 8 7 6 5 4 3 2 1

The paper in this book is certified under a sustainable forestry program.

Human Kinetics
1607 N. Market Street
Champaign, IL 61820
USA

United States and International
Website: **US.HumanKinetics.com**
Email: info@hkusa.com
Phone: 1-800-747-4457

Human Kinetics' authorized representative for product safety in the EU is Mare Nostrum Group B.V., Mauritskade 21D, 1091 GC Amsterdam, The Netherlands.

Email: gpsr@mare-nostrum.co.uk

L1118

At one time or another in our lives, we come in contact with someone whom we consider our teacher, supervisor, mentor, or role model. In many instances, such a person may also become our friend. I was fortunate during my professional preparation to have all of these and more in one individual. During my professional preparation, this person emphasized the importance of knowing human anatomy because he believed it was a keystone to understanding athletic performance and preventing, recognizing, treating, and rehabilitating athletic trauma. This emphasis has inspired me throughout the preparation of this book. I'd like to thank Robert Nicolette, former head athletic trainer (1957–1969) at the University of Illinois, on behalf of all of us who were fortunate enough to have known him. He has touched everyone we work with through our association with him. I dedicate this book to him to express how much I appreciate him. Rest in peace, Nick!—Robert S. Behnke

CONTENTS

Preface **ix** | Acknowledgments **xiii**

PART I General Concepts of Anatomy **1**

Chapter 1 Structures **3**

Bones 4
Joints 8
Muscles 12
Levers 17
Nerves 19
Blood Vessels 21
Other Tissues 21
Motor Unit 24

Chapter 2 Movement **29**

Anatomical Locations and Directional Terms 30
Planes and Axes 30
Fundamental Movements 31
Kinetic Chain 34

PART II Upper Extremity **37**

Chapter 3 The Shoulder **39**

Bones of the Shoulder Girdle 39
Bones of the Shoulder Joint 42
Joints and Ligaments of the Shoulder Girdle 43
Ligaments of the Shoulder Joint 46
Other Ligaments of the Shoulder 46
Fundamental Movements and Muscles of the Shoulder Girdle 47
Fundamental Movements and Muscles of the Shoulder Joint 51
Combined Actions of the Shoulder Girdle and Shoulder Joint 60

Chapter 4 **The Elbow and Forearm 61**

Bones of the Elbow and Forearm 61
Joints and Ligaments of the Elbow and Forearm 64
Fundamental Movements and Muscles of the Elbow and Forearm 65

Chapter 5 **The Wrist and Hand 73**

Bones of the Wrist and Hand 74
Joints and Ligaments of the Wrist and Hand 75
Fundamental Movements of the Wrist and Hand 77
Extrinsic Muscles of the Wrist and Hand 78
Intrinsic Muscles of the Hand 83
Muscles of the Thumb 85

Chapter 6 **Nerves and Blood Vessels of the Upper Extremity 89**

Nerves of the Brachial Plexus 91
Major Arteries of the Upper Extremity 95
Major Veins of the Upper Extremity 97

▶ PART II Summary Tables 100

Articulations of the Upper Extremity 100
Muscles, Nerves, and Blood Supply of the Upper Extremity 103

PART III The Head, Spinal Column, Thorax, and Pelvis 109

Chapter 7 **The Head 111**

Bones of the Head 111
Joints of the Head 115
Ligaments of the Head 117
Sinuses 117
Fundamental Movements and Muscles of the Head 118

Chapter 8 **The Spinal Column and Pelvis 123**

Bones of the Spinal Column 125
Ligaments of the Spinal Column 126
Fundamental Movements and Muscles of the Spinal Column 130
Bones of the Pelvis 139
Ligaments of the Pelvis 141
Fundamental Movements and Muscles of the Pelvis 143

Chapter 9 The Thorax 145

Bones of the Thorax 145
Joints and Ligaments of the Thorax 147
Fundamental Movements and Muscles of the Thorax 149
Structures Within the Thorax 153

Chapter 10 Nerves and Blood Vessels of the Head, Spinal Column, Thorax, Heart, and Lungs 159

The Brain 159
Peripheral Nervous System 164
Arteries of the Head and Brain 166
Veins of the Head and Brain 170
Nerves of the Thorax and Trunk 173
Arteries of the Thorax and Trunk 175
Veins of the Thorax and Trunk 177
Nerves, Arteries, and Veins of the Heart and Lungs 177

▶ PART III Summary Tables 183

Articulations of the Head, Spinal Column, Thorax, and Pelvis 183
Muscles, Nerves, and Blood Supply of the Head, Thorax, and Spinal Column 185

PART IV Lower Extremity 199

Chapter 11 The Hip and Thigh 201

Bones of the Hip Joint and Thigh 201
Ligaments of the Hip Joint 203
Fundamental Movements of the Hip Joint 205
Muscles of the Hip Joint and Upper Leg (Thigh) 205

Chapter 12 The Knee 215

Bones of the Knee 215
Ligaments of the Knee 219
Menisci of the Knee 224
Fundamental Movements of the Knee and Lower Leg 224
Muscles of the Knee and Lower Leg 225

Chapter 13 The Lower Leg, Ankle, and Foot 231

Bones of the Lower Leg 231
Bones of the Foot 233
Joints and Ligaments of the Ankle and Foot 236
Fundamental Movements of the Lower Leg, Ankle, and Foot 241
Muscles of the Lower Leg, Ankle, and Foot 243

Chapter 14 **Nerves and Blood Vessels of the Lower Extremity 255**

Nerves of the Lumbosacral Plexus 255
Major Arteries of the Lower Extremity 261
Major Veins of the Lower Extremity 265

▶ PART IV Summary Tables 268

Articulations of the Lower Extremity 268
Muscles, Nerves, and Blood Supply of the Lower Extremity. 271

Glossary **277** | Answers to Pop Quiz Questions **279** | Suggested Readings **281**

Index **283** | About the Authors **297**

Earn Continuing Education Credits/Units **298**

PREFACE

The human body is the most fascinating machine ever created. Science has long studied it and attempted to improve it through various methods, even going so far as trying to make parts interchangeable and create new synthetic parts. Learning about oneself through the study of the human body could lead to a longer and healthier life; it can begin with a basic understanding of the various elements that make up the human body.

For a first-time experience of the study of human anatomy, *Kinetic Anatomy* provides a good overview of the human body's various structures. For seeking further understanding of human anatomy, *Kinetic Anatomy* provides the basics that facilitate more in-depth study, including the human body's physiological functions associated with the anatomical structures presented in this text.

Goals of the Text

The goals of *Kinetic Anatomy* are to (1) familiarize the reader with the vocabulary of human anatomy, (2) describe the essentials of human anatomy for movement, and (3) provide the reader with the knowledge needed to pursue healthy living.

Having a firm understanding of the vocabulary of human anatomy allows the reader to communicate effectively with colleagues, physicians, therapists, educators, coaches, allied health personnel, and others using a universal language of human anatomy.

This text also provides a firm understanding of how the human body is constructed and how it moves by discussing bones, tying the bones together with ligaments to make articulations (joints), placing muscles on the bones (crossing joints), and then observing how the joints move when the muscles contract. The book also discusses the nerves (including the central nervous system's brain and spinal nerves and the peripheral nervous system) and blood vessels (including the heart), as well as the lungs, all of which provide elements essential for skeletal movement. But the main emphasis is on putting together the human body for the purpose of studying movement. Knowing what structures are involved and how they should function allows the reader to identify problems and correct them to enhance physical activity.

Finally, this book imparts knowledge that allows the reader to pursue a healthy life. Knowing about your body can alert you to potential problems and, with other acquired information, help you prevent or resolve them and lead a healthy lifestyle.

Updates to the Fifth Edition

The fifth edition of *Kinetic Anatomy* includes additions to the text and, more extensively, the ancillary materials. Where appropriate, new aspects of movement content have been added to expand the depth of current subjects. The new edition still offers the following instructor ancillaries: an instructor guide, a presentation package, an image bank, a test package, and chapter quizzes. In this edition, we continue to offer valuable study aids that are available to students, including anatomy coloring and labeling exercises, study questions, suggested learning activities, and functional movement exercises.

Organization of the Text

The text and illustrations are devoted to the structures that play a primary role in moving the human body: bones, ligaments, joints, muscles, and the nerves and blood vessels supplying innervations and circulation to those structures. Unique to this text is content on anatomical structures not often considered when studying the anatomy of movement: the brain, the heart, and the lungs. This content provides entry-level readers with further understanding of anatomical structures involved in movement. Although the bones, ligaments, muscles, nerves, and blood vessels are the primary structures that create motion in the human body, other structures of the nervous system (brain and peripheral nervous system), the heart, and the respiratory system are presented to illustrate how they contribute to human movement.

The text is organized into four parts. Part I presents the basic concepts of anatomy. The remainder of the text, like many textbooks in the areas of kinesiology and biomechanics, divides the body into the upper extremity (part II of this text); the head (brain), spinal column, thorax (heart and lungs), and pelvis (part III); and the lower extremity (part IV). Each anatomical chapter in parts II, III, and IV follows the same format: bones; joints and ligaments; muscles; and, where appropriate, the inclusion of major organs essential for movement (the brain, the heart, and the lungs). Parts II, III, and IV also include summary tables for muscles, bones, joints, ligaments, movements, nerves, and blood vessels. These tables have been reviewed extensively, and while some of the terms for the structures may differ elsewhere, the terminology used in this edition of *Kinetic Anatomy* is the most commonly used for the structures in the current study of human anatomy.

Participation in organized sports, informal recreational activities, and personal fitness programs has produced an increased interest in an unfortunate aspect of this participation: head trauma and other types of injuries. The more common of these injuries are discussed in relation to the particular anatomical structures involved. *Kinetic Anatomy* looks at the anatomy of the head and brain, including the central and peripheral nervous systems, as well as the blood vessels of the circulatory system. The vast network of blood vessels (numerous arteries and veins with multiple branches) is discussed, with identification of names and anatomical areas. Sprains, strains, and overuse conditions are defined and explained at appropriate instances in the text. In-depth investigation of both the nervous and circulatory systems and trauma to the bones, ligaments and joints, and muscles is encouraged and will require advanced anatomical study beyond the entry-level information provided in *Kinetic Anatomy*.

Further discussion is presented regarding joint strength and movement, the function of muscles (agonists, antagonists, fixers or stabilizers, and synergists), levers, and exercise. These additions, at the end of appropriate chapters, are presented to enhance your understanding of muscle function if you desire future study in kinesiology and human biomechanics.

These updates to *Kinetic Anatomy* make it a more inclusive entry-level text for undergraduate and secondary students and others seeking basic information about the anatomical structures of the human body in relation to movement.

Key Features of the Text

When one studies human anatomy, many devices are available to supplement learning. Human cadavers; audiovisual aids, including photos, illustrations, models, and software programs; and numerous other means can assist learning. *Kinetic Anatomy* additionally facilitates learning by providing a cost-free and readily available aid for comprehending various aspects of human anatomy to allow movement: the hands-on experience. Throughout the book, readers will find Hands On sidebars that provide instructions for feeling specific anatomical structures either on themselves or on a partner. In a very basic way, this makes the study of human anatomy a personal and practical experience available to everyone.

Detailed anatomical illustrations show readers the key structures that contribute to human movement in the anatomical areas discussed in any particular chapter. The artist has made every effort to accurately present these structures as they appear in the human body.

To enhance understanding, the text has a feature called Pop Quiz. Photographs of real-life sports activities appear with accompanying questions about what structures create the movement and what stresses are placed on the structures during the activity shown. The photographs help readers further understand what the structures being discussed do when they create movement. The answers to the questions appear at the end of the book.

Focus On sidebars are presented throughout the book to illustrate circumstances in everyday activity that relate to the specific anatomical structures in the text. Health conditions commonly mentioned in everyday life are discussed to hopefully advance readers' understanding of these conditions. References to these various conditions in the print and electronic media should be more meaningful to readers as a result of these boxes.

Finally, parts II, III, and IV end with summary tables. These summary tables provide a quick resource when seeking the components of a particular joint, its type, bones, ligaments, and movements, as well as the components of a muscle, including its origin, insertion, action, nerve supply, and blood supply. Whether readers are answering questions, preparing a paper, or making a presentation on a particular anatomical structure or human movement, the summary tables can serve as a quick reference.

Instructor Resources

Instructors have access to a full array of ancillary materials that support the text.

- **Instructor guide.** The instructor guide includes many valuable tools to help instructors build a lecture. For each chapter, instructors will find a chapter overview, chapter objectives, a lecture outline, and lecture aids (additional items that would be useful to have on hand when covering a chapter's content).
- **Presentation package.** The presentation package has more than 200 slides that highlight the content in each chapter. Select figures are also included.
- **Image bank.** The image bank includes most of the figures, tables, and photos from the text. Instructors can use these images to supplement lecture slides, create handouts, or develop other teaching materials for their classes.
- **Test package.** The test package includes more than 600 multiple-choice, true-false, and fill-in-the-blank questions. Instructors can use these questions to create or supplement tests or quizzes.

Instructor ancillaries are free to adopting instructors, including an ebook version of the text that allows instructors to add highlights, annotations, and bookmarks.

Student Resources

Students have access to a full array of ancillary materials that support the text.

- **Anatomy coloring and labeling exercises.** Coloring and labeling anatomical structures are effective ways to study the structure and function of human anatomy. These exercises, which are adapted from the anatomical drawings in the text, provide realistic drawings to color and label to help students memorize the content.
- **Study questions.** Students can use learning aids such as true-false, multiple-choice, and fill-in-the-blank questions to ensure they have a firm grasp of the key points of the chapter content and to prepare for tests and quizzes.
- **Suggested learning activities.** These activities enable students to apply material from the chapter through common physical activities.
- **Functional movement exercises.** These exercises ask students to apply the anatomical structures to the movements they perform, thereby reinforcing what structures do and how they do it.
- **Quizzes.** Instructors may assign students short quizzes to demonstrate mastery of each chapter's content.

To access the web resource, visit **https://ancillaries.humankinetics.com/KineticAnatomy5E**. You will not be prompted to enter a username or password to access this content.

ACKNOWLEDGMENTS

Several people at Human Kinetics have been responsible for the guidance needed to bring this edition of *Kinetic Anatomy* to completion. Dr. Loarn Robertson, former senior acquisitions editor, was responsible for deciding that *Kinetic Anatomy* would be a worthy addition to the entry-level study of human anatomy. Since his retirement, this project has been supervised by several acquisitions and developmental editors. This fifth edition of *Kinetic Anatomy* has been wonderfully handled by Brittany Phillips (acquisitions editor) and the excellent editorial team at Human Kinetics.

Again, we must thank Dr. Rainer Martens, founder of Human Kinetics, for approving the project that resulted in the creation of *Kinetic Anatomy*. His contributions in the areas of sport, physical education, health education, and recreation have received worldwide recognition and appreciation from authors, teachers, and students. His thoughts and actions in the publishing business opened avenues in these areas at a time when they were sorely needed and are now so widely accepted. Dr. Behnke not only appreciates all his efforts but also appreciates having had the opportunity to call him a colleague during his academic career.

Thanks again to all who have made working with Human Kinetics a pleasure and have produced a publication that hopefully will make the study of human anatomy enjoyable for anyone interested in learning about the human body and how it moves.

Dr. Behnke would like to thank his wife, Carol; his children, Mike, Jim, Kelly, Susan, and Tom; and his grandchildren and great grandchildren for their love, support, and patience.

Dr. Rachel Koldenhoven would like to thank her husband, Zach; her children, Riley and Sadie; and her mom, Carol, for always supporting her academic pursuits, for their love, and for their encouragement, which made this new edition possible. She would also like to thank Dr. Duane Knudson for recommending her to become the new author for the fifth edition of *Kinetic Anatomy*.

PART I

General Concepts of Anatomy

CHAPTER 1

Structures

Human anatomy has been defined simply as the structure of organisms pertaining to humankind. A structure is, by one definition, something composed of interrelated parts to form an organism, and an organism is simply defined as a *living thing*. The body is made up of four types of tissues (a collection of similar types of cells).

1. Connective tissue makes up bone, cartilage, and soft tissue, such as skin, fascia, tendons, and ligaments.
2. Muscle tissue is divided into three types: skeletal, which moves the parts of the skeleton; cardiac, which causes the pumping action of the heart; and smooth, which lines arterial walls and other organs of the body.
3. Nerve tissue is divided into neurons, which conduct impulses involving the brain, the spinal cord, spinal nerves, and cranial nerves, and neuroglia, which are specifically involved in the cellular processes that metabolically and physically support the neurons.
4. Epithelial tissue is divided into four varieties, and all of them are involved with the structures of the respiratory, gastrointestinal, urinary, and reproductive systems.

The study of human anatomy as it pertains to movement concentrates on the bones, joints (and associated ligaments), and muscles responsible for the human body's movement. Additionally, the role of the nervous system in stimulating muscle tissue; the role of the vascular system in providing the muscle tissue with energy and removing by-products; the bone, joint, and muscle components of the body's lever systems; and the effects of exercise need to be studied. *Kinetic Anatomy* presents brief overviews of the respiratory, circulatory, and autonomic nervous systems. Although human anatomy also includes other structures, such as the endocrine, digestive, reproductive, and urinary systems and sensory organs, this text concentrates specifically on those anatomical structures chiefly responsible for producing movement of the human organism.

Proper vocabulary is extremely important when discussing anatomy. Common terms make communication with others (physicians, coaches, therapists, athletic trainers) much easier, and it is essential that a student of human anatomy become familiar with standard terminology presented in this chapter. Knowledge of the structures and common terms used to describe movement anatomically also facilitates the use of specific coaching principles; the use of therapeutic techniques involving human movement for prevention, treatment, and rehabilitation of various physical conditions; and the application of scientific principles to human movement.

Although all systems of the human organism can be said to contribute in some unique ways to movement, this text emphasizes those systems (skeletal, articular, muscular, nervous, and circulatory) that directly accomplish movement. Primary concentration is on the following structures: bones, ligaments, joints, and muscles producing movement, with additional comments about the nerves and blood vessels in each specific anatomical area.

Bones

The body contains 206 bones. Bones have several functions, such as support, protection, movement, mineral storage, and blood cell formation. Arrangements of bones that form joints and the muscular attachments to those bones determine movement. Bones are classified by their shapes into five groups: long, short, flat, sesamoid, and irregular bones (figure 1.1). The bones that provide the framework for the body and make movement possible are classified as long bones (figure 1.2). A long bone has a shaft, known as the *diaphysis*, and two large prominences at either end of the diaphysis, known as the *epiphyses*. Early in life, the epiphysis is separated from the diaphysis by a cartilaginous structure known as the *epiphyseal plate*. The bone grows from these epiphyseal plates at both ends of the diaphysis; thus, this area is often referred to as the *growth plate*. Once a bone has reached its maximum length (maturity), the epiphyseal plate "closes" (bone tissue has totally replaced the cartilaginous tissue), and the epiphysis and diaphysis become one continuous structure. Around the entire bone

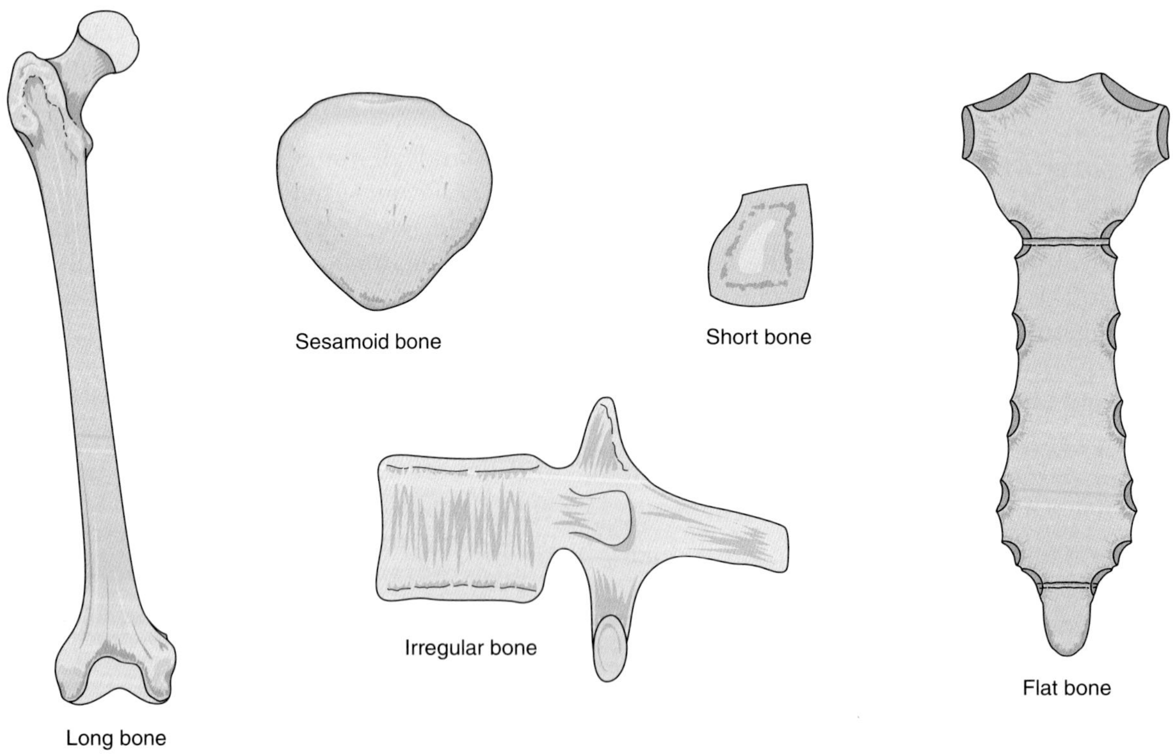

Figure 1.1 Five types of bones.

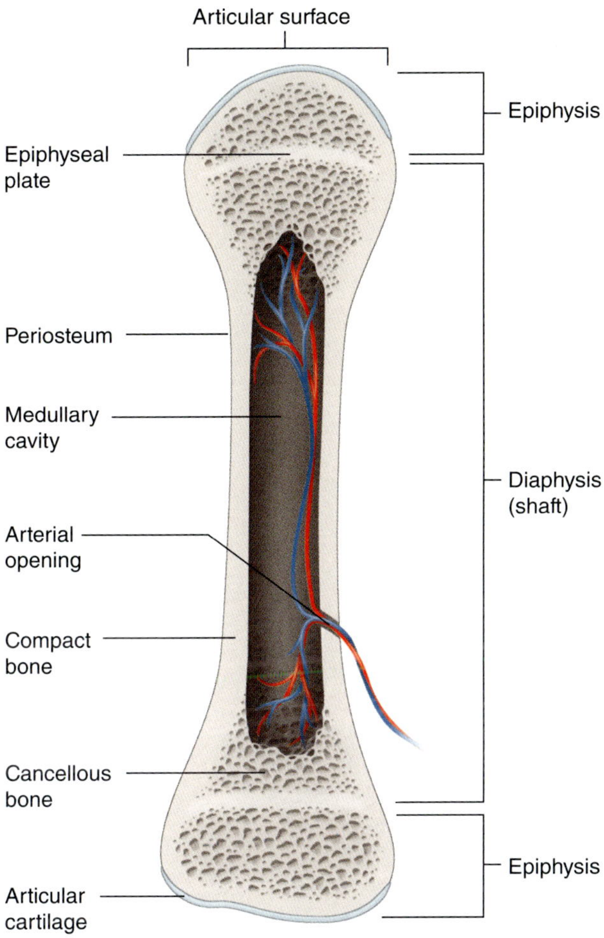

Figure 1.2 Structure of a long bone.

is a layer of tissue known as the *periosteum*, where bone cells are produced. Additionally, the very ends of each bone's epiphyses are covered with a material known as *articular cartilage*. This covering provides for smooth movement between the bones that make up a joint and protects the ends of the bones from wear and tear.

Short bones differ from long bones in that they possess no diaphysis and are fairly symmetrical. Bones in the wrist and ankle are examples of short bones. Flat bones, such as the bones of the head, chest, and shoulder, get their name from their flat shape. Irregular bones are simply bones that cannot be classified as long, short, or flat. The best example of an irregular bone is a vertebra of the spinal column. Sesamoid (sesame seed–shaped) bones are small, nodular bones embedded in a tendon. The kneecap (patella) is the largest sesamoid bone in the body; other sesamoid bones are found in the hand and foot.

Hands On

Sit comfortably on the floor with your legs extended and relaxed. Use your hands to gently palpate (feel) your patella (kneecap). Notice how the patella moves slightly under your fingers. Feel the tendons above (the quadriceps tendon) and below (the patellar tendon) the patella. These structures surround the patella, making it a sesamoid bone.

Bones, like all other tissues of the human body, except neurological tissue, respond to stresses placed on them. This response is known as *Wolff's law*. Examples of bones shaped in response to stress are the femur, which is slightly curved to adapt to compressive forces on the medial side and distraction forces on the lateral side, and the scapula, which is shaped relative to the lines of pull from numerous muscles attached to it.

Several terms are commonly used to describe the features of bones. These features are usually referred to as *anatomical landmarks* and are basic to one's anatomical vocabulary.

- A process is a projection from a bone (figure 1.3).
- A tuberosity on a bone is a large bump (see figure 1.3).
- A tubercle is a smaller bump (see figures 1.3 and 1.4).

All three of these bony prominences usually serve as the attachment for other structures.

- A spine, or spinous process, is typically a longer and thinner projection of bone, unlike any of the previously mentioned prominences (figure 1.5).

FOCUS ON

Osteoporosis

The loss of calcium and other minerals resulting from the natural aging process can cause bones to become porous and brittle. This condition is known as *osteoporosis*, which can lead to broken bones and postural disfigurement. Approximately 50% of people older than 60 have this condition.

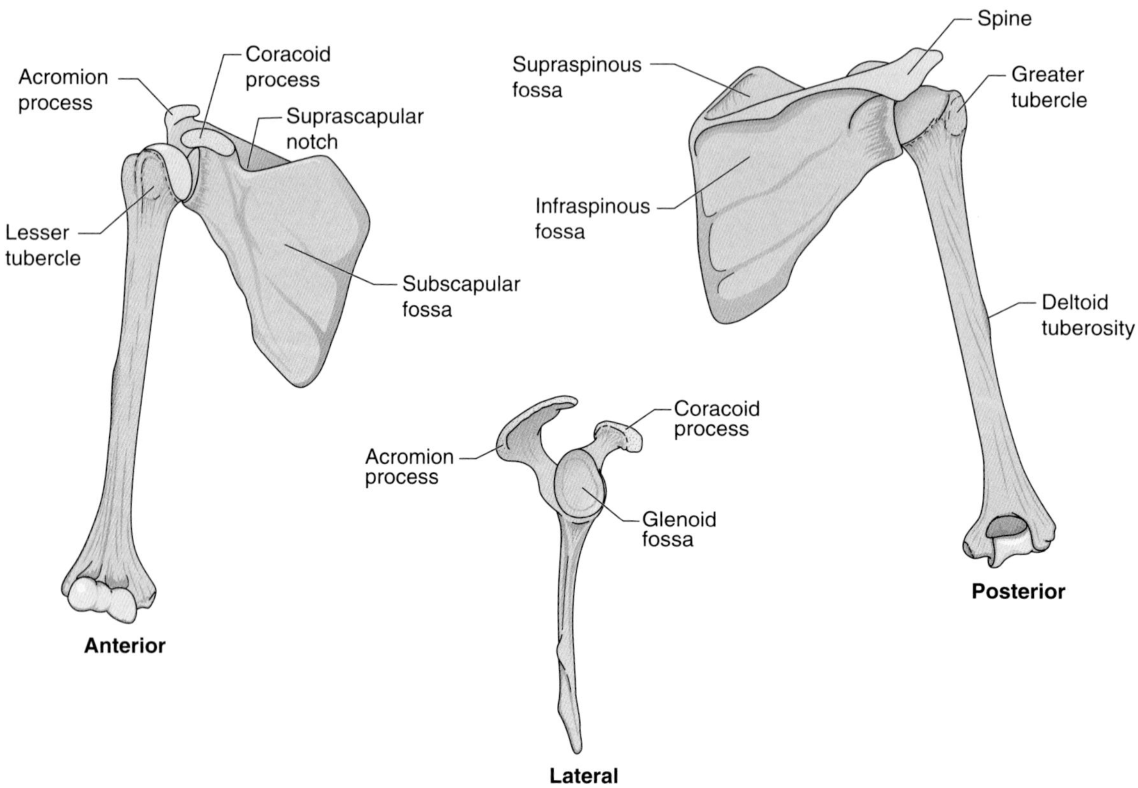

Figure 1.3 Landmarks of the shoulder bones: anterior (front), lateral (side), and posterior (back) views.

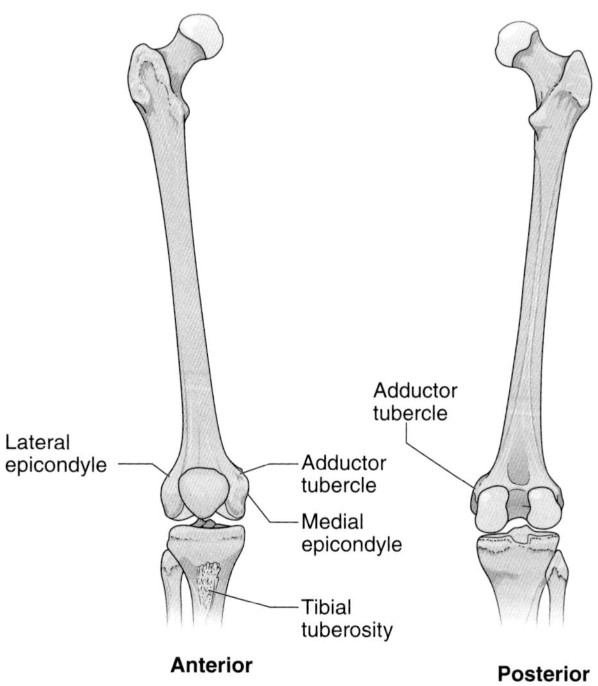

Figure 1.4 Landmarks of the thigh and leg bones.

- The large bony knobs at either end of a long bone are the condyles (figure 1.6).
- The part of the condyle that articulates (joins) with another bone is the articular surface (see figure 1.2).
- Smaller bony knobs that sometimes appear just above the condyles of a bone are epicondyles (see figure 1.4).
- A **fossa** is a smooth, hollow surface on a bone that usually functions as a source of attachment for other structures (see figure 1.3).
- A smaller and flatter smooth surface is a facet (figure 1.7). Facets also serve as attachments for other structures.
- A notch is an area on a bone that appears to be cut out and allows for the passage of other structures, such as blood vessels or nerves (see figure 1.8).
- Similar in function to a notch but appearing as a hole in a bone is a foramen (see figure 1.5).

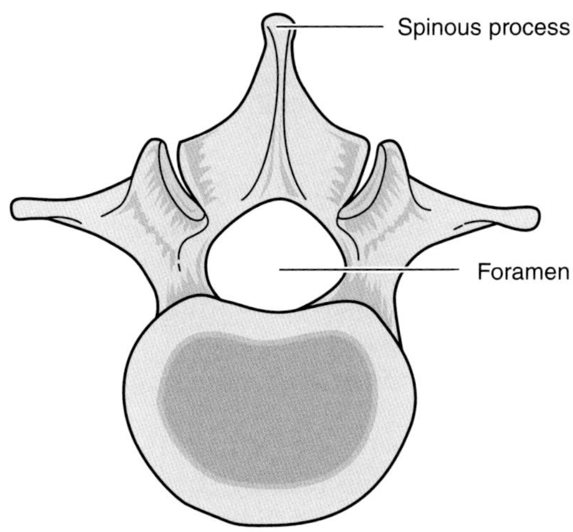

Figure 1.5 Superior (from above) view of a typical vertebra.

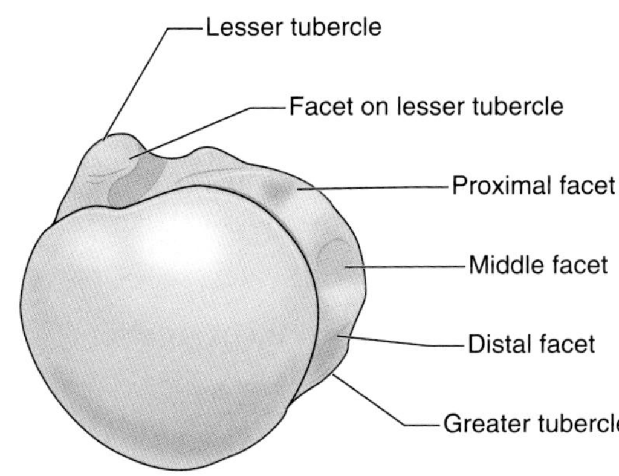

Figure 1.7 Superior view of the humerus.

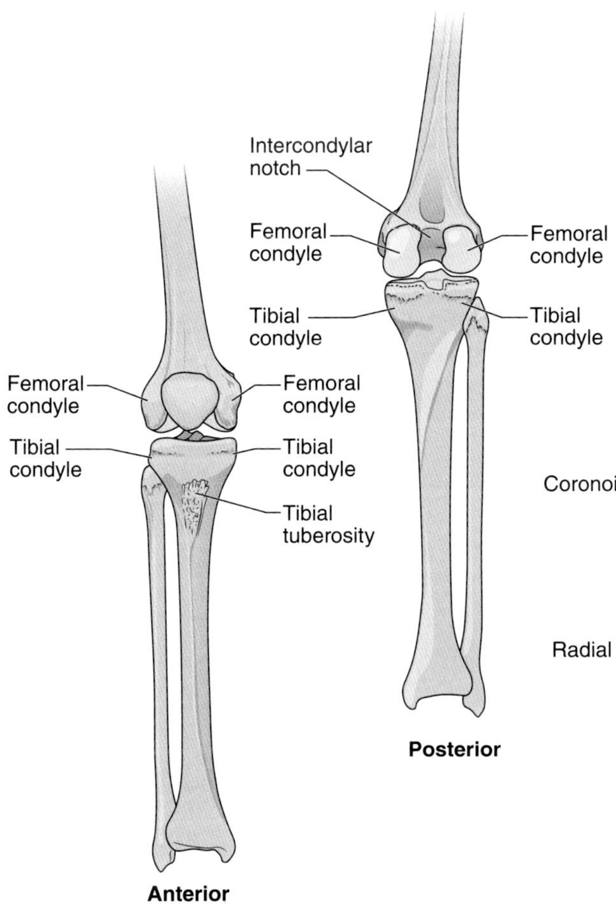

Figure 1.6 The knee.

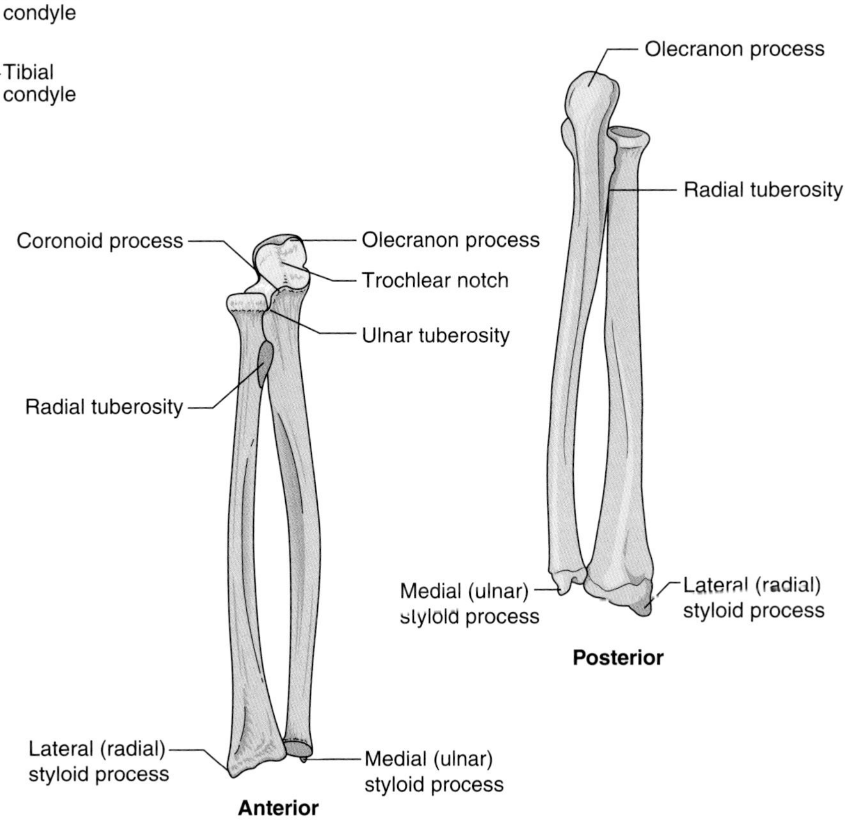

Figure 1.8 Landmarks of the elbow bones.

Joints

The place where two or more bones join together anatomically is referred to as an *articulation*. The terms *joint* and *articulation* are interchangeable, and the study of joints is known as *arthrology*. As bones move through a joint's range of motion, movement will occur between the surfaces of the bones.

- Rolling is a movement in which the joint surface of one bone rolls on the joint surface of another bone, and new points of contact are made on each bone throughout the joint's range of motion. Examples of rolling include a ball rolling along a surface and a bicycle's tire rolling along a road.
- Gliding is a linear movement of a joint in which one point on a bone's joint surface contacts new points on the adjacent bone's joint surface. An example is a hockey puck gliding across the ice; the flat surface of the puck contacts different points on the ice as it glides across it.
- Spin is the rotation of one joint surface relative to another joint surface in which the same point on each surface remains in contact with each other during the movement. An example of spin is a figure skater spinning on the ice in the same place.

Distraction occurs when joint surfaces are separated, and compression occurs when joint surfaces are pressed together.

Tying bones together at articulations are structures of dense, fibrous connective tissue known as *ligaments* (figure 1.9). A ligament is a cord, band, or sheet of strong, fibrous connective tissue that unites the articular ends of bones, ties them together, and facilitates or limits movements

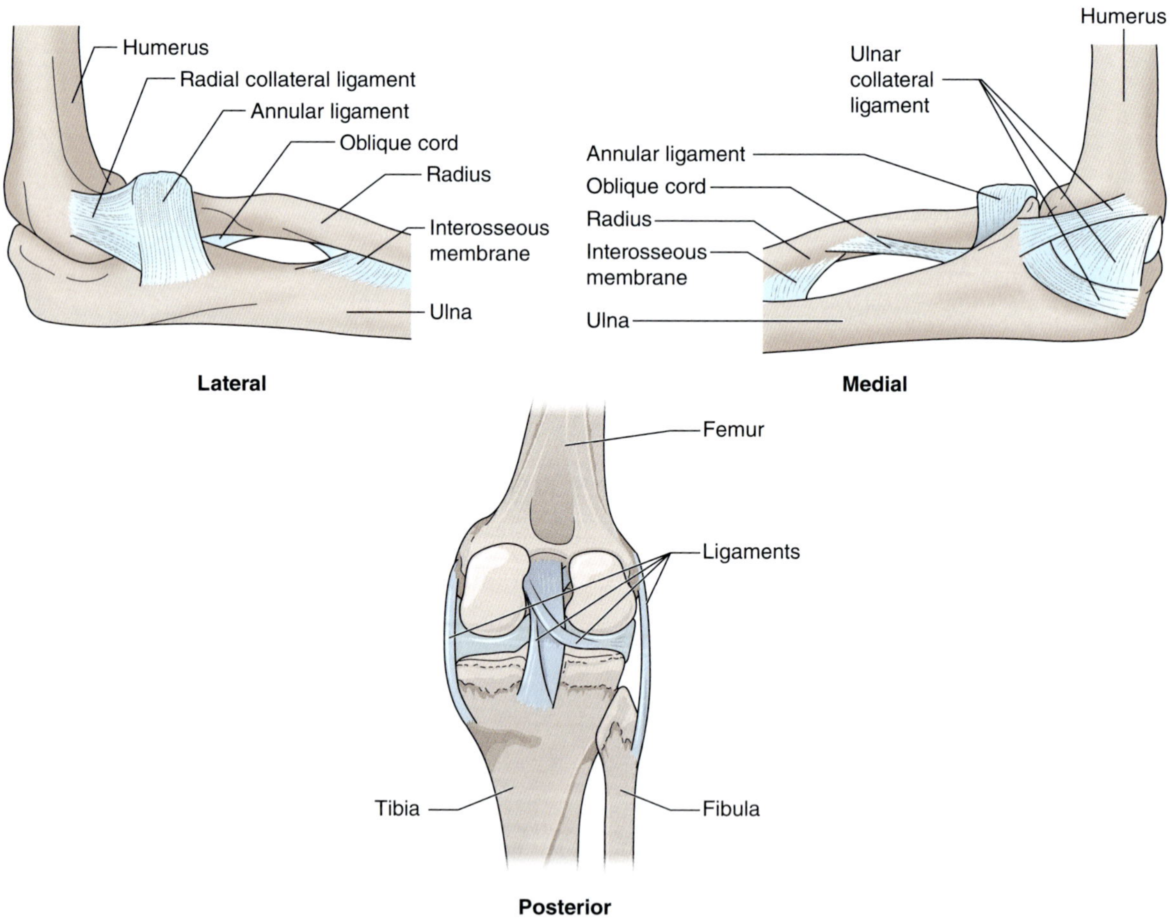

Figure 1.9 Major ligaments of the elbow and the ligaments of the knee.

between the bones. Ligaments are not the sole support for the stability of joints. The muscles that cross the joint and the actual formations of the articulating bones also contribute to joint stability.

There are two major forms of joints: diarthrodial and synarthrodial. **Diarthrodial joints** are distinguished by a separation of the bones and the presence of a joint cavity. These joints are divided by their shape into six subdivisions (figure 1.10).

1. The **hinge joint** has one concave surface, with the other surface looking like a spool of thread. The elbow joint is an example of a hinge-type diarthrodial joint.
2. The **ball-and-socket** type of diarthrodial joint consists of the rounded head of one bone fitting into the cuplike cavity of another bone. Both the hip and shoulder joints are examples of the ball-and-socket type of diarthrodial joint.
3. The **irregular joint** consists of irregularly shaped surfaces that are typically either flat or slightly rounded. These are also known as *plane joints* or *gliding joints*. The joints between the bones of the wrist (carpals) are an example of this type of joint. Gliding movement occurs between the carpal bones.
4. The **condyloid joint** consists of one oval-shaped surface fitting into an oval-shaped socket or depression. These joints are also known as *ellipsoidal joints*. Although the description of the condyloid joint is similar to that of the ball-and-socket joint, the difference is that the condyloid joint is capable of movement in only two planes about two axes, whereas the ball-and-socket joint is capable of movement in three planes about three axes. (Note: Planes and axes are discussed in chapter 2.) An example of a condyloid joint is where the metacarpal bones of the hand meet the phalanges of the fingers.
5. The **saddle joint** is often considered a modification of the condyloid joint. Both bones have a surface that is convex in one direction and concave in the opposite direction, like a saddle. These joints are rare, and the best

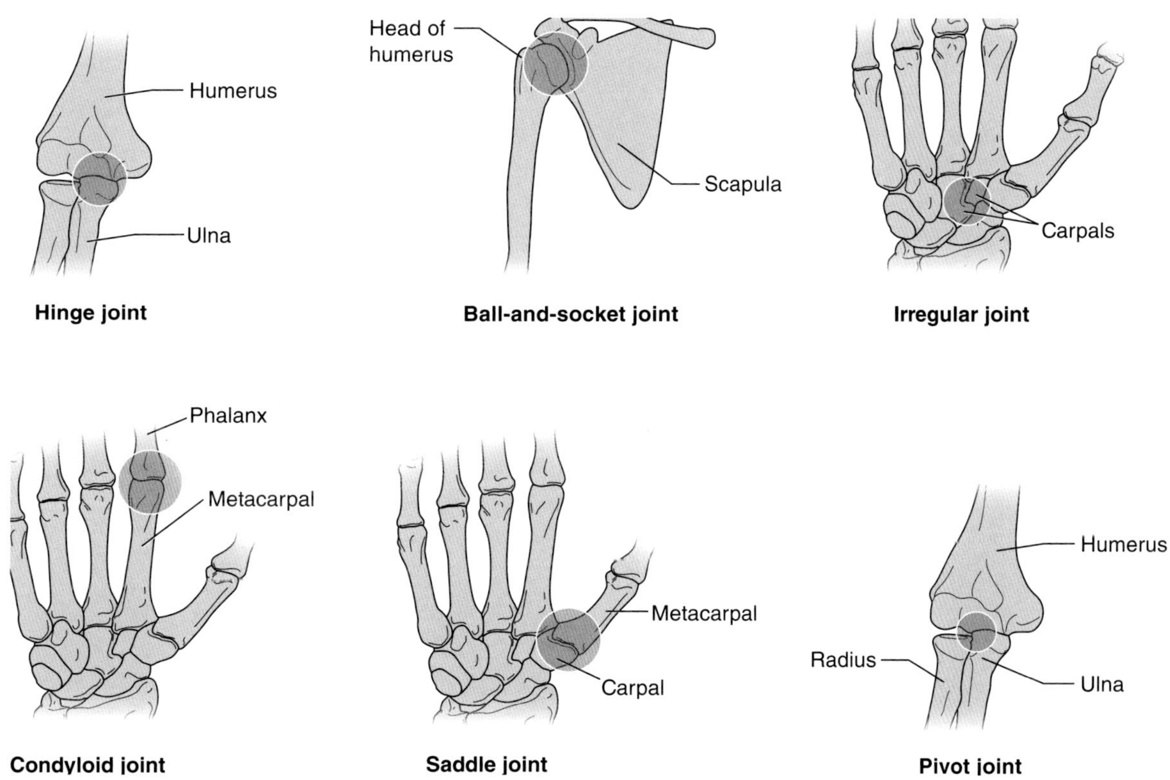

Figure 1.10 Six types of diarthrodial joints.

example is the joint between the wrist and the thumb (carpometacarpal joint).

6. In the **pivot joint**, one bone rotates about the other bone. The radius bone (of the forearm) rotating on the humerus (upper-arm bone) is an example of a pivot joint.

All the diarthrodial joints are considered synovial joints. The synovial joints are where the greatest amount of movement occurs. They are characterized by a space between the articulating surfaces (figure 1.11); a synovial membrane lining the joint secretes synovial fluid for lubrication and provides nutrients to joint structures. Synovial joints are surrounded by a joint (articular) capsule. These joints are classified into four categories by the type of movement they permit in planes and about axes (figure 1.12).

1. Joints between bones that allow only a gliding type of movement over each other are known as **nonaxial joints**, such as those found in the wrist and foot.
2. **Uniaxial joint**, such as the elbow joint, permits movement in only one plane about one axis.

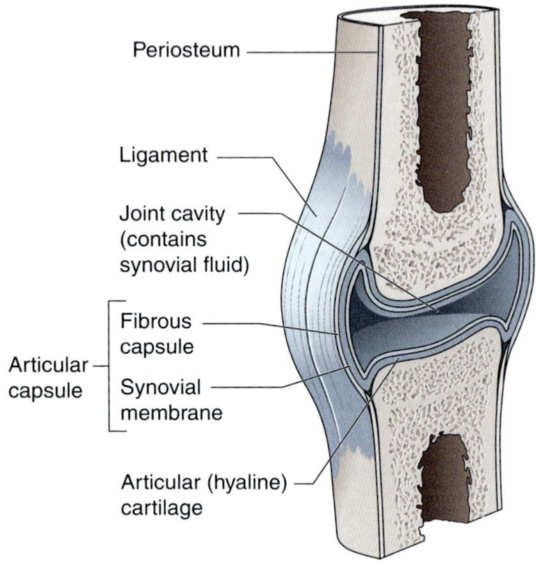

Figure 1.11 A diarthrodial (synovial) joint.

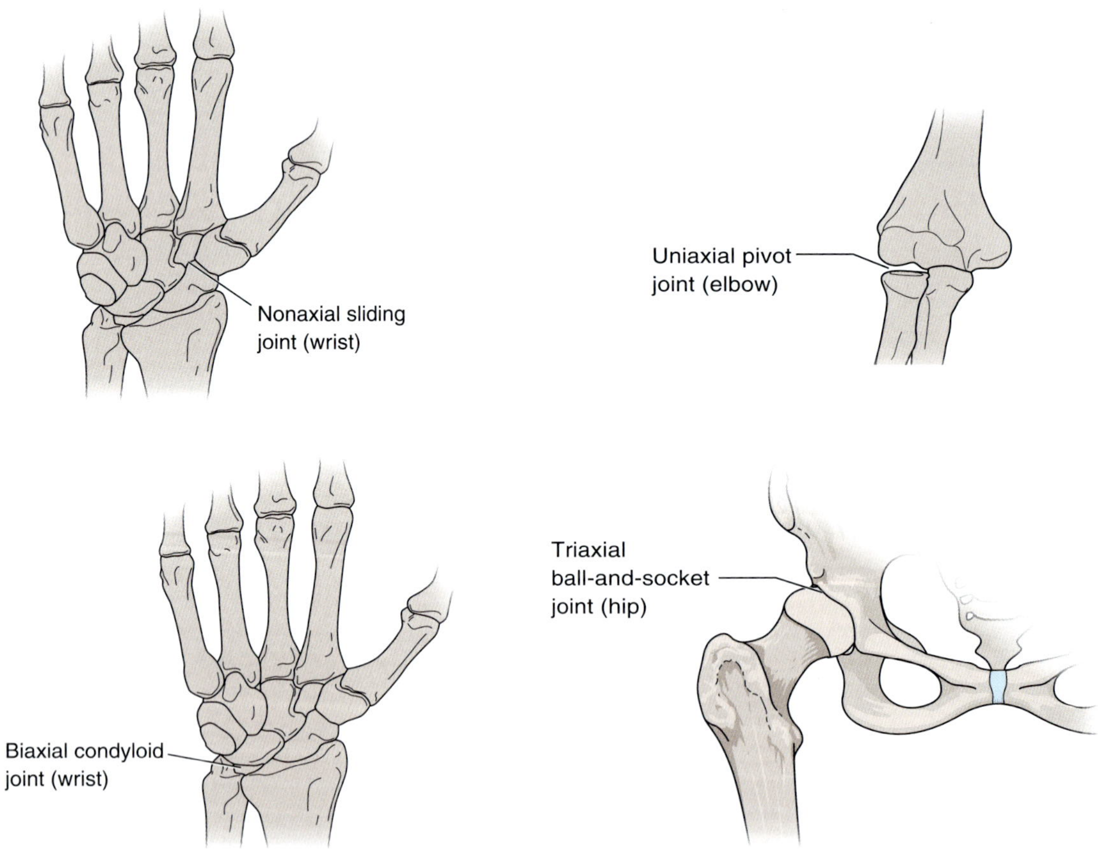

Figure 1.12 Four types of synovial joints.

3. A **biaxial joint**, such as the wrist, permits movement in two planes, about two axes.
4. A **triaxial joint** allows movement in three planes, about three axes, illustrated by the movements of the shoulder and hip joints, which are both ball-and-socket joints.

Synarthrodial joints have no separation or joint cavity, unlike the diarthrodial joints. There are three subdivisions of synarthrodial joints (figure 1.13): suture, cartilaginous, and ligamentous.

1. The suture joint has no detectable movement and appears to be sewn (sutured) together like a seam in clothing. The bones of the skull are the classic example of suture joints. There is no movement in these joints.
2. Cartilaginous joints allow some movement, but other than the joints found in the spinal column, they do not play a major role in movement. A cartilaginous joint contains fibrocartilage that deforms to allow movement between the bones and also acts as a shock absorber between them. Examples include the intervertebral joints, the pubic symphysis, and the sacroiliac joints.
3. The ligamentous joints tie together bones that have very limited or no movement. The joints between two structures of the same bone (e.g., the coracoid and acromion processes of the scapula) and between the shafts of the forearm and lower-leg bones are examples of the ligamentous form of a synarthrodial joint.

Joint strength is determined by several factors:

- The physical structure of the bones contributes to joint strength. (For example, both the shoulder and hip joints are classified as ball-and-socket joints, but the deeper structure of the hip joint makes it stronger than the shoulder joint.)
- The strength, number, and anatomical position of ligaments also determine joint strength, as do the number and strength of muscles and tendons crossing a joint. The strength of the muscles and tendons is extremely important for injury prevention and rehabilitation.
- To a lesser extent, other structures (blood vessels, nerves, skin, and fascia) crossing a joint also contribute to joint strength.

The degree of movement in joints varies not only from person to person but also within a particular person. The following factors influence the degree of movement in joints:

- The bones involved (joint structure can limit range of motion; consider elbow movement versus hip movement)
- The thickness and laxity of the ligamentous structures
- The amount of fat and muscular tissue around a joint
- The strength and flexibility of the muscle tissue crossing the joint

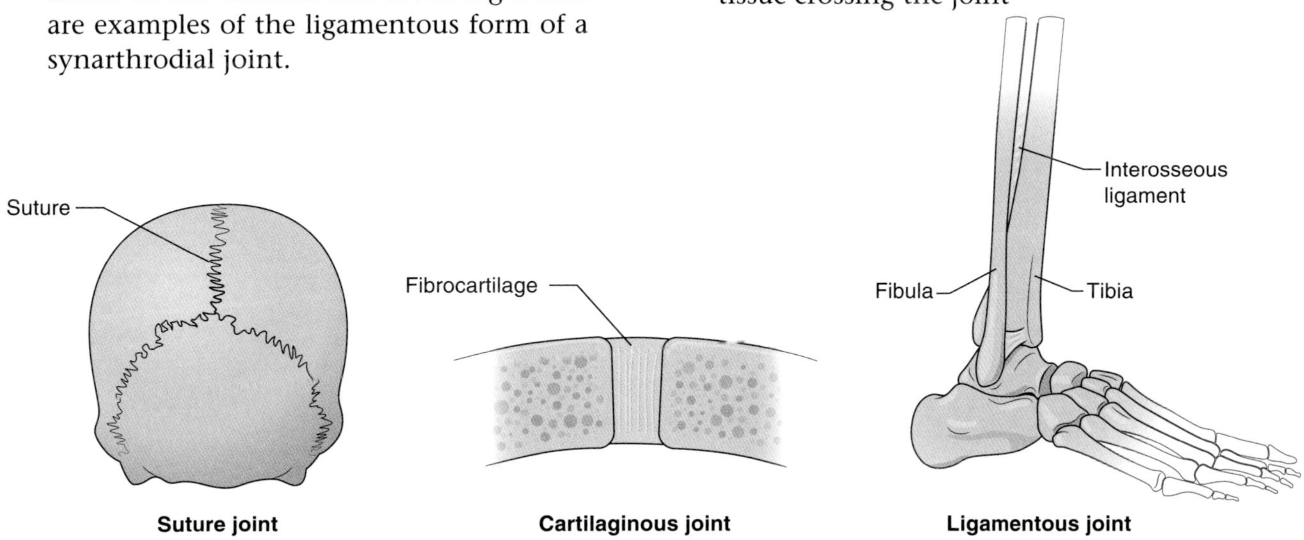

Figure 1.13 Three types of synarthrodial joints.

- The resistance of other structures (mentioned previously as structures contributing to joint strength: blood vessels, nerves, skin, and fascia crossing the joint)

Muscles

The study of muscle is known as **myology**. Chemically, muscles consist of water and solids; those solids are proteins, carbohydrates, inorganic salts (including calcium, chloride, iron, magnesium, phosphorus, potassium, and sodium), enzymes, fat globules, nitrogenous extractives (e.g., creatine, uric acid), and nonnitrogenous extractives (e.g., lactic acid, glycogen). Further analysis of the chemical composition of muscle is found in the study of human physiology.

The number of muscles in the human body depends on several factors. Not all people have exactly the same number of muscles. Some muscles may appear on one side of the body and not the other (e.g., psoas minor). Some muscles are totally absent in some people (e.g., palmaris longus). Various texts may list certain muscles separately, while others might consider certain muscles as part of a larger muscle (e.g., flexor hallucis brevis and flexor digitorum brevis). Most authorities agree on the number 680 as the total muscles in the human body, with approximately 240 having separate names. Muscles are named using the following criteria:

- Action, such as flexor or extensor
- Attachment to bones, such as sternocleidomastoid
- Direction of pull, such as oblique or rectus
- Location, such as tibialis or ulnar
- Size, such as maximus or minor
- Shape, such as teres or trapezius
- Structure, such as triceps or quadriceps
- Some combination of these criteria, such as flexor digitorum brevis

Muscle tissue is often categorized into three types:

1. Smooth, which occurs in various internal organs and vessels
2. Cardiac, which is unique to the heart
3. Skeletal, which causes movement of the bones and their joints

Unlike skeletal muscles, smooth muscles (figure 1.14*a*) are not organized into motor units, and they receive nervous control via the autonomic nervous system. Smooth muscles are typically found in blood vessels, where they increase and decrease the lumen (opening) of the vessels to assist in the flow of blood through the circulatory system; in hollow organs, such as the stomach and bladder; and in the alimentary (digestive) tract, where they create a rhythmic type of contraction. In the intestines, this rhythmic movement is known as *peristalsis*. Cardiac muscle fibers (figure 1.14*b*) are associated with the heart. This type of muscle creates a rhythmic contraction consisting of two phases, systole and diastole. Systole is the contraction of cardiac muscle, and diastole is the period when the cardiac muscle relaxes. Both smooth and

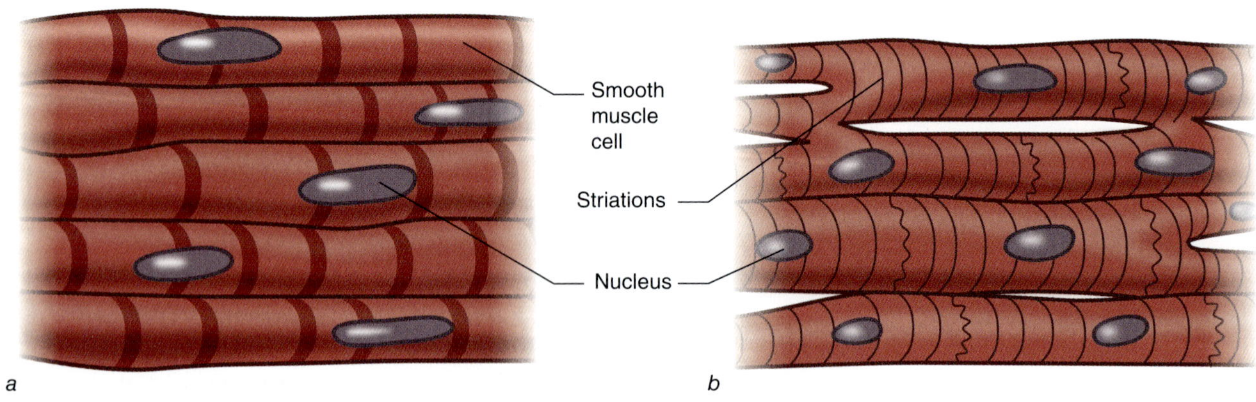

Figure 1.14 *(a)* Smooth and *(b)* cardiac muscle cells.

cardiac muscle are structurally different from skeletal muscle.

For the purpose of looking at anatomy and movement, this text concentrates on the skeletal muscle. Skeletal muscle has the ability to produce force and movement about a joint through four properties:

1. **Extensibility**: The ability of a muscle to be stretched beyond its normal resting length.
2. **Elasticity**: The ability of a muscle to return to its original length when stretching ceases.
3. **Contractility**: The ability of a muscle to shorten and develop tension when stimulated.
4. **Excitability (irritability)**: The ability of a muscle to be sensitive or responsive to chemical, electrical, or mechanical stimuli.

The various forms of skeletal muscle are fusiform, quadrate, triangular, unipennate, bipennate, multipennate, and longitudinal (figure 1.15). Most skeletal muscles are either fusiform or pennate. Fusiform muscles are formed by long, parallel fibers and typically are involved in movements over a large range of motion. Pennate muscles consist of short, diagonal fibers and are involved in movements that require great strength over a limited range of motion.

The strength a particular muscle can generate is dependent on several factors:

- *Cross-sectional area*: All else being equal, a larger cross section means more muscle fibers and therefore greater strength.
- *Length*: Longer fibers typically generate a force creating more motion since they shorten over a greater distance.
- *Texture*: A muscle with less noncontractile tissue will have more muscle fibers per area and, therefore, is capable of generating a greater force.
- *Specificity*: The chemical structure of muscle (e.g., viscosity of fluids present, amount of sarcoplasm, number of amino acids) can affect muscle actions.
- *Tension*: A muscle generates more force when it is placed on stretch, or under tension.
- *Coordination*: Poor coordination between muscles can create friction between muscle fibers, reducing the force that can be generated.

The fibers of a muscle form the muscle belly. At either end of the belly, a unique form of connective tissue, a tendon, attaches the muscle to the bones. Tendons are extensible and elastic, like skeletal muscle, but they are not contractile. They are similar to ligaments in that both are dense, fibrous connective tissues. The main difference is that tendinous tissue does not have as much elasticity as ligamentous tissue. Tendons of skeletal muscles are usually defined as either tendons of origin or tendons of insertion (figure 1.16). Tendons of origin are usually longer and attached to the proximal (closest to the center of the body) bone of a joint, which is typically the less mobile (fixed) of the two bones of a joint. The origin

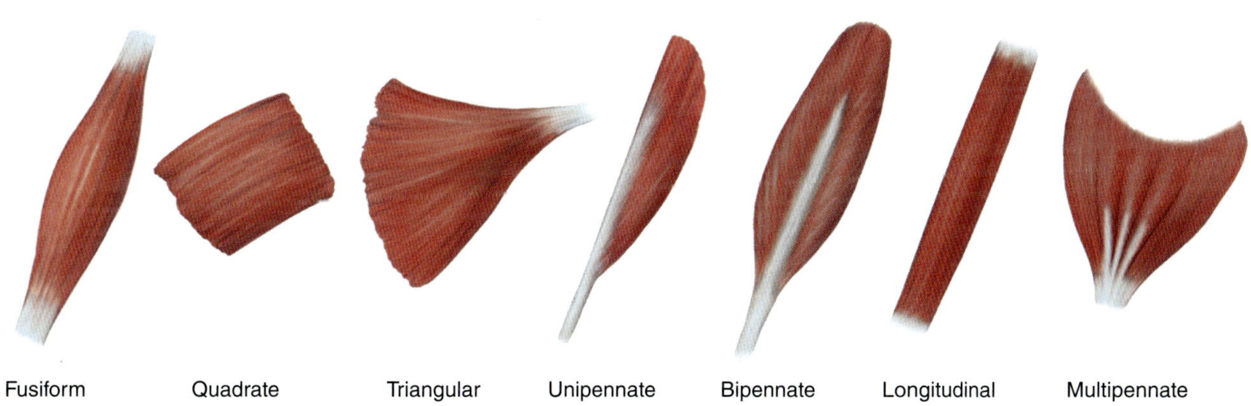

Figure 1.15 Various fiber arrangements of skeletal muscles.

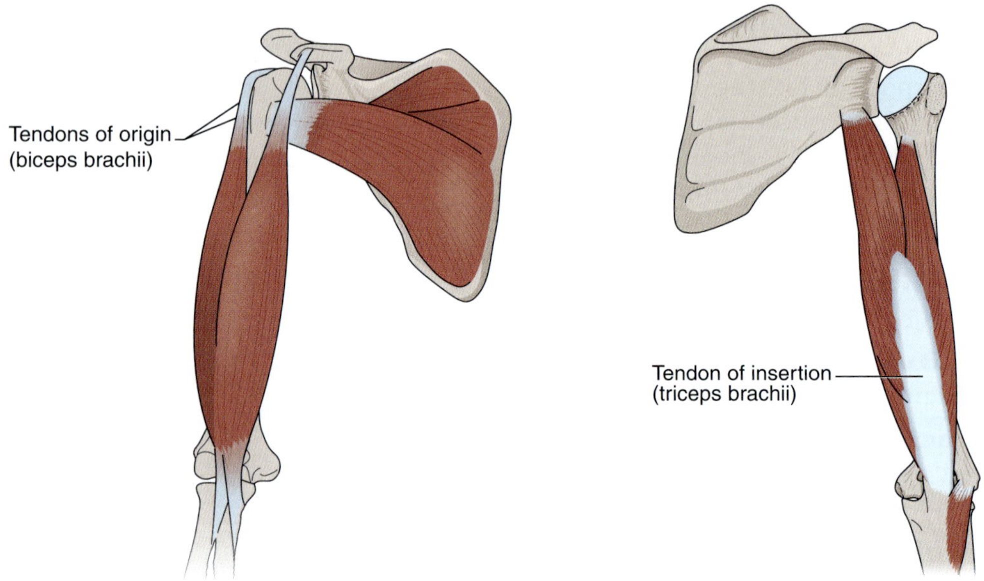

Figure 1.16 Tendons of origin and insertion.

of a muscle is usually the most stable attachment. Tendons of insertion are often shorter and attached to the more distal (farther from the center of the body) bone of a joint, which is typically the more movable (unstable) of the two bones of a joint. While most tendons of muscles are attached to bone, they may also be found attached to other tendons, fascia, ligaments, or even skin. Additionally, because tendons cross bony areas or need to be confined to certain areas, the tendons are covered by connective tissue known as *tendon sheaths* (figure 1.17) to protect them from wear and tear from the bony structures they cross.

Contraction of a skeletal muscle usually results in movement of bones in some direction. The movement (action) is typically described as one or more of the following:

- Flexion
- Extension
- Abduction
- Adduction
- Lateral rotation
- Medial rotation
- Pronation
- Supination
- Circumduction

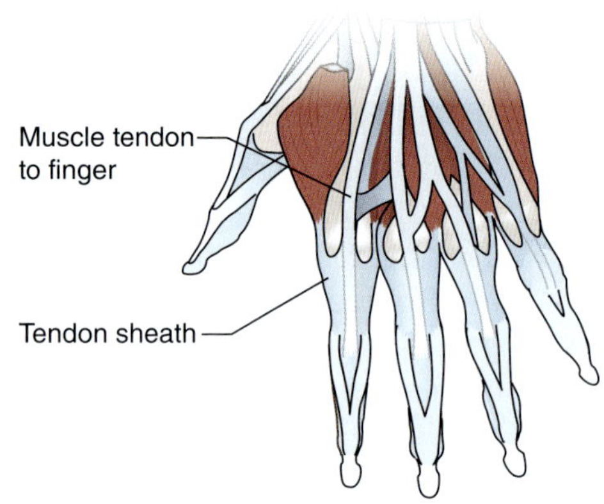

Figure 1.17 Tendon sheaths.

These actions are described in more detail in chapter 2.

In almost every movement, more than one muscle is involved in producing the motion. The interrelationship of the skeletal muscles involved in a movement can be described by the actual function that a particular muscle provides. The muscle identified as the main muscle producing a particular movement is known as the *prime mover*, or *agonist*. Any muscle helping a prime mover accomplish its action is known as a *synergist*. The prime mover is typically opposed by another muscle, which is called the *antagonist*.

At the elbow joint, the triceps brachii is the antagonist to the biceps brachii (agonist) during elbow flexion. A muscle may also simply hold a bone in place while another muscle performs its function. This action is known as *fixation*, and the muscle is called a *fixator*.

As a muscle moves a joint through a motion, the line of pull of the muscle may change the joint action. A muscle's line of pull is relative to its orientation around a joint. As a muscle contracts, its line of pull may change, which can result in a muscle having an action different from its original action. Further study of muscle actions will be left to studies in kinesiology and biomechanics.

Skeletal muscle has several functions.

1. The obvious function of skeletal muscle is to provide movement.
2. Skeletal muscle provides protection from external trauma and acts as a shock absorber for underlying bones and internal organs.
3. Support to joints through the tension of muscles and tendons is another function of skeletal muscle. A prime example of joint support by muscles and tendons is body posture. During sitting, standing, walking, running, and almost every activity of the body, muscles and tendons crossing joints provide support.
4. An often-overlooked function of skeletal muscle is heat production through contraction, a good example of which is the simple act of shivering.

Skeletal muscle has a distinctive property in that it can have the ability to contract and generate tension **concentrically**, **eccentrically**, or **isometrically**. Development of tension within skeletal muscle causes it to pull on either one or both of its attachments. When a muscle generates tension and its attachments move closer together, a **concentric contraction** occurs. When a muscle generates tension and its attachments move farther apart, an **eccentric contraction** occurs. When a muscle generates tension and there is no movement, an **isometric contraction** occurs.

The human body has the ability to adapt to the demands imposed upon it, which is known as the *SAID principle*: specific adaptation to imposed demands. Muscle has the ability to change in size and strength relative to the type of demands placed upon it. Muscle can increase in size and strength when periodically stimulated at a greater intensity than it is accustomed to in its current state.

The general structure of skeletal muscle is shown in figure 1.18. Skeletal muscle is encased by a form of connective tissue known as the *epimysium*. Within the epimysium are numerous bundles of muscle fibers that are individually wrapped in a fibrous sheath known as the *perimysium*. Within the perimysium, the muscle fibers are in turn enclosed in a connective sheath known as the *endomysium*. A muscle fiber consists of a number of myofibrils, which are the contractile elements of muscle. Some muscle fibers are large enough to be seen, while others are visible only through a microscope. Individual myofibrils are enclosed by a viscous material known as *sarcoplasm* and wrapped in a membrane known as the *sarcolemma*. Lengthwise, myofibrils consist of bands of alternating dark and light filaments of contractile proteins known as *actin* and *myosin* (see figures 1.18 and 1.19). This alternating pattern produces a striped (striated) appearance when viewed under a microscope.

A myofibril is divided into a series of sarcomeres, which are considered the functional units of skeletal muscle (see figure 1.19). Sarcomeres contain an I-band (isotropic band), the light-colored portion made up mostly of the protein filament actin, and an A-band (anisotropic band), the dark-colored area made up mostly of the protein filament myosin (see figure 1.19). A sarcomere is that portion of a myofibril that appears between two Z-lines (Z-lines bisect I-bands). Actin is also found in A-bands. As the actin filaments extend into the A-band, they overlap with the myosin filaments, contributing to the darker appearance at the edges of the A-band. The lighter-colored central portion of the A-band is known as the *H-zone*. This region is lighter in color because actin does not extend into this area, and the myosin filament is thinner in the middle than at its outer edges. The two protein filaments, actin and myosin, are the site of muscular movement (contraction). The myosin filament has cross-bridges (small extensions) that

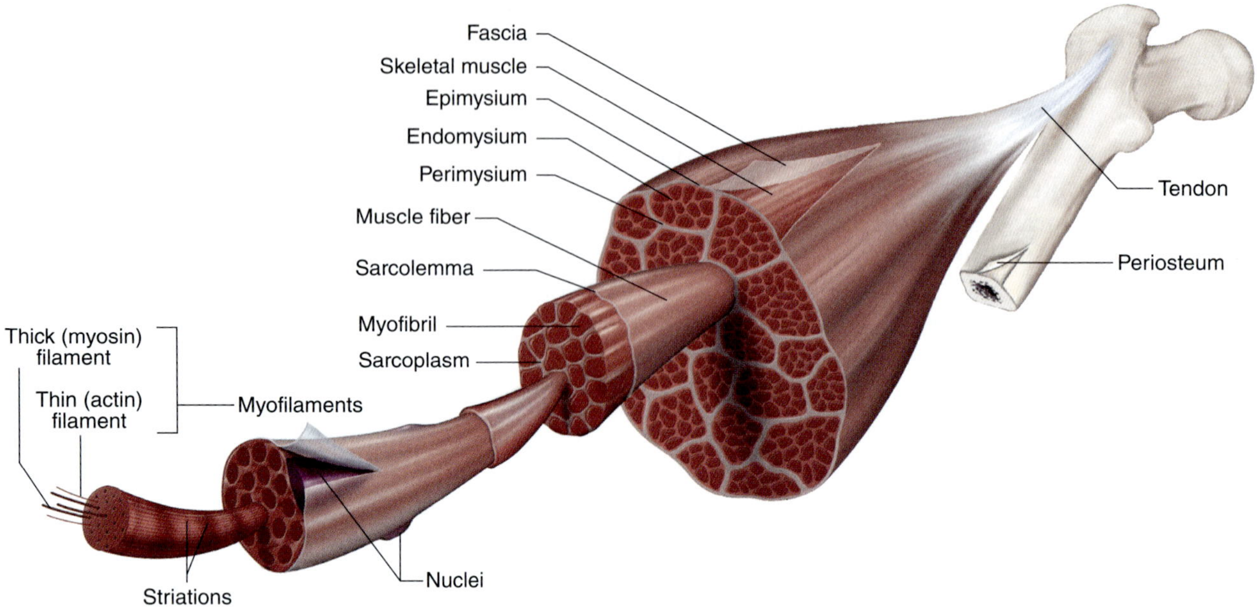

Figure 1.18 Structure of a skeletal muscle.

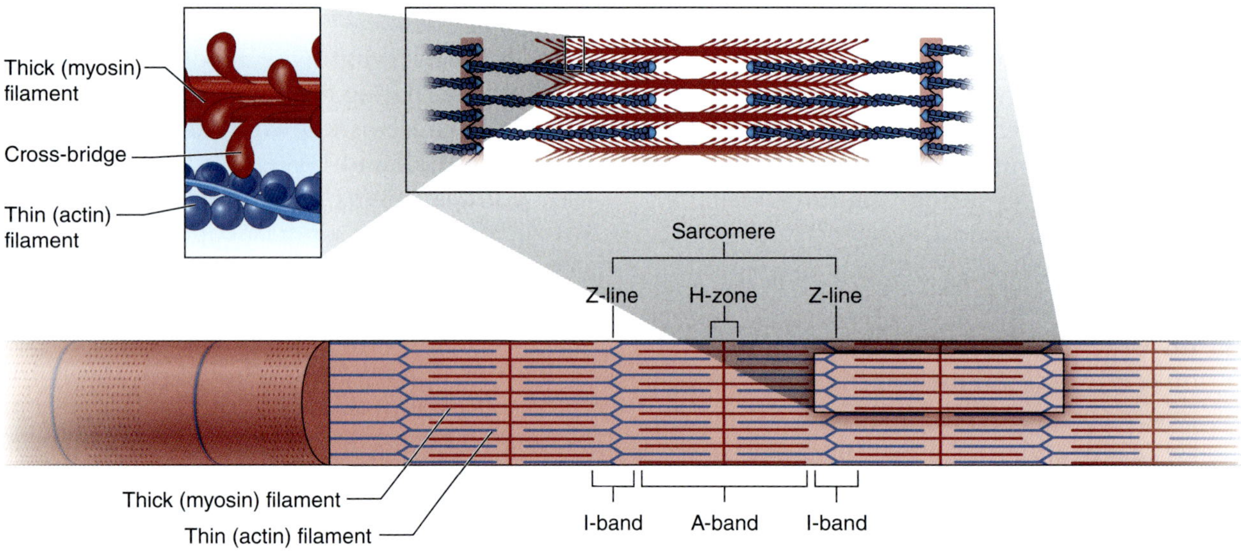

Figure 1.19 A muscle fiber and its myofibrils.

reach out at an angle toward the actin filaments (see figure 1.19).

There are two primary types of skeletal muscle fibers, commonly known as *fast-twitch* and *slow-twitch fibers* (figure 1.20). Most muscles contain both types of fibers, but, depending on heredity, function, and, to a lesser degree, training, some muscles contain more of one type of fiber than the other. Fast-twitch fibers are large and white and appear in muscles used to perform strength

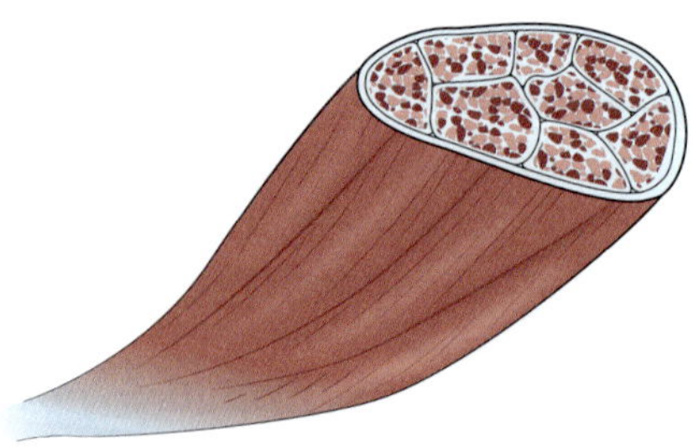

Figure 1.20 Fast-twitch (light) and slow-twitch (dark) muscle fibers.

activities. The slow-twitch fibers are smaller and darker (red) than the fast-twitch fibers (primarily because they have a greater supply of myoglobin). Slow-twitch fibers are slow to fatigue and are prevalent in muscles involved in performing endurance activities. A runner with a higher percentage of slow-twitch fibers in lower-extremity muscles is more likely to develop into a distance runner, whereas a runner with a higher percentage of fast-twitch fibers in lower-extremity muscles is more likely to become a sprinter. Some of the major factors distinguishing *red* muscle fibers from *white* muscle fibers are that red muscles (1) are typically smaller, (2) have more endurance, (3) contract more slowly, (4) contain more myoglobin, (5) have greater density, and (6) have a greater amount of sarcoplasm.

Various forms of exercise have numerous effects on muscles. Exercise increases muscle size by increasing the size of fibers and thickening the sarcolemma and connective tissue. It increases muscle strength through this increase in fiber size and by training the body to activate more muscle fibers. Over a period of time, endurance training activities will increase blood flow to the muscle tissue being exercised, resulting in an increased ability to continue working (endurance). While this training does not alter the composition of the muscle, it does enhance what is already there. Exercise improves neuromuscular control by improving the transmission of nerve impulses to muscles, resulting in increases in both endurance and strength. Through the process of training, exercise improves muscular coordination. It also increases blood flow to muscles, which increases the amount of myoglobin present. The combination of increased blood flow and a greater presence of myoglobin in the muscle increases the amount of nutrients and oxygen present for the muscle to produce work.

Before discussing the innervation of muscles, which allows them to function, a simple explanation is needed of how the structures presented so far—bones, joints, and muscles—combine to produce movement through a system of levers.

Levers

Bones, ligaments, and muscles are the structures that form levers in the body to create human movement. In simple terms, a joint (where two or more bones join together) forms the axis (or fulcrum), and the muscles crossing the joint apply the force to move a weight or resistance. Levers are typically labeled as either first, second, or third class. All three types are found in the body, but most levers in the human body are third class.

A first-class lever has the axis (fulcrum) located between the weight (resistance) and the force (figure 1.21*a*). An example of a first-class lever is a pair of pliers or scissors. First-class levers in

POP QUIZ

Which of these two runners is likely to be a better sprinter than a distance runner? Briefly explain your reasoning.

Tetra Images/Getty Images

Westend61/Getty Images

FOCUS ON

Muscle Viscosity

Viscosity is more easily understood when you consider motor oil used in an automobile. The viscosity (thickness) of the oil depends on the temperature: The oil either thins (as the temperature increases) or thickens (as the temperature decreases). Viscosity resists molecular rearrangement caused when a muscle contracts, and, therefore, more energy is needed to move quickly (and overcome resistance) than slowly because of muscle viscosity. Viscosity tends to slow the speed of muscle contraction, so depending on the activity, lowering muscle viscosity can affect performance. Some authorities believe that one of the benefits of a warm-up before physical activity is that muscle viscosity changes with the increased temperature within the muscle tissue, which makes the muscle more able to endure the stress of the physical activity.

The study of muscle physiology and exercise physiology reveals other values of warming up. Anyone involved in prescribing physical activity (physicians, coaches, therapists, athletic trainers, and personal trainers) needs to understand the physiological factors involved in the warm-up, including the effect on muscle viscosity.

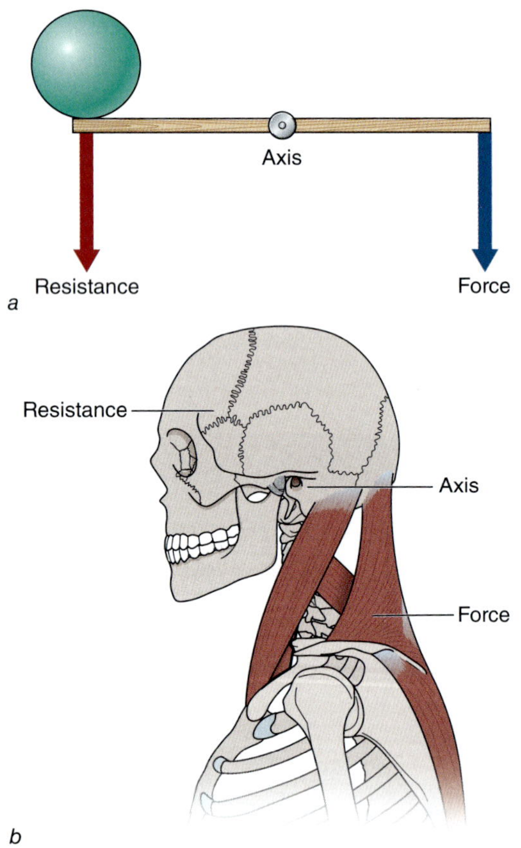

Figure 1.21 *(a)* A first-class lever in *(b)* the human body.

the human body are rare. One example is the joint between the head and the first vertebra (the atlantooccipital joint: figure 1.21*b*). The weight (resistance) is the head, the axis is the joint, and the muscular action (force) comes from any of the posterior muscles attaching to the skull, such as the trapezius.

In a second-class lever, the weight (resistance) is located between the axis (fulcrum) and the force (figure 1.22*a*). The most obvious example is a wheelbarrow, where a weight is placed in the bed of the wheelbarrow between the wheel (axis) and the hands of the person using the wheelbarrow (force). In the human body, an example of a second-class lever is found in the lower leg when someone stands on tiptoes (figure 1.22*b*). The axis is formed by the metatarsophalangeal joints, the resistance is the weight of the body, and the force is applied to the calcaneus bone (heel) by the gastrocnemius and soleus muscles through the Achilles tendon.

In a third-class lever, the most common in the human body, force is applied between the resistance (weight) and the axis (fulcrum; figure 1.23*a*). Picture someone using a shovel to pick up an object. The axis is the end of the handle where the person grips with one hand.

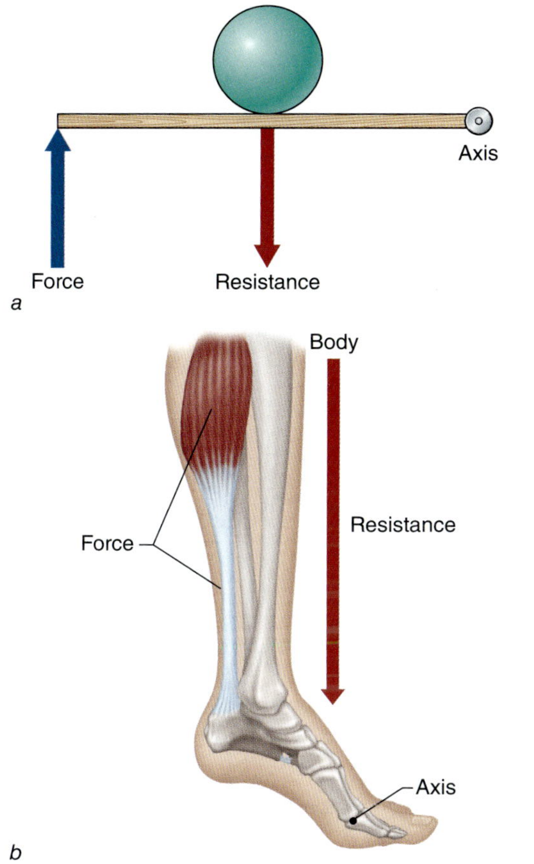

Figure 1.22 *(a)* A second-class lever in *(b)* the human body.

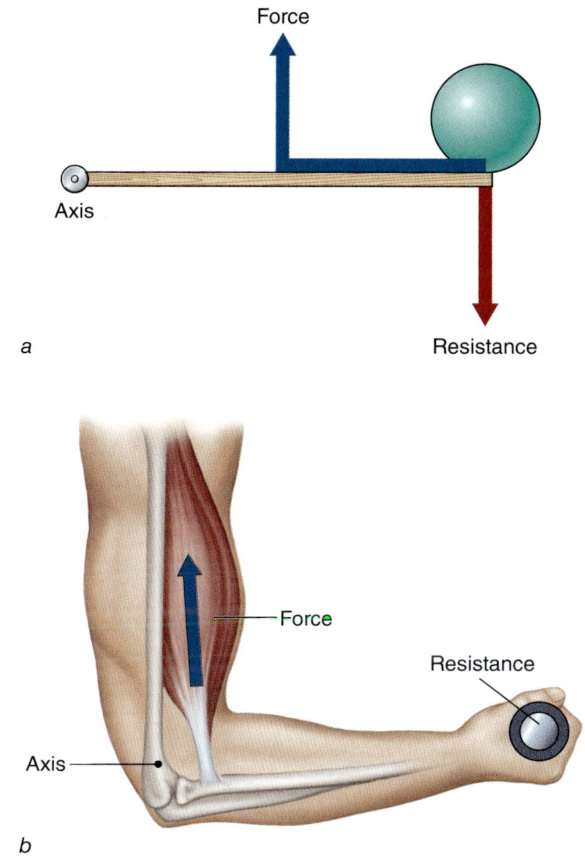

Figure 1.23 *(a)* A third-class lever in *(b)* the human body.

The other hand, placed somewhere along the shaft of the handle, applies force. At the other end of the shovel (the bed), a resistance (weight) is present. There are numerous third-class levers in the human body; one example can be illustrated in the elbow joint (figure 1.23*b*). The joint is the axis (fulcrum). The resistance (weight) is the forearm, wrist, and hand. The force is the biceps muscle when the elbow is flexed.

Nerves

The body has three main nervous systems: an autonomic nervous system, a central nervous system, and a peripheral nervous system.

• The autonomic nervous system controls the glands and smooth muscle of the body. This system is often divided into the parasympathetic system (cranial and sacral portions) and the sympathetic system (thoracic and lumbar portions; see chapter 10).

• The central nervous system consists of the brain and spinal cord. The brain is divided into the cerebrum (frontal, parietal, occipital, and temporal lobes), the brain stem, and the cerebellum. The outer layer of the cerebrum, which contains many cell bodies and dendrites, is often referred to as the *gray matter*; the inner portion of the cerebrum (the white matter) contains primarily axons (figure 1.24).

The peripheral nervous system consists of 12 pairs of cranial nerves and 31 pairs of spinal nerves. The cranial nerves are both sensory and motor in nature and typically receive a particular sensory stimulus (externally, such as smell, sight, temperature, pain, or pressure, and internally,

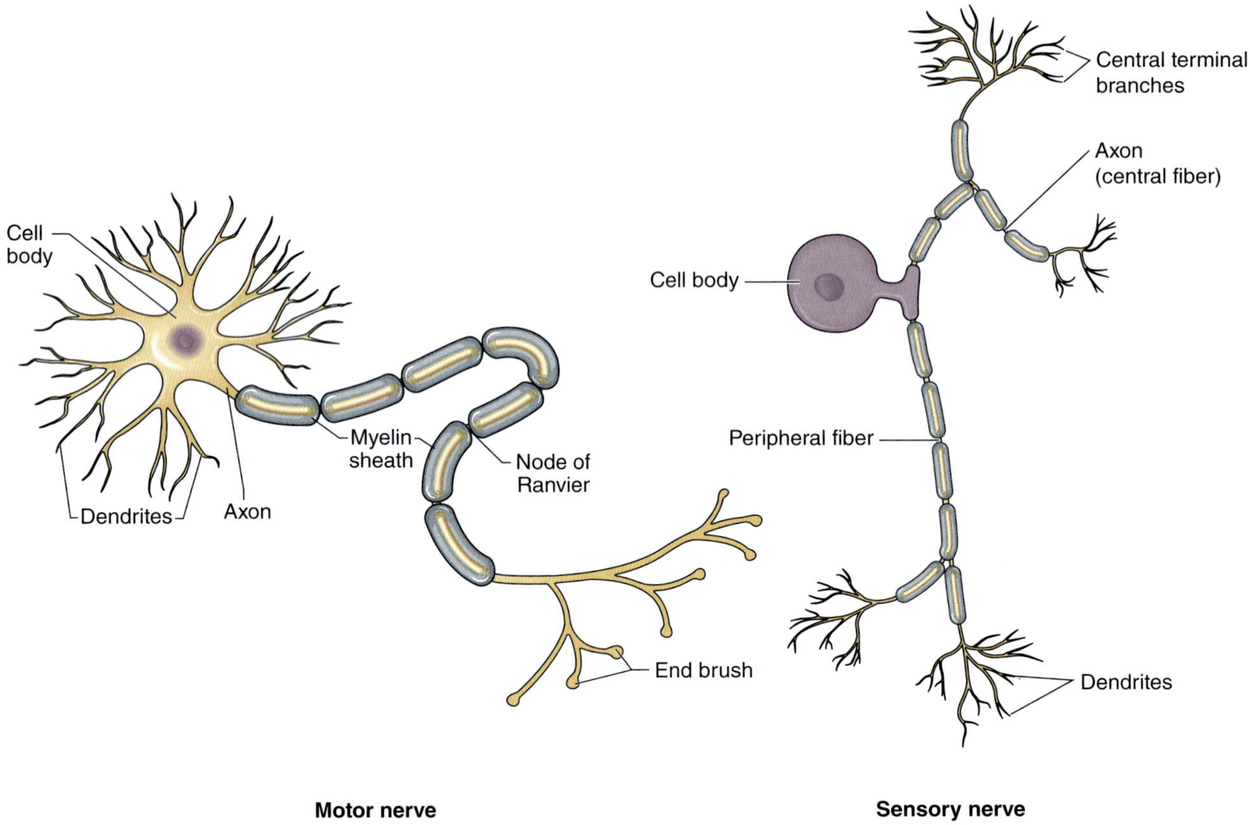

Figure 1.24 A motor (efferent) nerve and a sensory (afferent) nerve.

such as hunger, thirst, fatigue, or balance) and convert it into a nerve impulse, which can result in an appropriate effect (response). The spinal nerves, divided into plexuses (networks of peripheral nerves), innervate (stimulate) the muscles to create movement. The major plexuses are the cervical, brachial, lumbar, sacral, and, to a limited extent, pudendal (coccygeal, discussed in chapter 10, see figure 10.12). Spinal column levels are typically referred to by a specific vertebra. For example, C5 is the fifth cervical vertebra, T8 is the eighth thoracic vertebra, and L2 is the second lumbar vertebra.

The nerve (see figure 1.24), or neuron, consists of a nerve cell body and projections from it, which are known respectively as the *axon* and the *dendrite*. In a motor nerve, the dendrite receives information from surrounding tissue and conducts the nerve impulse *to* the nerve's cell body (responsible for neuron nutrition), and the axon conducts the nerve impulse *from* the cell body to the muscle fibers. A nerve innervating a muscle is referred to as a *motor neuron*, and a motor neuron plus all the muscle fibers it innervates is known as a *motor unit*. Another structural component of a motor nerve is the myelin sheath, which insulates the axon. The gaps in the myelin sheath are known as the *nodes of Ranvier*; impulses "leap" along the myelin sheath (from node to node), allowing the impulses to travel at higher speeds than they would across an unmyelinated axon. At the end of the axon is a structure known as the *motor end plate*, which consists of end brushes (terminal branches) that are in very close proximity to the muscle fibers. This connection between the nerve fibers and the muscle fibers is referred to as the *myoneural junction*, also known as the **neuromuscular junction**. Motor nerves carry impulses *away* from the central nervous system to the muscle tissues, whereas sensory nerves (not discussed in this chapter) carry impulses from muscles, ligaments, tendons, and other tissues

FOCUS ON

Palpation

Knowing the textural differences between structures will facilitate hands-on applications of human anatomy.

- The skin's texture, thickness, and flexibility will vary throughout the body. For example, the skin on the dorsum (top) of the foot is thin and pliable, whereas the skin on the plantar (bottom) aspect of the foot is thicker and firmer.
- Bones have a constant, hard, rigid feel and do not change shape or move when palpated.
- Skeletal muscle has a striated texture that feels like connected bands of tissue oriented in a particular direction relative to the muscle's action. Skeletal muscle feels firm when it is contracted and soft and pliable when it is relaxed.
- Although tendons vary in size and shape, they have a smooth and flexible cord-like texture.
- Ligaments may or may not be pliable, depending on their depth within the body, but, if palpable, they are situated between two adjacent bones and feel dense and taut.
- Nerves are tube shaped, mobile, and tender when depressed. When palpated, a nerve may create a sensation of tingling, burning, or numbness down its distribution pathway.

to the central nervous system. Motor nerves are also referred to as *efferent nerves*; sensory nerves are also referred to as *afferent nerves*. All skeletal motor nerves innervate muscles and are connected indirectly or directly with the motor area of the brain known as the *cerebral cortex* (see chapter 10).

Blood Vessels

The blood vessels bring nutrients to the muscle tissue and carry away the waste products produced as the muscle tissues expend energy. When the heart pumps, blood moves out of the heart into a huge vascular tree consisting of arteries, arterioles (smaller arteries), capillaries, veins, and venules (small veins). There are three tissue layers (tunics) of the walls of arteries, veins, and capillaries (tunica intima, tunica media, and tunica adventitia). The middle layer (tunica media) contains various quantities of smooth muscle fibers, depending on the type of vessel. The arteries and arterioles (figure 1.25) distribute blood to the tissues, where capillaries provide the blood directly to the cells. The veins and venules (figure 1.26) collect the blood from the capillaries and return it to the heart. The middle wall of the arteries contains a vast amount of smooth muscle that contracts with the heart to pump the blood throughout the body. The veins contain small valves that permit blood to flow in only one direction (toward the heart). All three layers of tissue in the veins are much thinner compared to the tissue in the arteries. As a result, smooth muscle fibers are either absent or minimal in veins, with only a few thin fibers found in the tunica media. For this reason, skeletal muscles assist in returning blood to the heart as they contract and squeeze the veins between muscles or between muscles and bones (figure 1.27). The skeletal muscles act as muscular venous pumps that squeeze blood upward past each valve. Gravity also assists venous return in veins that are found above the heart. There are more valves in the veins of the extremities, where upward blood flow is opposed by gravity.

Other Tissues

Other types of tissues associated with bones, joints, and muscles are fascia and bursa. Fascia is another form of fibrous connective tissue of

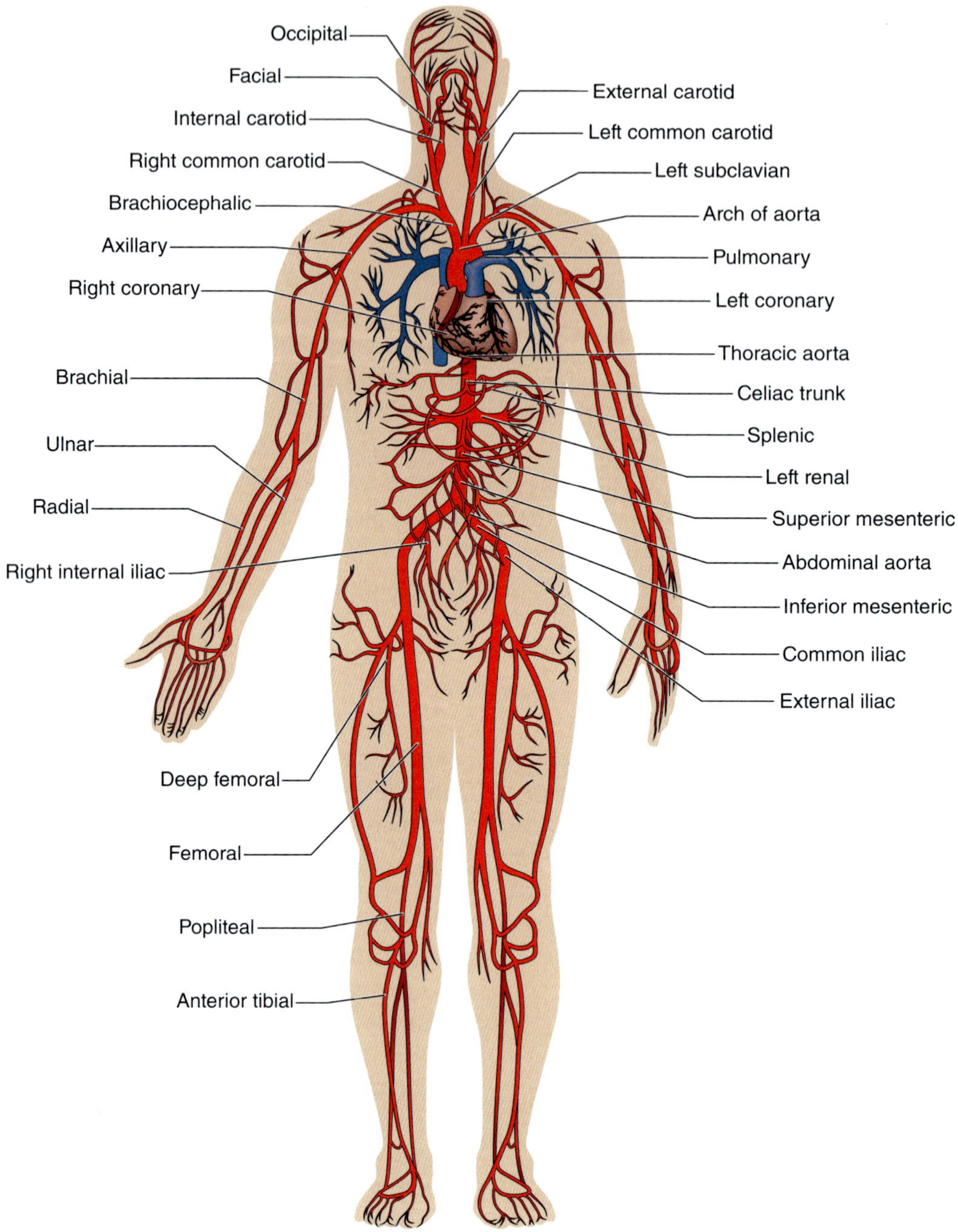

Figure 1.25 Major arteries of the body.

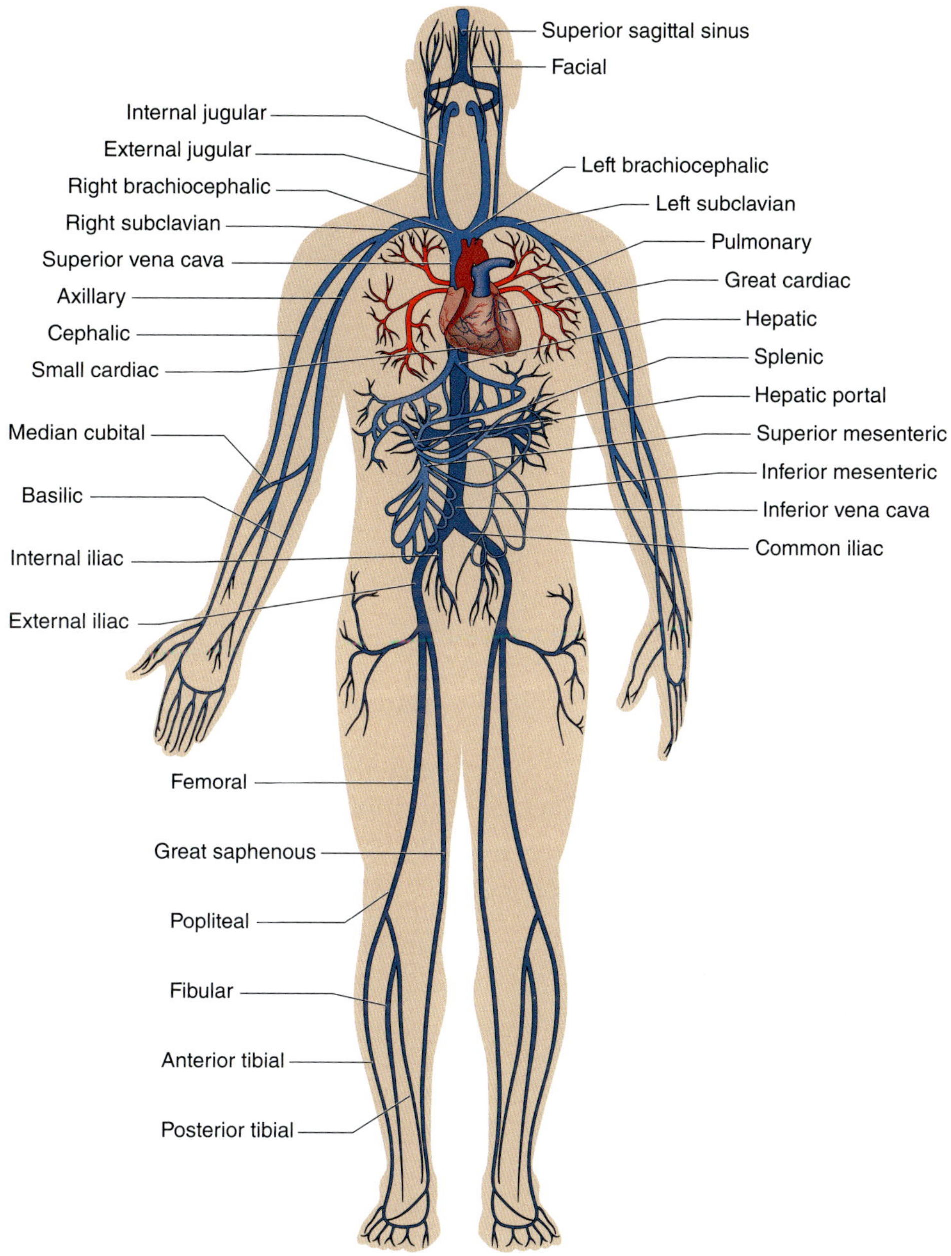

Figure 1.26 Major veins of the body.

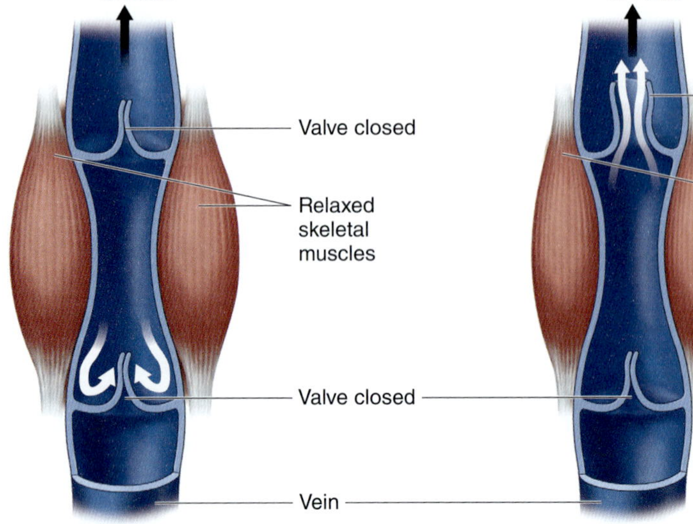

Figure 1.27 A vein's valve action.

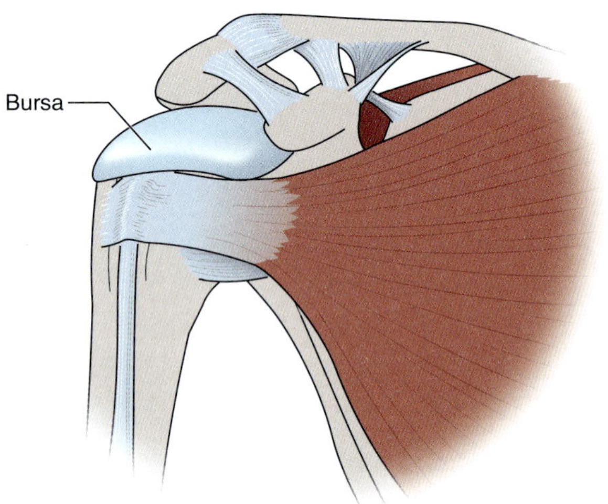

Figure 1.28 A typical bursa.

the body that covers, connects, or supports other tissues. One form of fascia, the sarcolemma of muscle, has already been discussed. A bursa (figure 1.28) is a saclike structure that contains bursa fluid and protects muscles, tendons, ligaments, and other tissues as they cross the bony prominences described earlier. Bursae provide lubricated surfaces to allow movements of muscles and tendons directly over structures, such as bone, without being worn away over time from friction. Trauma to a bursa can inflame it and create a condition known as *bursitis*. This trauma can be from an infection, pressure, or a direct blow to the area. Bursae are identified by their position in the body. There are subfascial bursae, located beneath the fascia; subcutaneous bursae, beneath the skin; submuscular bursae, beneath and between muscles; and subtendinous bursae, beneath and between tendons.

Motor Unit

We have now discussed the bones, the ligaments that connect the bones to form articulations (joints), the muscles that cross joints and create movement, the nerves that innervate the muscles, and the blood vessels that supply all these structures—and all are considered essential to movement. We now look more closely at the motor unit. Some textbooks examine the physiology of nerve impulses that generate muscular contractions and cause movement. In this book, we concentrate on the anatomy of the structures that actually produce movement.

A **motor unit** is defined as a motor nerve and all the muscle fibers it supplies (figures 1.29 and 1.30). The structural parts of the motor unit are the motor nerve and the muscle fiber. All the motor units together are referred to as the body's *neuromuscular system*.

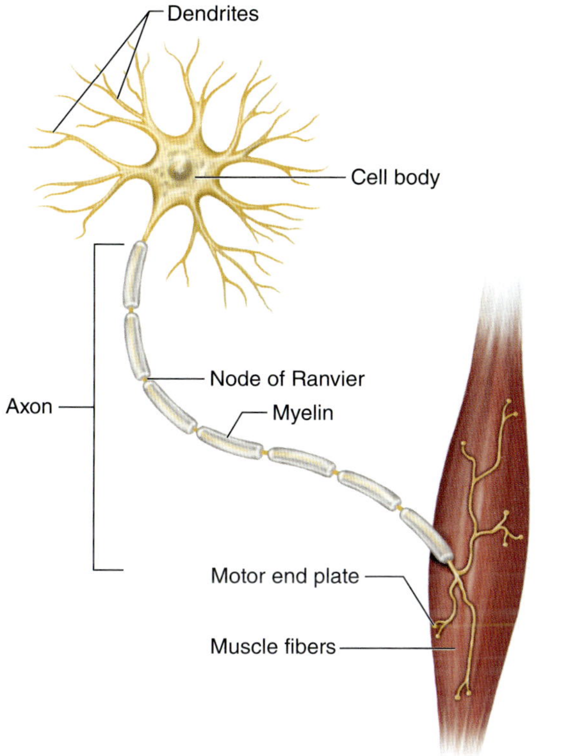

Figure 1.29 The motor unit, made up of a motor neuron and muscle fibers.

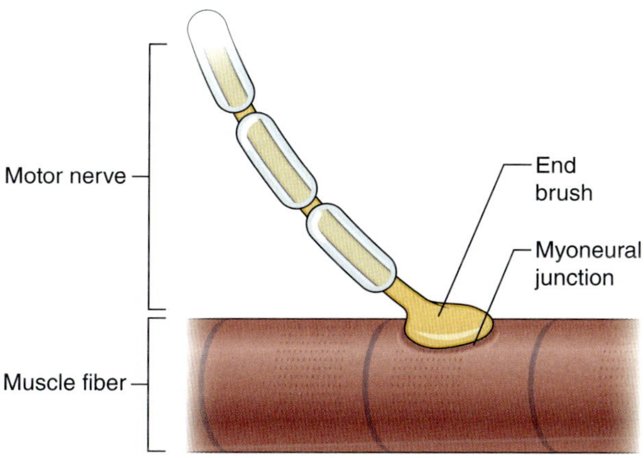

Figure 1.30 Schematic of a motor unit.

The space between the end brushes of the motor end plate and the muscle fibers is known as the *myoneural junction* or *neuromuscular junction*. Although the end brushes never actually come in direct contact with the muscle fibers, it is at the myoneural junction that the synapse (connection) occurs. At the synapse (figure 1.31), the end brushes of the motor nerve's axon release a

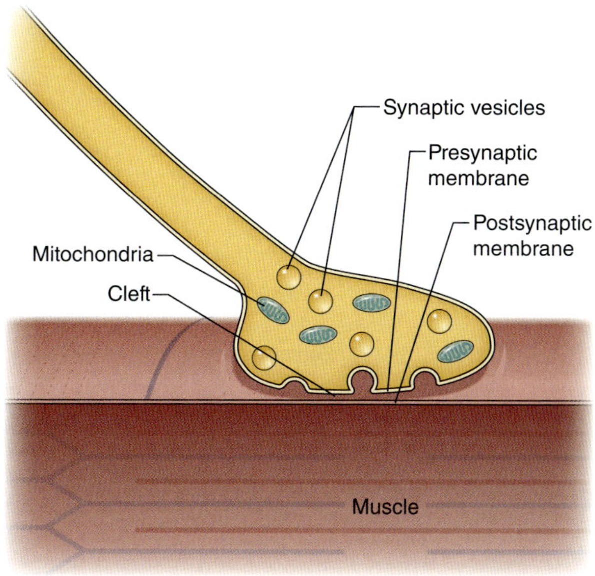

Figure 1.31 A neuromuscular synapse and its related structures.

chemical known as *acetylcholine*. This chemical stimulates the outer covering (sarcolemma) of the muscle fiber, which allows the impulse to continue to the muscle fiber and causes it to contract if the impulse is great enough to reach the muscle fiber's threshold. The space between the axon's end brush and the muscle's sarcolemma, where acetylcholine passes from the nerve to the muscle tissue, is known as the *synaptic cleft*. In summary, a nerve impulse travels from the spinal cord (or brain) to a dendrite of a spinal nerve, from the dendrite to the nerve's cell body, and from the cell body over the nerve's axon to the axon's end brushes (motor end plate), where a chemical is released at the synapse.

Now that we have covered the anatomical structures responsible for transmitting the nerve impulse to the muscle, let us look at the component that makes up the other half of the motor unit: the muscle. As the impulse passes from nerve to muscle, calcium is released from the sarcoplasmic reticulum and transverse tubules (figure 1.32) within the muscle fibers, two structures closely allied with the actin and myosin protein filaments. The calcium release causes the myosin cross-bridges to wiggle or swivel in such a fashion that they contact the actin filaments surrounding them and cause the actin to move

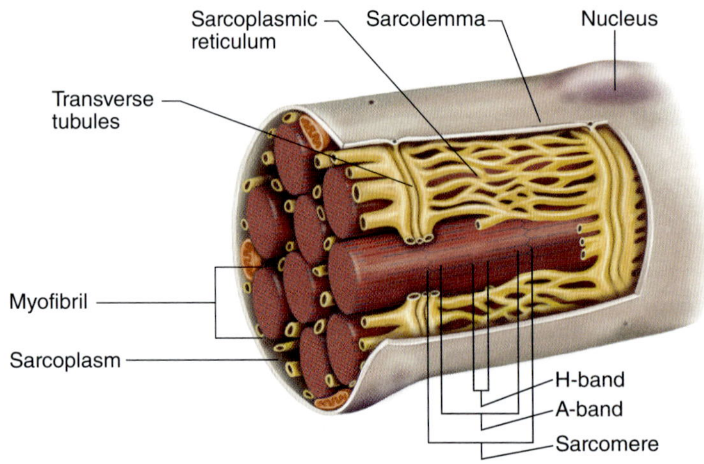

Figure 1.32 Sarcoplasmic reticulum and transverse tubules.

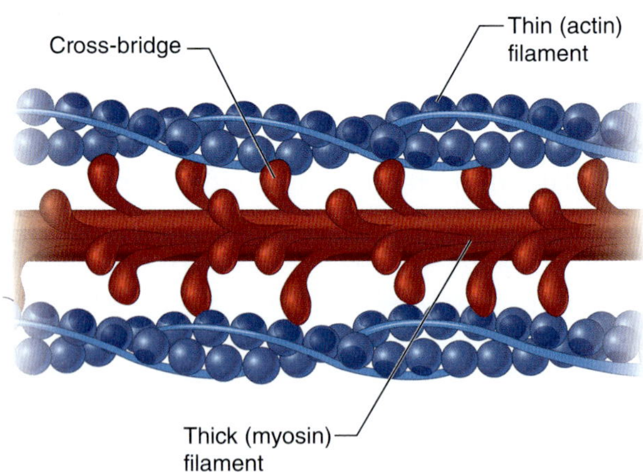

Figure 1.33 Cross-bridge formation and movement.

toward the center of the sarcomere (figure 1.33). This chemical communication within muscle fibers is responsible for the action known as the *sliding filament mechanism*.

In review, the impulse from the motor nerve crosses the synapse at the myoneural junction and activates the release of calcium through the sarcoplasmic reticulum and transverse tubules, causing the cross-bridges of the myosin protein filament to contact the actin protein filaments and produce movement of the actin filaments toward the center of the sarcomere, thus shortening the sarcomere.

Motor units differ widely in the number of muscle fibers innervated by one motor nerve. The ratio of muscle fibers per motor nerve can range from as low as 10 muscle fibers to as high as 2,000 muscle fibers per motor nerve. The lower the ratio of fibers per nerve, the more motor units are required to innervate all the fibers in a muscle. This is the case in muscles required to perform fine movements, such as those in the hand or the eye. By comparison, the biceps muscle performs elbow flexion and forearm supination, which are definitely not considered fine movements, and therefore has a very high ratio of muscle fibers per motor nerve.

If a stimulus from a nerve is intense enough to reach the threshold of a muscle fiber, all muscle fibers innervated by that nerve contract completely. There is no such thing as a partial contraction of a muscle fiber. This is referred to as the **all-or-none law of muscle contraction**. Depending on the effort required (e.g., lifting a piece of paper versus a 50 lb [23 kg] weight), various degrees of muscle contraction (gradation of strength) are called on to perform the activity. The gradation of a muscular contraction depends on two major factors: (1) the number of motor units recruited and (2) the frequency at which they are stimulated. As the force required increases, more motor units are called into action. Additionally, they are stimulated more frequently. If impulses are sent rapidly enough to muscle fibers that they contract before totally relaxing from the previous contraction, a greater force of contraction can occur (up to a point). Once the muscle is receiving impulses at such a rate that it cannot relax, it reaches a state of continuous contraction known as *tetanus* (figure 1.34). The application of the term **tetanus** to this state of continuous contraction—the result of physical effort—should not be confused with another usage of this term, an infectious disease that can cause involuntary muscular contractions.

Courses in biology, human physiology, exercise physiology, kinesiology, biomechanics, and other fields examine the generation of a nerve impulse and the resulting contraction of muscle in detail. The preceding material should be considered an introductory overview and by no means a detailed analysis of the neuromuscular system and its motor units.

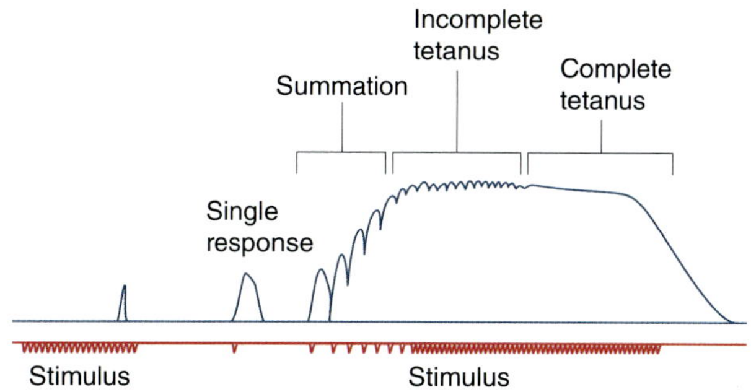

Figure 1.34 Muscular force increases with frequency of impulses.

Visit the web resource to complete the activities for this chapter.

CHAPTER 2

Movement

Now that you understand the structures involved in movement (bones, ligaments, and muscles) and the terms used to describe them, let us look at the universal language that describes the movements performed by these structures.

When we describe a human movement, there is an anatomical point that is universally accepted as being the position all movements start from: the anatomical position. In this position, all joints are considered to be in a neutral position, or at 0°, with no movement having yet occurred. Occasionally, you might also hear the term fundamental position. Carefully note that the only difference between the two positions is the position of the hands (figure 2.1). The anatomical position is preferred to the fundamental position for any discussion of human movement because the hand position in the

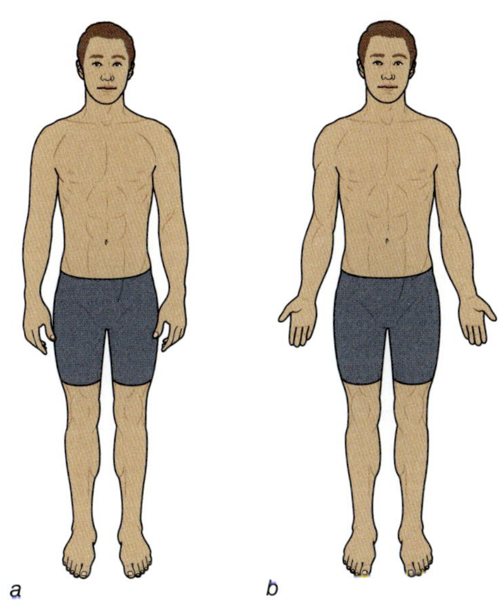

Figure 2.1 *(a)* Fundamental and *(b)* anatomical positions.

fundamental position makes certain upper-extremity movements impossible. In the following sections, the description of any movement starts from the anatomical position.

Anatomical Locations and Directional Terms

Several terms are considered universal for discussing the spatial relationship between one anatomical structure and another. The term *superior* refers to something that is above, or higher than, another structure (e.g., your head is superior to your chest). The opposite term, *inferior*, means something is below, or lower than, another structure (e.g., your chest is inferior to your head). *Lateral* refers to something farther away from the midline of the body than another structure (e.g., your arms are lateral to your spinal column). *Medial* means a structure is closer to the midline of your body than another structure (e.g., your nose is medial to your ears). *Anterior* refers to a structure that is in front of another structure (e.g., your abdomen is anterior to your spinal column). *Posterior* refers to a structure that is behind another structure (e.g., your spinal column is posterior to your abdomen).

The terms *proximal* (close) and *distal* (far) are usually used in reference to structures of the extremities (arms and legs). Proximal means closer to the trunk, and distal means farther from the trunk (e.g., your knee is proximal to your ankle, and your hand is distal to your wrist). The term *dorsal* indicates the top side of an animal, such as the dorsal fin on the top of a fish, or the posterior of the human body. (The dorsal aspect of your hand is commonly called the *back of your hand.*) The term *volar* refers to the underside, or bottom aspect, of a structure. (The volar aspect of your wrist or hand is also referred to as the *palmar aspect*, whereas the volar aspect, or sole, of the foot is referred to as the *plantar aspect.*)

Planes and Axes

Human movement that takes place from the starting (anatomical) position is described as taking place in a plane (a flat surface) about an axis (a straight line around which an object rotates). Muscles create movements of body segments in one or more of the three planes that divide the body into different parts. These three planes are perpendicular (at right angles) to each other (figure 2.2). The sagittal plane (also known as the *anteroposterior plane*) passes from the front

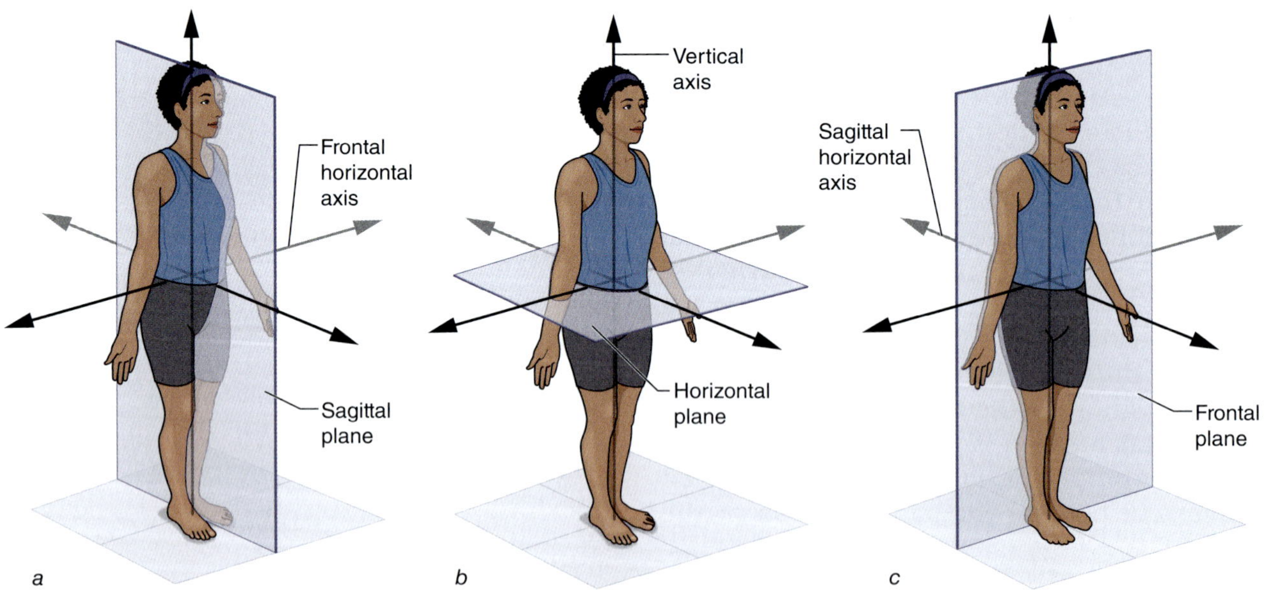

Figure 2.2 The cardinal planes and axes: *(a)* sagittal, *(b)* horizontal, and *(c)* frontal planes.

through the back of the body, creating a left side and a right side of the body. There could be any number of sagittal planes; however, there is only one cardinal sagittal plane. The term *cardinal* refers to the one plane that divides the body into equal segments, with exactly one half of the body on either side of the cardinal plane. Therefore, the cardinal sagittal plane divides the body into two equal halves on the left and right. The term *cardinal plane* appears in some texts as the *principal plane*. The terms are interchangeable.

The horizontal plane (also known as the *transverse plane*) passes through the body horizontally to create top and bottom segments of the body. There could be any number of horizontal planes, but there is only one cardinal horizontal plane, which divides the body into equal top and bottom halves.

The frontal plane (also known as the *lateral plane*) passes from one side of the body to the other, creating a front side and a back side of the body. Again, there could be any number of frontal planes, but there is only one cardinal frontal plane, which divides the body into equal front and back halves.

The point at the intersection of all three cardinal planes is the body's center of gravity. When all segments of the body are combined and the body is considered one solid structure in the anatomical position, the center of gravity lies approximately in the low-back area of the spinal column. As body parts move from the anatomical position or as weight shifts through weight gain or loss or by carrying loads, the center of gravity also shifts. No matter what the body's position or weight distribution, however, half of the weight of the body (and its load) will always be to the left and right, in the front and back, and above and below the center of gravity. The center of gravity of the body constantly changes with each movement, each change in weight distribution, or both.

Earlier, we defined an axis as a straight line around which an object rotates. In the human body, we picture joints as axes, and bones as the objects that rotate about them in a plane perpendicular to the axis. There are three main axes, and rotation is described as occurring in a plane about the axis that is perpendicular to the plane (figure 2.2). The sagittal plane rotates about a frontal horizontal axis (figure 2.2*a*).

Hands On

The knee joint is a frontal horizontal axis, and the lower leg is the object that moves in the sagittal plane when you bend your knee.

The horizontal plane rotates about a vertical (longitudinal) axis (figure 2.2*b*).

Hands On

As you turn your head to the left and right as if to silently say no, your head rotates in a horizontal plane about the vertical axis created by your spinal column.

The frontal plane rotates about the sagittal horizontal axis (figure 2.2*c*).

Hands On

When you raise your arm to the side, your shoulder joint is the sagittal horizontal axis, and your arm is the object moving in the frontal plane.

For a summary of the relationship between anatomical planes and associated axes, see table 2.1.

Fundamental Movements

Again, remember that movement takes place in a plane about an axis. There are three planes and three axes, with two fundamental movements possible in each plane. In the sagittal plane, the fundamental movements are known as *flexion* and *extension*. Flexion is defined as a reduction of the angle formed by the bones of the joint (figure 2.3). In flexion of the elbow joint, the angle between the forearm and upper arm decreases. Extension is defined as an increase in the joint angle (figure 2.4). Returning a joint in

Table 2.1 Planes, Axes, and Fundamental Movements

Plane	Axis	Movements
Sagittal (anteroposterior)	Frontal horizontal	Flexion and extension
Frontal (lateral)	Sagittal horizontal	Abduction and adduction
Horizontal (transverse)	Vertical	Rotation

Figure 2.3 Flexion at the shoulder, hip, and knee.

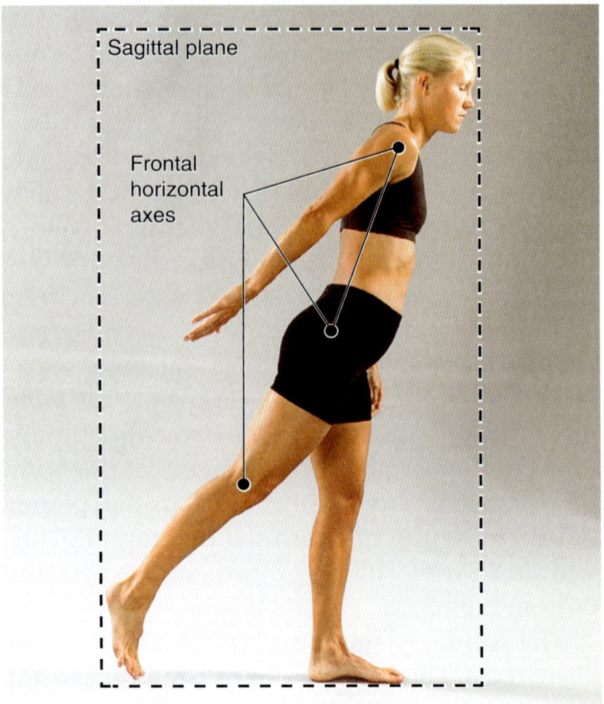

Figure 2.4 Extension at the shoulder, hip, and knee.

flexion to the anatomical position is considered extension. Further extension beyond the anatomical position is referred to as *hyperextension*. At the ankle, we call these motions dorsiflexion and plantarflexion. Dorsiflexion is when the foot moves upward toward the anterior leg, and plantarflexion is when the foot moves downward. For example, when you press down on the gas pedal of a car, you plantarflex your ankle. Fundamental movements in the frontal plane are known as *abduction* and *adduction*. Abduction is defined as movement away from the midline of the body (figure 2.5). As you move your arm away from the side of your body in the frontal plane, you are abducting the shoulder joint. Movement toward the midline of the body, for example, returning your arm from an abducted shoulder position to the anatomical position, is defined as *adduction* (figure 2.6). At the ankle, we call these motions inversion and eversion. Medial (inward) movement of the ankle is known as *inversion*. Lateral (outward) movement of the ankle is known as *eversion*. The fundamental movement in the horizontal plane is simply defined as *rotation* (figure 2.7). The earlier example of shaking your head "no" is a rotation of the head. For describing movement in the upper (arm) and lower (leg) extremities, the terms *external rotation* and *internal rotation* are often used (figure 2.8). When the anterior (front) surface of the arm or leg rotates laterally (away from the midline of the body), the movement is defined as *external rotation* (or lateral rotation). When the anterior surface of the arm or leg rotates medially (toward the midline of the body), the movement is defined as *internal rotation* (or medial rotation). Two terms refer to actions of the forearm and ankle. The term *pronation* refers to the turning of the forearm toward the body, resulting in the volar, or palmar, surface of the hand facing the body, or, if the elbow is flexed, palm down. At the ankle, pronation involves combined motions of dorsiflexion, eversion, and external rotation. *Supination*, the reverse of pronation, refers to turning the forearm outward and palm upward from the pronated position. At the ankle, supination involves combined motions of plantarflexion, inversion, and internal rotation.

Figure 2.5 Abduction at the shoulder and hip.

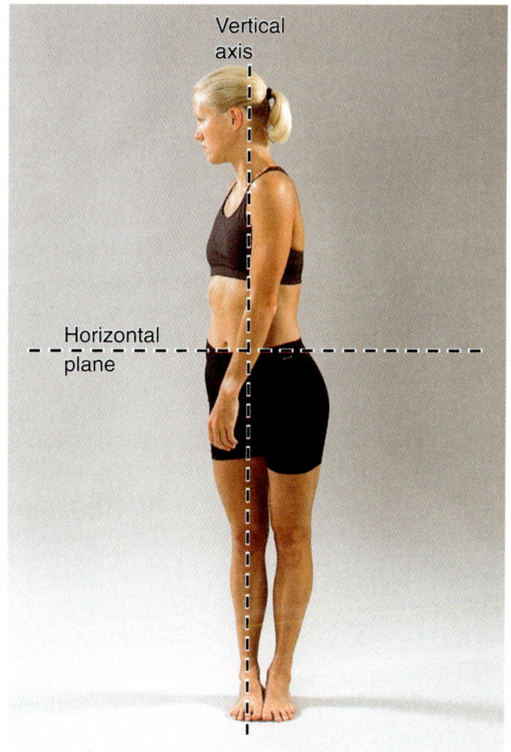

Figure 2.7 Rotation as a twisting movement of the spine.

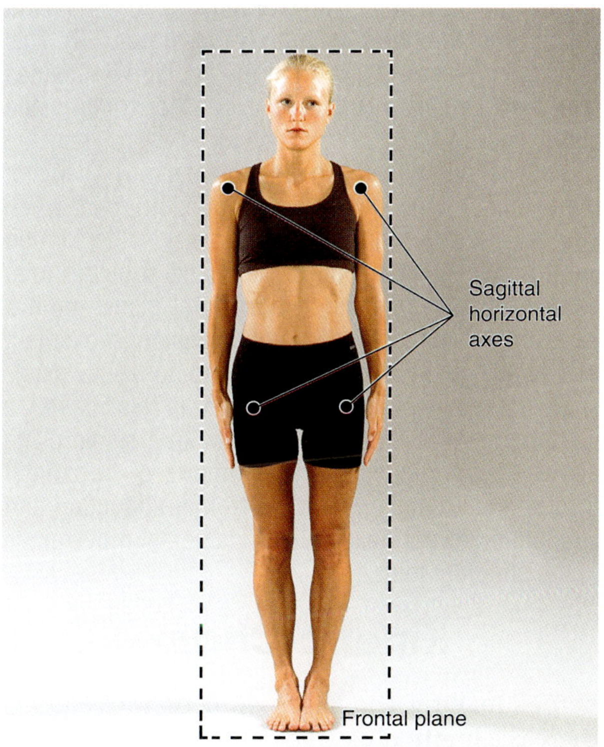

Figure 2.6 Adduction at the shoulder and hip as a return from the abducted position in figure 2.5.

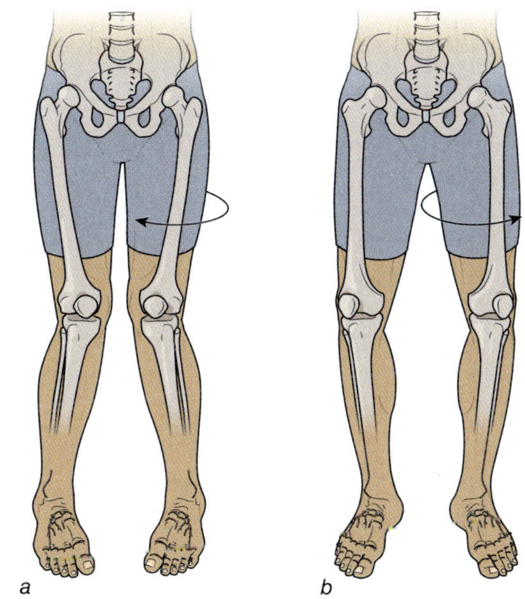

Figure 2.8 *(a)* Internal and *(b)* external rotation of the lower extremity.

FOCUS ON

The Subtalar Joint: A Triplanar Joint in Action

The subtalar joint is the articulation between the talus and the calcaneus. Due to the unique shapes of the articulating surfaces, the subtalar joint moves around an oblique axis that runs diagonally through the three planes. This makes the subtalar joint a triplanar joint despite it being a uniaxial joint. The oblique axis allows the motions of pronation and supination to occur. Pronation requires the combination of dorsiflexion, abduction, and eversion. Supination requires the combination of plantarflexion, adduction, and inversion.

Figure 2.9 Circumduction at the shoulder.

When a body segment (e.g., a forearm) moves through an arc of motion (e.g., at the elbow), the movement takes place in a sagittal plane about a horizontal (frontal) axis, and that arc of motion is referred to as that joint's *range of motion* (ROM). Every diarthrodial joint in the human body that is capable of movement has a ROM. A joint's ROM varies, depending on several factors, such as bone structure, ligamentous attachments, muscle health, trauma, and external restrictions (e.g., casts and bracing), but all unimpaired joints have a typical ROM.

Joints capable of creating movement in two (biaxial) or three (triaxial) planes are also capable of another movement, circumduction, which, because it combines two or more fundamental movements, is *not* considered a fundamental movement of any joint. When movement occurs in two or three planes in a sequential order, the joint is said to be circumducting. Moving your arm at the shoulder joint in a "windmill" motion is an example of circumduction (figure 2.9).

There are two other movements of the shoulder joint that some may consider as fundamental movements, but, by the definition of fundamental movements presented earlier in this chapter, they are not: horizontal flexion and extension, and horizontal abduction and adduction. The following "Hands On" example should demonstrate why.

Hands On

Stand in the fundamental position. Raise your arm in front of you to shoulder height (see figure 2.3). This movement is defined as *shoulder flexion* (movement in a sagittal plane about a frontal and horizontal axis).

From this position, move your arm to a position aligned with the side of your body (see figure 2.5). This movement places your arm in the position defined as *shoulder abduction* (movement in a horizontal plane about a vertical axis).

You just performed two fundamental movements of the shoulder joint (flexion and abduction), and, when you reversed the initial movements, you performed another two fundamental movements: shoulder joint adduction (movement in a horizontal plane about a vertical axis) and shoulder joint extension (movement in a sagittal plane about a frontal and horizontal axis).

By the definition presented earlier in this chapter, a fundamental movement takes place in a single plane about a single axis. The movements of horizontal flexion and extension, and horizontal abduction and adduction, do not qualify as fundamental movements of the shoulder joint.

Kinetic Chain

Before leaving the discussion of movement, there is a method used to describe movement as the kinetic chain (KC). Human movement can be described as an open kinetic chain (OKC) or as a closed kinetic chain (CKC). These forms of human movement are studied further in courses such as kinesiology, biomechanics, exercise physiology, and in other areas of study (e.g., exercise therapy and rehabilitation). For the purpose of this text, a brief explanation and an example of OKC and CKC terms should familiarize one who is studying human movement.

In a CKC exercise, the body part is farthest away from the rest of the body and attached to an object (e.g., floor, wall, or ground). This form of exercise typically involves more than movement at a single joint. Think of the classic push-up. The hands are located farthest away from the rest of the body and are attached to the floor. Among the joints involved are the elbow, the shoulder, and other joints. OKC exercise emphasizes movement of the body part farthest away from the rest of the body, and the body part is *not* attached to a fixed object. Most OKC exercises involve a single joint movement. As an example, picture the classic biceps curl. Note that the body part farthest away from the body is not fixed to an immovable object, and the movement occurs about a single joint: the elbow.

POP QUIZ

Which of these activities best demonstrates an example of an OKC exercise?

Visit the web resource to complete the activities for this chapter.

PART II

Upper Extremity

CHAPTER 3

The Shoulder

Any discussion of the shoulder must start with the fact that the shoulder is actually two distinct anatomical structures: the shoulder girdle and the shoulder joint. The shoulder girdle consists of the clavicle and scapula bones, and the shoulder joint is formed by the scapula and the humerus bones. The primary function of the shoulder girdle is to position itself to accommodate movements of the shoulder joint.

Bones of the Shoulder Girdle

Two bones make up the structure known as the *shoulder girdle*: the clavicle and the scapula (figure 3.1). The clavicle is a long, slender, S-shaped bone that attaches to the sternum (breastbone) at the medial end and to the scapula at the lateral end. The clavicle is often referred to as the *collarbone*. It is the only bony attachment that the upper extremity has to the trunk. Because of its shape and the fact that it is held in place at either end by strong, unyielding ligaments and has little protection from external forces (with the exception of skin), the clavicle is an often-fractured bone.

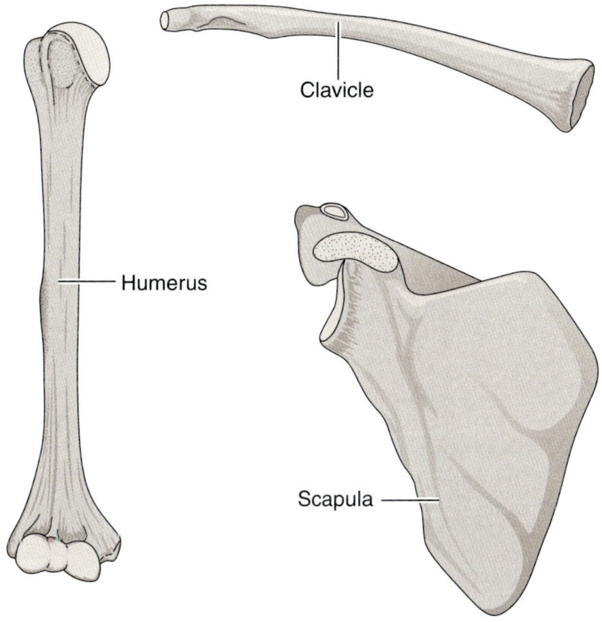

Figure 3.1 Anterior view of the bones of the shoulder.

The lateral end of the clavicle is referred to as the *acromial end*, and the medial end is referred to as the *sternal end* (figure 3.2). Prominent bony landmarks are observed on the superior and inferior views (figure 3.3) and include the deltoid tubercle, the conoid tubercle, the trapezoid line, the costal tuberosity, and the subclavian groove. These structures are important as places of attachment for soft tissue.

The scapula is the large, triangular, winglike bone in the upper posterior portion of the trunk. This bone, also referred to as the *shoulder blade*, is considered a bone of both the shoulder girdle and the shoulder joint. Figure 3.4 illustrates the many

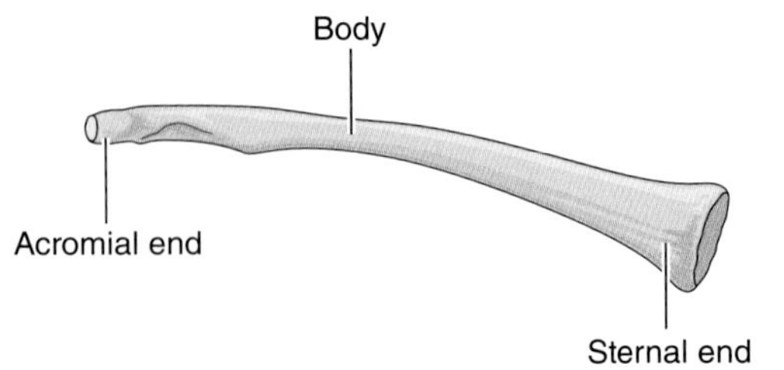

Figure 3.2 Sternal and acromial ends of the clavicle.

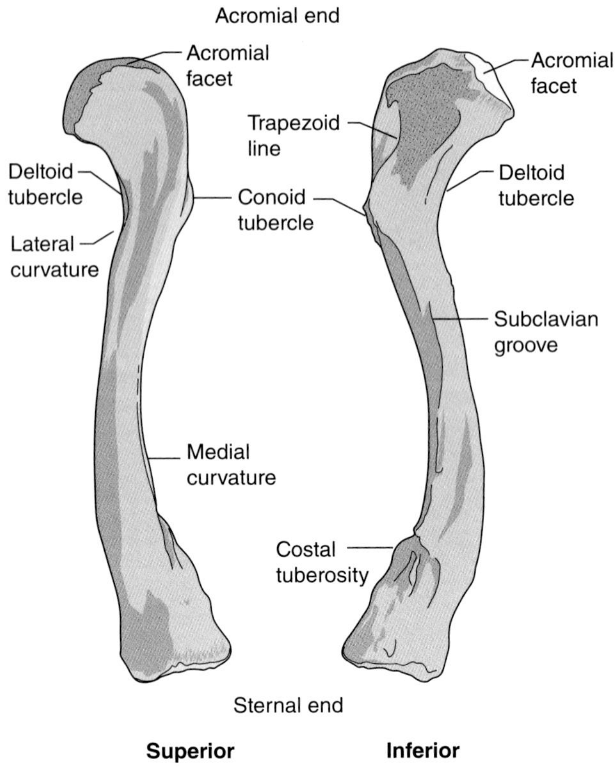

Figure 3.3 Prominent bony landmarks of the clavicle.

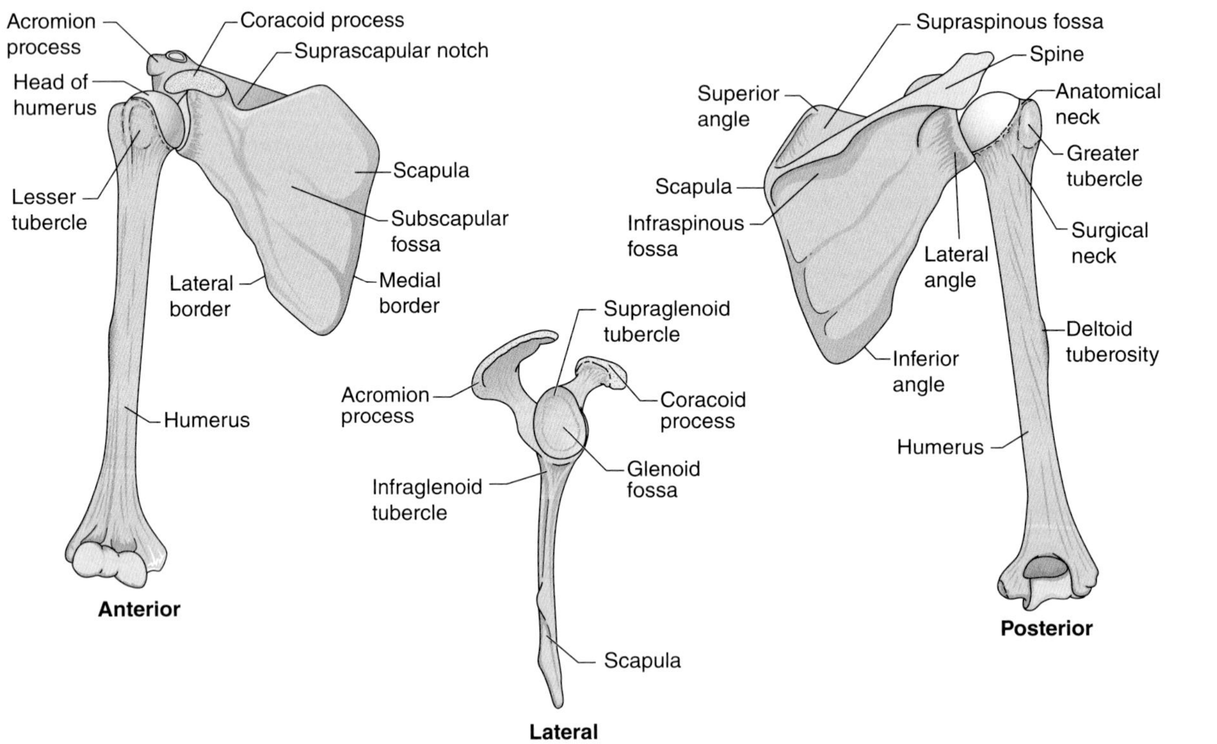

Figure 3.4 Landmarks of the shoulder bones.

bony prominences of the scapula, including the lateral and medial borders and the inferior angle at the junction of the two borders.

Hands On

You can locate the lateral border, medial border, and inferior angle of the scapula on a partner (figure 3.5).

Additionally, the superior border at the medial end of the scapula becomes the superior angle, and at its lateral end has a notch known as the *scapular notch*. Two large bony prominences at the superior lateral portion of the scapula are known as the *coracoid process* (anterior) and the *acromion process* (posterior).

Hands On

Either on yourself or a partner, palpate (touch) both the coracoid and acromion processes (figures 3.6 and 3.7).

The acromion process is the lateral expansion of a ridge of bone approximately one-third of the way down the posterior aspect of the scapula. This ridge of bone is known as the *spine of the scapula*.

Hands On

Apply pressure with your index and middle fingers along the upper third of your partner's scapula, and you will feel this spine (figure 3.8).

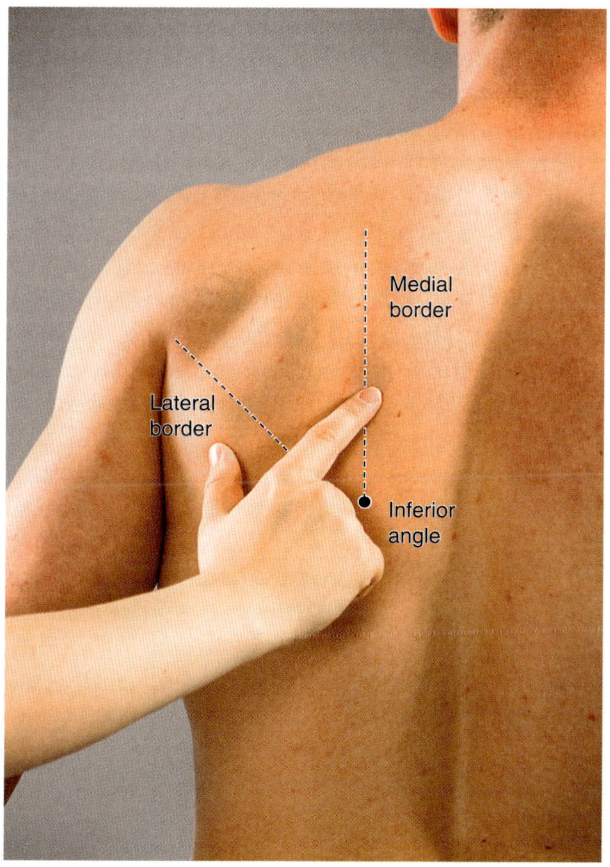

Figure 3.5 Locating the lateral border, medial border, and inferior angle of the scapula.

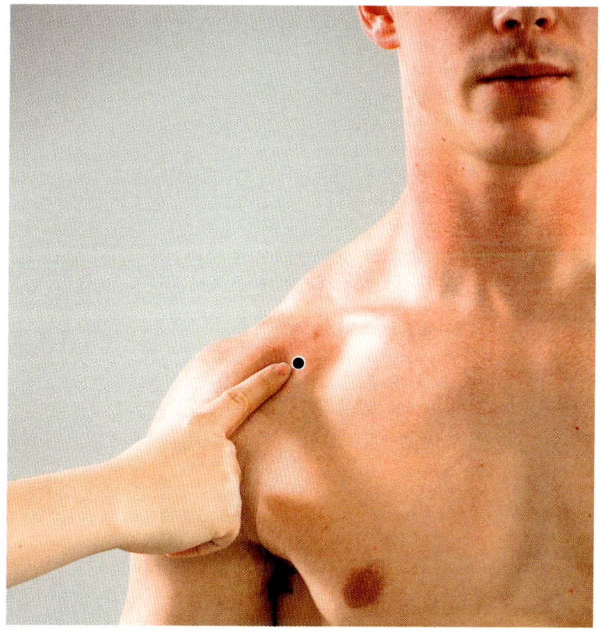

Figure 3.6 Locating the coracoid process.

Figure 3.7 Locating the acromion process.

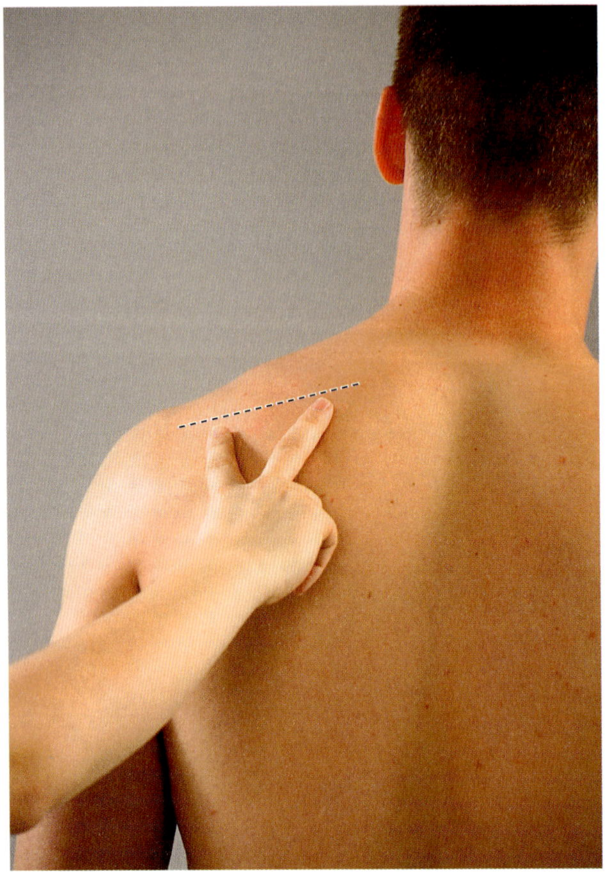

Figure 3.8 Locating the spine of the scapula.

Most laterally, the scapula forms a smooth, round, slightly depressed surface known as the *glenoid fossa*. This cavity forms the socket for the shoulder joint. Above and below the glenoid fossa are two bony prominences known, respectively, as the supraglenoid and infraglenoid tubercles. The smooth area of bone between the lateral and medial borders of the scapula on the anterior surface is known as the *subscapular fossa*. On the posterior surface of the scapula, the smooth bony surfaces above and below the spine are known, respectively, as the supraspinous fossa and infraspinous fossa (see figure 3.4).

Bones of the Shoulder Joint

The shoulder joint is the articulation between the scapula and the humerus (bone of the upper arm). The joint is known as the *glenohumeral (GH) joint* because of the two articulating bony surfaces. The prominent structure of the scapula

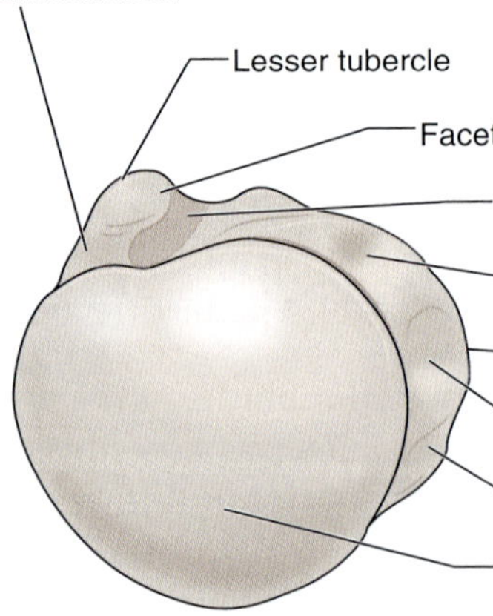

Figure 3.9 Landmarks of the proximal end of the humerus.

in terms of the shoulder joint is the anatomical area labeled the glenoid fossa. The shoulder joint is classified as a ball-and-socket joint, and the glenoid fossa, although somewhat shallow, is considered the socket of the joint.

The "ball" of the shoulder joint is the structure known as the *head of the humerus*. This chapter discusses the humerus only because it is part of the shoulder joint (the proximal end; figure 3.9). The humerus at its distal end is discussed further in the chapter on the elbow joint.

The head of the humerus is separated from the shaft of the bone by two necks. The **anatomical neck** is located between the head of the humerus and two bony prominences known as the *greater* and *lesser tubercles*, sometimes referred to as *tuberosities*. The **surgical neck** (see figure 3.4) is actually the upper portion of the shaft of the humerus. A groove known as the *intertubercular* (bicipital) *groove* is created by the greater and lesser tubercles. Atop both the lesser and greater tubercles appear four flat surfaces known as *facets*. The lesser tubercle has one facet, and the greater tubercle has three: a proximal, middle, and distal facet. Approximately halfway down the shaft of the humerus, on the lateral surface, is a bony prominence known as the *deltoid tuberosity* (see figure 3.4).

Hands On

Apply pressure to either your own or your partner's humerus, approximately halfway between the head and the distal end at the elbow, on the lateral aspect; you will feel the bump of the deltoid tuberosity (figure 3.10).

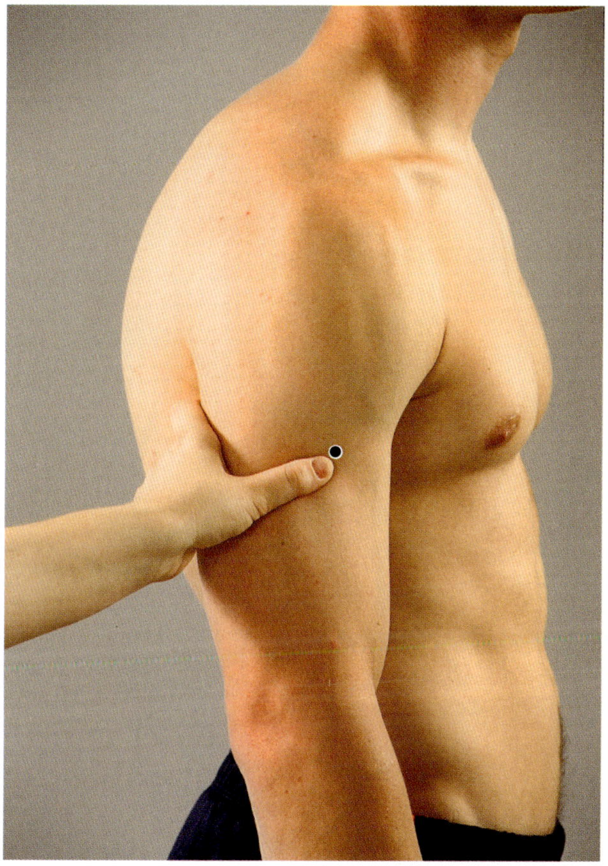

Figure 3.10 Locating the deltoid tuberosity.

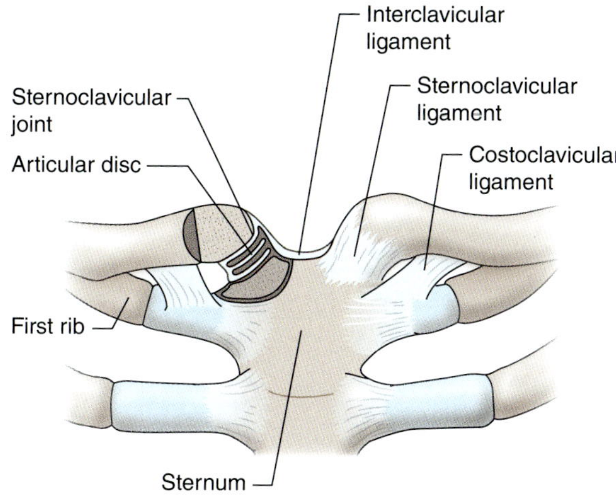

Figure 3.11 The sternoclavicular, costoclavicular, and interclavicular ligaments and the articular disc.

Joints and Ligaments of the Shoulder Girdle

The shoulder girdle has two joints, one at either end of the clavicle, known as the *acromioclavicular* (AC) and *sternoclavicular* (SC) *joints*. Another shoulder joint, known as the *scapulothoracic joint*, is the joint between the anterior side of the scapula and the posterior aspect of the thorax. The scapulothoracic joint is often described as a functional joint and not a true anatomical joint because there are no ligaments that attach the scapula to the thorax (chest). Movement in the SC joint is slight in all directions and of a gliding, rotational type. The joint derives its stability from its bony arrangement, because the sternal end of the clavicle fits into the clavicular notch of the manubrium of the sternum (see chapter 9), and from the ligamentous arrangement that ties the clavicle and sternum together.

Three primary ligaments are responsible for the SC articulation (figure 3.11). The SC ligament, with anterior, superior, and posterior fibers, and two other ligaments, help to stabilize the SC articulation: the costoclavicular ligament, which secures the sternal end of the clavicle to the first rib, and the interclavicular ligament, which secures the sternal ends of both clavicles into the clavicular notch of the manubrium of the sternum. Also, an articular disc is present between the sternal end of the clavicle and the clavicular notch of the manubrium of the sternum.

Hands On

Place your index finger in the space between each of your clavicles; you will feel the clavicular notch formed between both SC joints (figure 3.12).

The joint between the lateral (acromial) end of the clavicle and the scapula is divided into two areas: the AC joint and the coracoclavicular joint. Simply described, the lateral end of the clavicle articulates with both the acromion process and the coracoid process of the scapula (figure 3.13).

Hands On

Palpate the area at the lateral end of the clavicle (figure 3.14). You should both feel and see the bump that is the AC joint.

The AC joint is the articulation between the acromion process of the scapula and the acromial end of the clavicle. There is a slight gliding type of movement between the two bones of this joint when elevation and depression of the acromial

Figure 3.12 Locating the sternoclavicular joint.

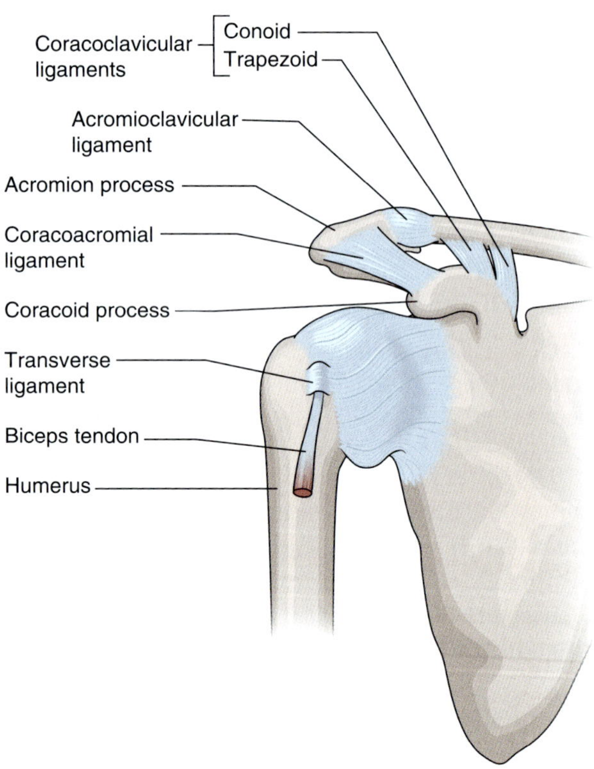

Figure 3.13 Anterior view of the acromioclavicular and coracoclavicular ligaments.

Figure 3.14 Locating the acromioclavicular joint.

end of the clavicle and the acromion process of the scapula take place. The AC ligament functions as the joint capsule, tying together and totally surrounding the lateral end of the clavicle and the acromion process of the scapula.

The other shoulder girdle joint, the coracoclavicular joint, is sometimes considered a component of the AC joint and sometimes treated as a separate joint. The joint is the articulation between the lateral (acromial) end of the clavicle and the coracoid process of the scapula. Two ligaments run between the coracoid process of the scapula and the inferior surface of the clavicle. The two ligaments, the conoid and the trapezoid, are often referred to as a single ligament, the *coracoclavicular ligament*. Although some people do not consider the coracoclavicular joint a true joint, slight movement occurs in all directions in the articulation. The trapezoid ligament is the more lateral component of the coracoclavicular ligament and runs from the superior aspect of the coracoid process of the scapula to the anterior inferior aspect of the clavicle. It opposes forward, upward, and lateral movement of the lateral aspect of the clavicle. The conoid ligament is the medial component of the coracoclavicular ligament and runs from the superior aspect of

FOCUS ON

Shoulder Separation

A sprain (partial or complete tearing) of the AC and coracoclavicular ligaments results in a visible gap between the clavicle and the scapula, a classic illustration of what is known as a *shoulder separation* (figure 3.15). The separation is actually a widening of the space between the lateral end of the clavicle and the acromion process of the scapula. The weight of the upper extremity often reveals this gap when the affected shoulder girdle is compared with the unaffected one. When you hear the term *shoulder separation,* know that it more correctly refers to the AC joint of the shoulder girdle.

Falling from a tree, being tackled in a football game, being taken down to the mat in a wrestling match, or slipping on ice and falling on the shoulder can lead to spraining the AC joint. To further study the topic of sprains, consult athletic training or sports medicine texts, and talk with athletic trainers and physicians who specialize in sports medicine. Knowledge of the basic anatomy of the shoulder will enhance your ability to understand the literature and communicate with these specialists.

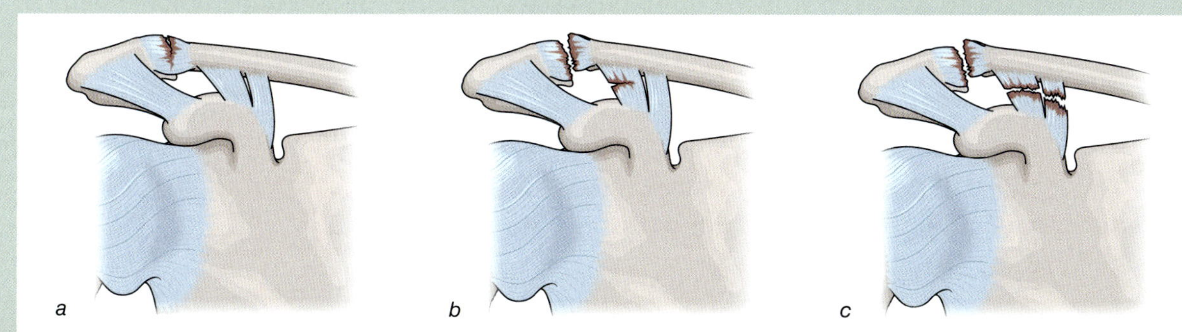

Figure 3.15 Three types of shoulder separation: *(a)* grade 1 (mild), *(b)* grade 2 (moderate), and *(c)* grade 3 (severe) sprains of the ligaments of the acromioclavicular joint as a result of trauma to the joint.

POP QUIZ

Based on this photo, it looks like this athlete could have sustained an injury to the shoulder girdle. Which of the following shoulder girdle injuries is most likely?

- **a.** Fractured clavicle
- **b.** Sprained AC joint
- **c.** Sprained SC joint

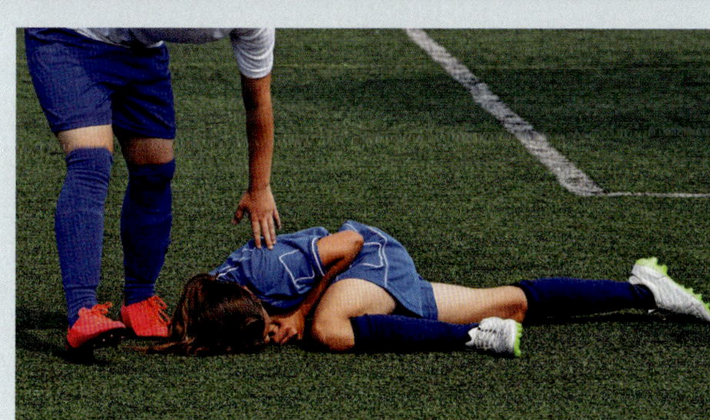

majorosl/iStock/Getty Images Plus

the coracoid process of the scapula to the posterior inferior aspect of the clavicle. It opposes backward, upward, and medial movement of the lateral aspect of the clavicle. The coracoclavicular ligament is a strong supporter of the AC ligament. Loss of these ligaments results in separation of the upper extremity from the trunk of the body.

Ligaments of the Shoulder Joint

The shoulder joint is the articulation between the head of the humerus and the glenoid fossa of the scapula, known as the *GH joint*. The ligaments of the GH joint (figure 3.16) include the capsular ligament, the GH ligament (superior, inferior, and middle sections), and the coracohumeral ligament. The capsular ligament attaches the anatomical neck of the humerus and the circumference of the glenoid of the scapula. The GH ligaments are located beneath the anterior surface of the joint capsule and reinforce the capsule. The superior GH ligament runs between the upper surface of the lesser tubercle of the humerus and the superior edge of the glenoid of the scapula. The middle GH ligament runs between the anterior surface of the lesser tubercle of the humerus and the anterior edge of the glenoid of the scapula. The inferior GH ligament runs between the lower anterior surface of the lesser tubercle of the humerus and the lower anterior edge of the glenoid of the scapula. The coracohumeral ligament runs between the anatomical neck of the humerus, near the greater tubercle, and the lateral aspect of the coracoid process of the scapula.

Other Ligaments of the Shoulder

In addition to ligaments of the shoulder joint and girdle, other ligaments of the shoulder include those specific to the scapula and the humerus.

Ligaments of the Scapula

Although we typically think of ligaments as tying bones together to form articulations, some ligaments run from one aspect of a bone to another aspect of the same bone, serving a function other than forming joints. Four such ligaments are prominent on the scapula (figure 3.17). The superior transverse scapular ligament crosses the scapular notch, converting the notch into a

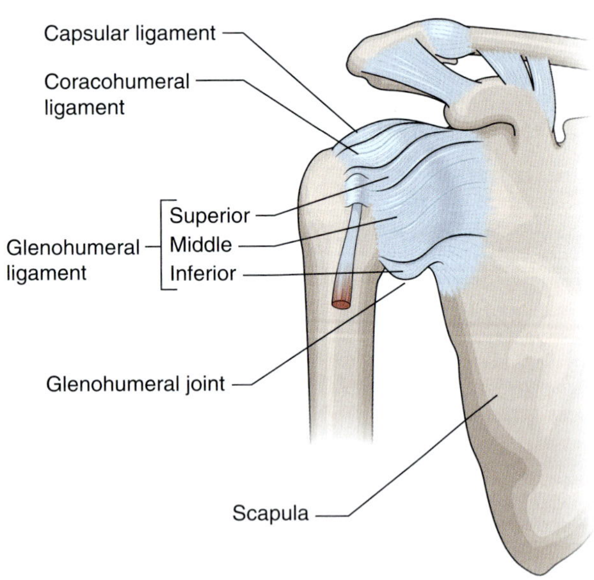

Figure 3.16 The capsular, coracohumeral, and glenohumeral ligaments.

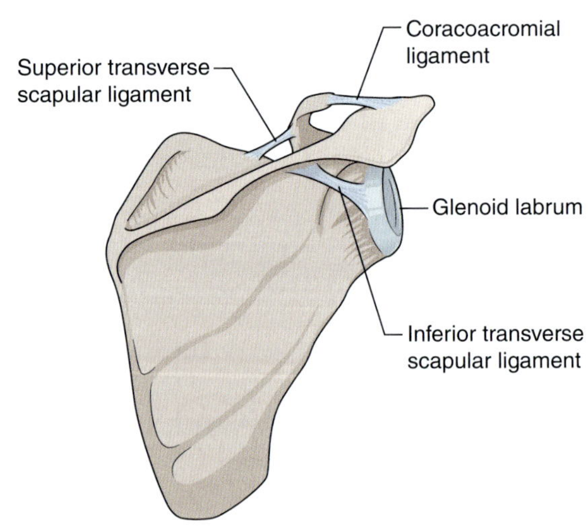

Figure 3.17 The superior and inferior transverse scapular ligaments, the coracoacromial ligament, and the glenoid labrum.

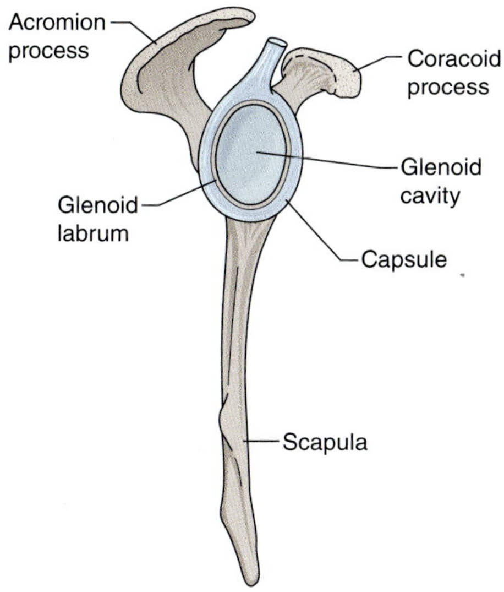

Figure 3.18 The glenoid labrum.

foramen through which the suprascapular nerve passes. The inferior transverse scapular ligament crosses from one edge to the other of the great scapular notch. This ligament forms a tunnel for the passage of the suprascapular nerve and the transverse scapular blood vessels, which, respectively, innervate (stimulate) and supply blood to the supraspinatus and infraspinatus muscles. The coracoacromial ligament crosses between the coracoid process and the acromion process of the scapula. Although a ligament of the scapula, this ligament limits the superior movement of the humeral head. The glenoid labrum (also known as the *glenoid lip*; figure 3.18) is a ligament that forms an edge around the entire circumference of the glenoid of the scapula. The glenoid labrum helps deepen the glenoid fossa for the head of the humerus to add to the stability of the GH joint.

Ligament of the Humerus

On the anterior surface of the proximal end of the humerus, two structures were discussed: the greater and lesser tubercles. Also mentioned was the intertubercular groove formed between these structures. Crossing the intertubercular groove is a ligament known as the *transverse humeral ligament* (figure 3.19). This ligament has one function: to hold the tendon of the long head of the biceps brachii muscle in the groove.

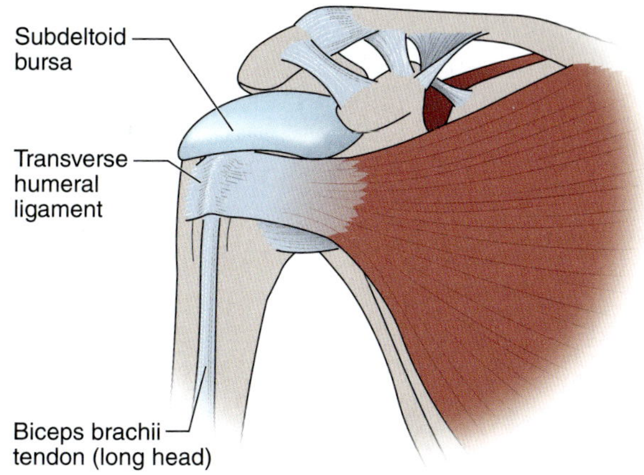

Figure 3.19 The transverse humeral ligament.

Fundamental Movements and Muscles of the Shoulder Girdle

In this chapter and all succeeding chapters, the fundamental movements involving the anatomical area under discussion are introduced first. A discussion of the muscles involved in those movements is then presented.

Movements of the Shoulder Girdle

Movements of the shoulder girdle are primarily for the purpose of accommodating shoulder joint movement through changing positions of the glenoid of the scapula. Although it was stated earlier that fundamental movements are those confined to a single plane about a single axis, the shoulder girdle is an exception. Remember that the starting position for describing all fundamental movements of any joint is the anatomical position. There are four fundamental movements of the shoulder girdle: elevation, depression, abduction, and adduction. Because of the relationships between the clavicle and scapula and between the scapula and the thorax, on which the scapula is positioned, movement exclusively in one plane about one axis is not always possible. Fundamental movements of the shoulder girdle are described relative to the direction in which the scapula moves.

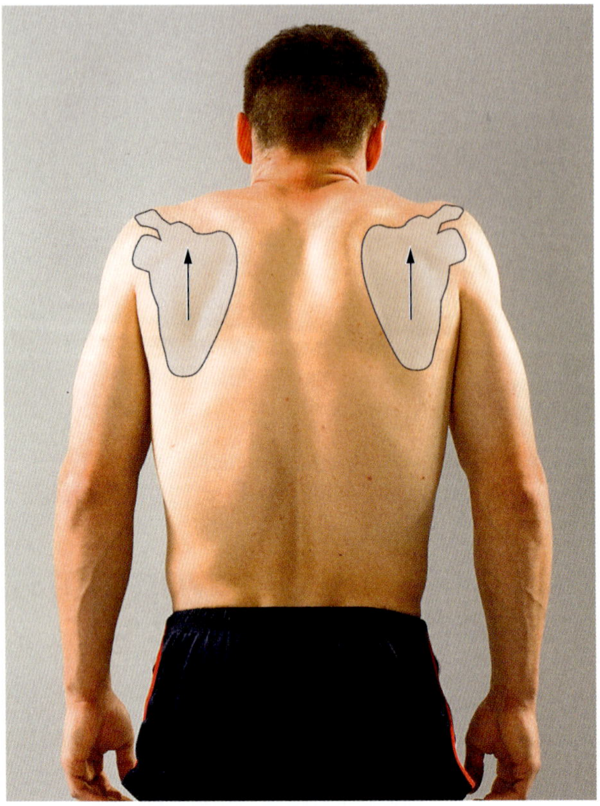

Figure 3.20 Elevation of the scapulae.

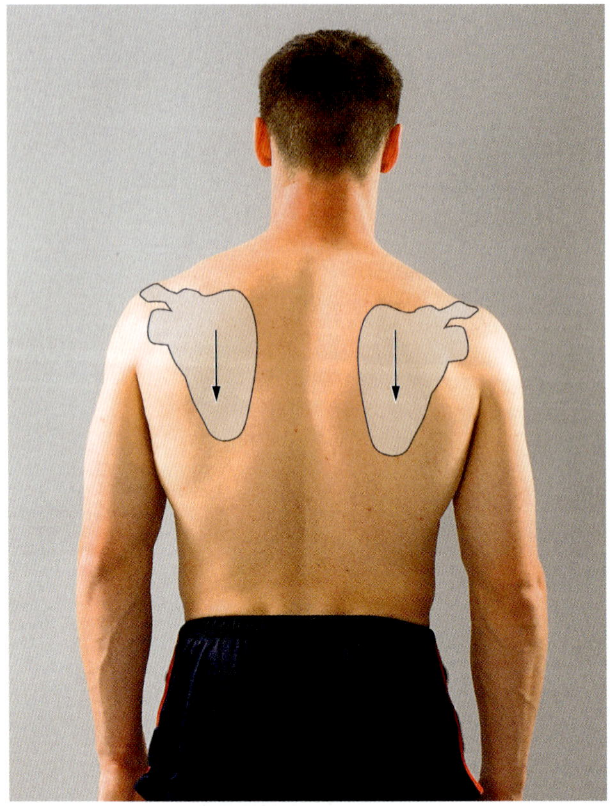

Figure 3.21 Depression of the scapulae.

Elevation of the shoulder girdle is defined as a superior (upward) movement of the scapula in the frontal plane (figure 3.20).

Depression of the shoulder girdle may be described as inferior (downward) movement of the scapula in the frontal plane (figure 3.21) but should more correctly be described as the return from elevation. Because of the anatomical starting position, depression of the shoulder girdle is not possible. Eccentric contraction (lengthening) of muscles that concentrically contracted (shortened) to cause elevation of the shoulder girdle results in depression of the shoulder girdle (return to the anatomical position).

Abduction of the shoulder girdle cannot be defined simply as movement away from the midline of the body by the scapula. Because the scapula is attached to the clavicle (at the AC joint) by ligaments and to the thorax (chest) by muscle tissue, the scapula cannot move purely laterally away from the midline of the body. The scapula must rotate about the distal end of the clavicle (at the AC joint) and tilt as it glides along the thorax (chest). Thus, abduction of the shoulder girdle is more correctly defined as upward rotation and lateral tilt of the scapula (figure 3.22). The upward rotation is defined as the upward movement of the glenoid of the scapula that accommodates shoulder joint movement. Additionally, the scapula tilts laterally as it glides along the curvature of the thorax (chest). This movement is also referred to as *protraction*. Hugging another person by placing your arms around that person requires you to protract both of your shoulder girdles.

Adduction of the shoulder girdle cannot be defined simply as movement toward the midline of the body by the scapula. Again, because the scapula is tied to the clavicle (at the AC joint) by ligaments and to the thorax (chest) by muscle tissue, the scapula cannot move purely medially toward the midline of the body. The scapula rotates about the distal end of the clavicle (at the AC joint) and tilts as it glides along the thorax (chest). Thus, adduction of the shoulder girdle is more correctly defined as downward rotation and medial tilt of the scapula (figure 3.23). The downward rotation is defined as the downward movement of the glenoid of the scapula that accommodates shoulder joint movement.

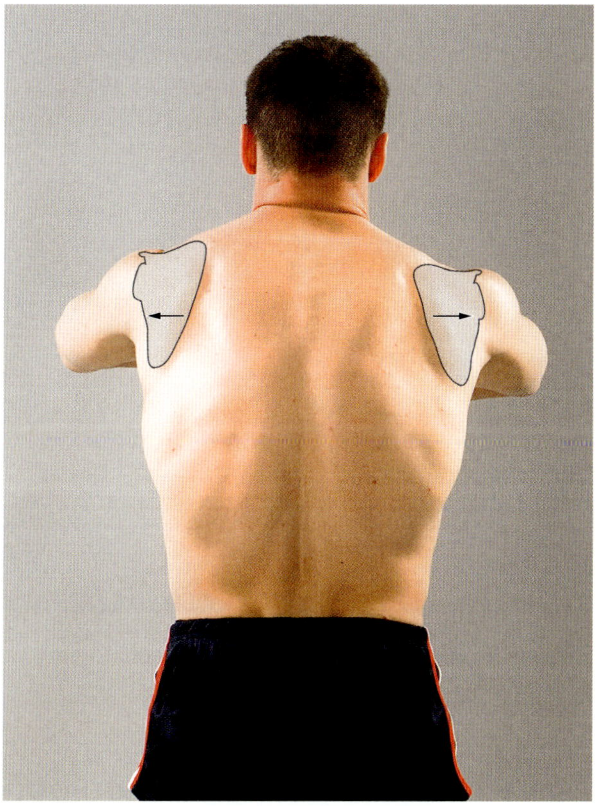

Figure 3.22 Abduction of the scapulae.

Additionally, the scapula tilts medially as it glides along the curvature of the thorax (chest). This movement is also referred to as *retraction*. Bringing your shoulders back, as when standing at attention, requires you to retract both of your shoulder girdles.

Anterior Muscles of the Shoulder Girdle

Six muscles are primarily involved in producing the fundamental movements of the shoulder girdle. Three muscles are anatomically anterior, and three are posterior to the shoulder girdle bones. The anterior muscles of the shoulder girdle include the pectoralis minor, serratus anterior, and subclavius.

- *Pectoralis minor*: The pectoralis minor originates from the third, fourth, and fifth ribs and inserts on the coracoid process of the scapula (figure 3.24). Because the ribs are the more stable attachment, contraction of the pectoralis minor results in the coracoid process of the scapula being pulled toward the ribs (shoulder girdle downward rotation, or adduction).
- *Serratus anterior:* Originating from the anterior lateral aspects of the upper nine ribs, the serratus anterior inserts on the anterior surface of the vertebral (medial) border of the scapula

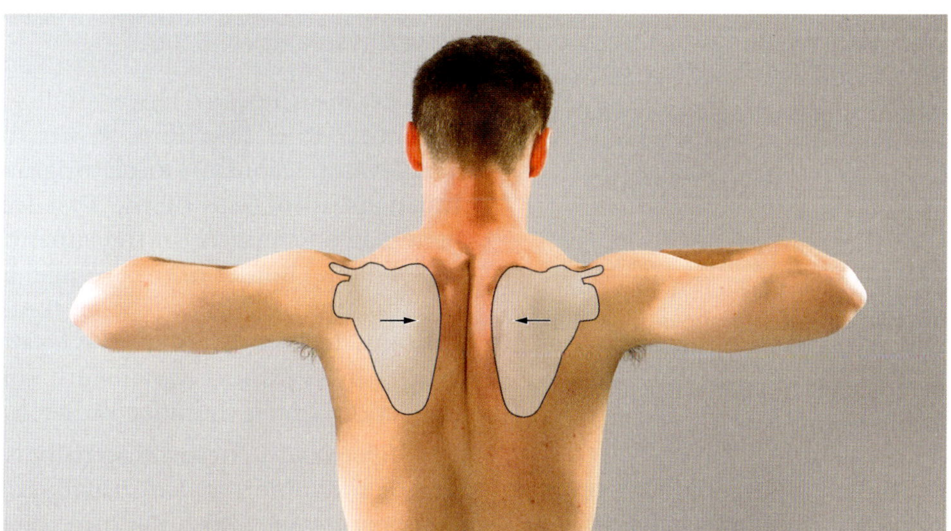

Figure 3.23 Adduction of the scapulae.

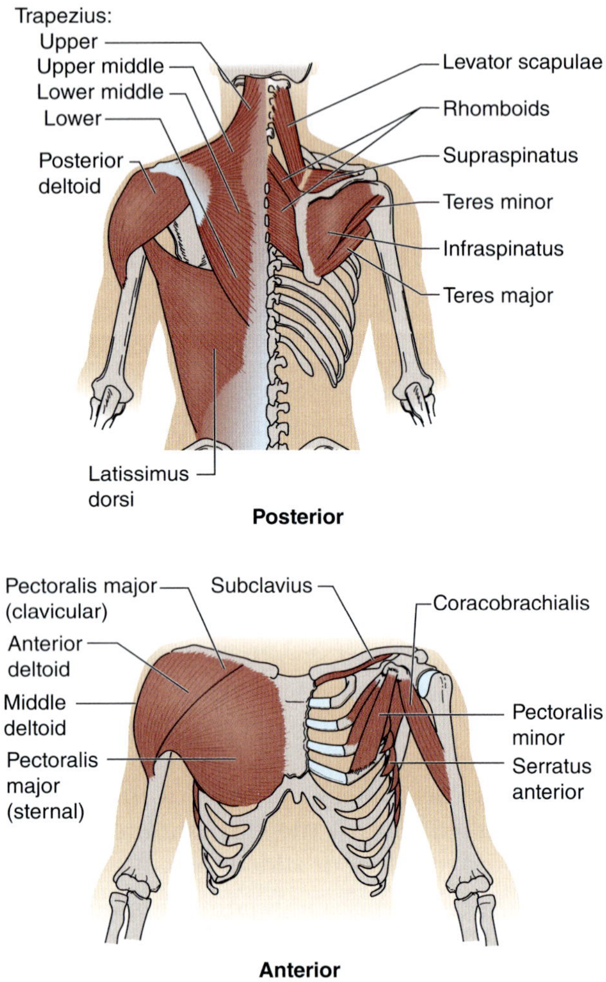

Figure 3.24 Muscles acting at the scapula and humerus.

(figure 3.24). Because the more stable attachment of the serratus anterior is on the ribs, contraction of the muscle results in abduction (upward rotation and lateral tilt) of the shoulder girdle.

Hands On

Have your partner raise their arm above their head, with the hand on the back of the head (figure 3.25). The serratus anterior should be prominently displayed laterally along the ribs.

- *Subclavius:* The subclavius originates on the first rib and inserts on the subclavian groove of the clavicle. Its main function is to assist the ligaments of the SC joint in providing stability to the joint (see figure 3.24).

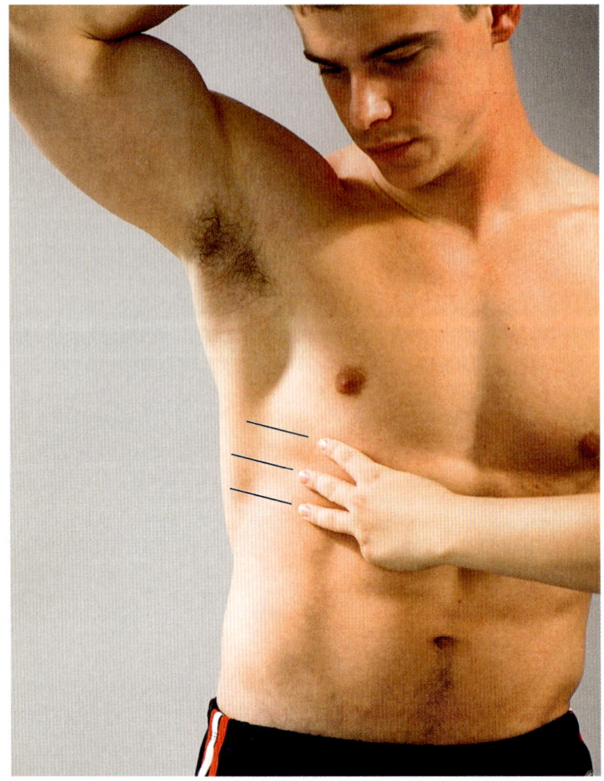

Figure 3.25 Locating the serratus anterior.

Posterior Muscles of the Shoulder Girdle

The posterior muscles of the shoulder girdle include the levator scapulae, rhomboids, and trapezius (see figure 3.24).

- *Levator scapulae:* This muscle's name tells the story of its function—to lift the scapula. The levator scapulae muscle originates on the transverse processes of the first four cervical vertebrae and inserts on the superior aspect of the vertebral border of the scapula. Because the cervical attachments are the more stable end of this muscle, contraction of the levator scapulae results in elevation (downward rotation and adduction) of the shoulder girdle.

- *Rhomboids*: These are actually two muscles (major and minor) that are usually considered as one because they both perform the same function. They originate on the spinous processes of the seventh cervical through the fifth thoracic vertebrae and insert on the vertebral (medial) border of the scapula. Because the spinous process attachments of the rhomboids are the most stable, contraction of the rhomboids results in elevation and adduction (downward rotation) of the shoulder girdle.

- *Trapezius:* This large, triangular muscle originates on the external occipital protuberance at the base of the skull and the spinous processes of all cervical and thoracic vertebrae and inserts on the spine of the scapula and posterior surface of the clavicle. Because of the size of the muscle, the angle of its various fibers, and its numerous functions, discussion of the muscle's action usually divides the muscle into four sections (upper, upper middle, lower middle, and lower). The upper fibers of the trapezius parallel to a great degree the levator scapulae muscle and therefore perform similar functions: elevation and adduction (downward rotation) of the shoulder girdle. The upper middle fibers of the trapezius also assist in elevation of the shoulder girdle but function to a greater degree in adduction of the shoulder girdle. The lower middle fibers of the trapezius function almost exclusively in adduction of the shoulder girdle. The lower fibers of the trapezius, because of the angle at which they run from the thoracic spine to the spine of the scapula, contribute to abduction (upward rotation) of the shoulder girdle.

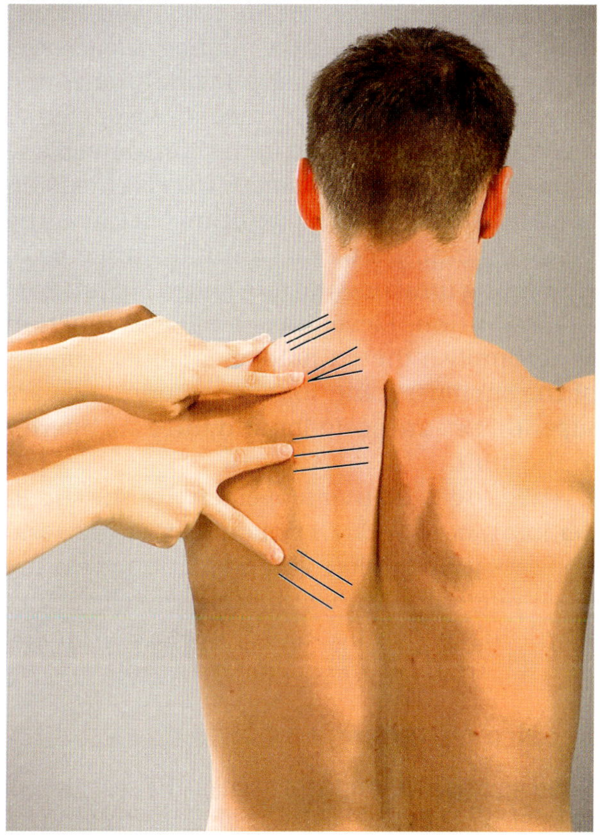

Figure 3.26 Locating the four parts of the trapezius.

Hands On

On your partner, locate the four portions of the trapezius muscle (figure 3.26).

Fundamental Movements and Muscles of the Shoulder Joint

We first observe the fundamental movements of the shoulder joint and then look at the muscles responsible for those movements.

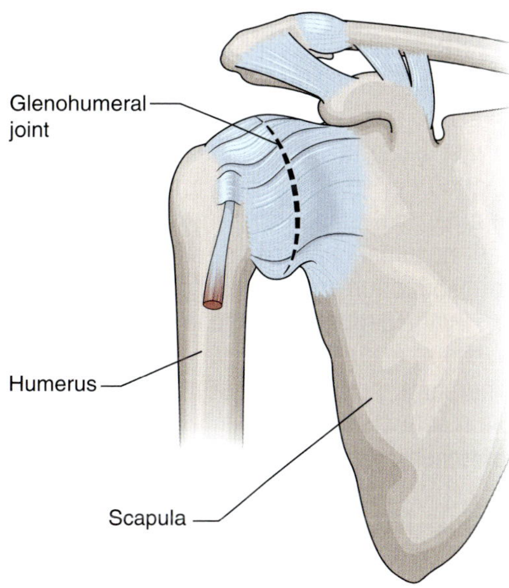

Figure 3.27 Anterior view of the glenohumeral joint.

Movements of the Shoulder Joint

The shoulder (GH) joint (figure 3.27) is classified as a triaxial joint because it is capable of movement in all three cardinal planes: flexion (anterior movement of the arm) and extension (return from flexion) in the sagittal plane about a frontal horizontal axis, abduction (movement away from the midline of the body) and adduction (movement toward the midline of the body) in the frontal plane about a sagittal horizontal axis, and internal (inward, medial) and external (outward, lateral) rotation in the horizontal plane about a vertical axis. *Hyperextension* is extension beyond the anatomical position. Because the shoulder joint is a triaxial joint, it is capable of combining fundamental movements to produce circumduction. Eleven major muscles function to accomplish the six fundamental movements of the shoulder joint: four are anterior, two are superior, two are posterior, and three are inferior to the joint.

There are two other movements of the shoulder joint, horizontal flexion and extension, and horizontal abduction and adduction, that are often referred to as *fundamental movements* but, based on the definition of fundamental movements introduced in chapter 2, are not.

Hands On

Stand up in the fundamental position. Raise your arm in front of you to shoulder height (see figure 2.3). This movement is defined as *shoulder flexion* (movement in a sagittal plane about a frontal and horizontal axis). From this position, move your arm to your side in a position aligned with the side of your body (see figure 2.5). This movement is defined as *shoulder abduction* (movement in a horizontal plane about a vertical axis).

You just performed two fundamental movements of the shoulder joint (flexion and abduction) and then another two fundamental movements when you reversed the initial movements (shoulder joint adduction from abduction and shoulder joint extension from flexion). By the definition from chapter 2, a fundamental movement occurs in a single plane about a single axis. Therefore, horizontal flexion and extension, and horizontal abduction and adduction, do not qualify as fundamental movements of the shoulder joint.

Anterior Muscles of the Shoulder Joint

The following muscles appear on the anterior aspect of the shoulder joint.

- *Pectoralis major:* The pectoralis major originates on the second to sixth ribs, the sternum, and the medial half of the clavicle and inserts on the anterior area of the surgical neck of the humerus just distal to the greater tubercle (see figure 3.24). The upper portion of the muscle is often referred to as the *clavicular part,* and the lower portion is referred to as the *sternal part.* Contraction of the pectoralis major muscle produces flexion, adduction, and internal rotation of the shoulder joint.

Hands On

Locate your partner's pectoralis major using figure 3.28 as a guide.

Figure 3.28 Locating the pectoralis major.

- *Coracobrachialis:* The coracobrachialis originates on the coracoid process of the scapula (where the tendon of origin is conjoined with the tendon of origin of the short head of the biceps brachii) and inserts on the middle of the medial side of the humerus opposite the deltoid tubercle on the lateral side (figures 3.24 and 3.29). The coracobrachialis flexes the shoulder joint and, because of its angle of pull, assists with adduction of the joint.

Hands On

Have your partner place her arm at a right angle to the body (abduction) as you palpate the coracobrachialis (figure 3.30).

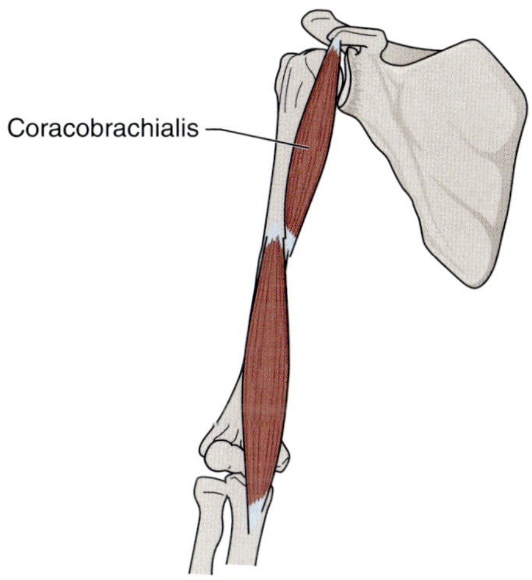

Figure 3.29 Anterior view of the coracobrachialis.

• *Biceps brachii:* Although the biceps brachii is frequently considered a flexor of the elbow (figure 3.31), both the long-head tendon and short-head tendon of the biceps brachii cross the shoulder joint. The long head originates on the supraglenoid tubercle on the superior edge of the glenoid of the scapula, and the short head originates on the coracoid process of the scapula (and is conjoined with the coracobrachialis tendon of origin). Both heads combine to form the belly of the muscle and insert on the tuberosity of the radius, which is one of the two forearm bones. Actions produced by contraction of this muscle at the shoulder joint include flexion and abduction by the long-head tendon and flexion, adduction, and internal rotation by the short-head tendon.

• *Subscapularis:* The subscapularis muscle (figure 3.31) is located on the anterior surface of the scapula between the scapula and the thorax. It originates on the large subscapular fossa on the anterior surface of the scapula and inserts on the lesser tubercle of the humerus. When it contracts, the subscapularis produces internal rotation and flexion at the shoulder joint. This muscle is one of four shoulder joint muscles that attach to a musculotendinous structure often referred to as the *rotator cuff* (discussed later in this chapter).

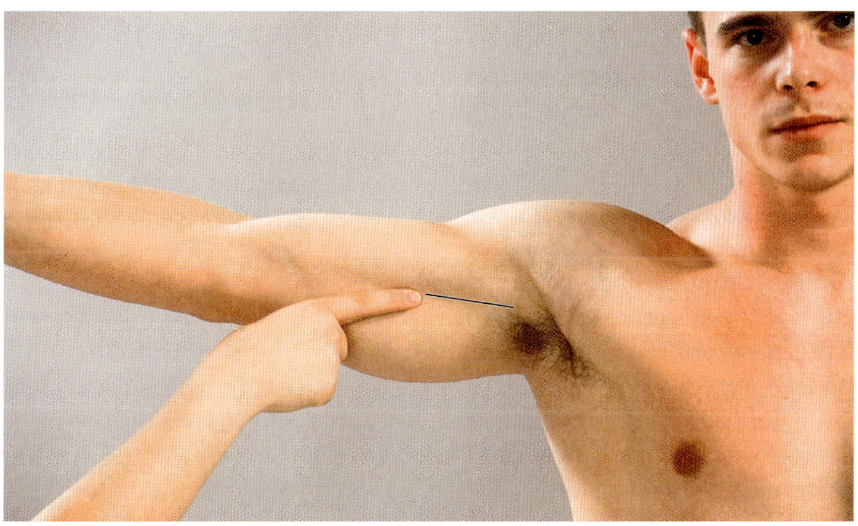

Figure 3.30 Locating the coracobrachialis.

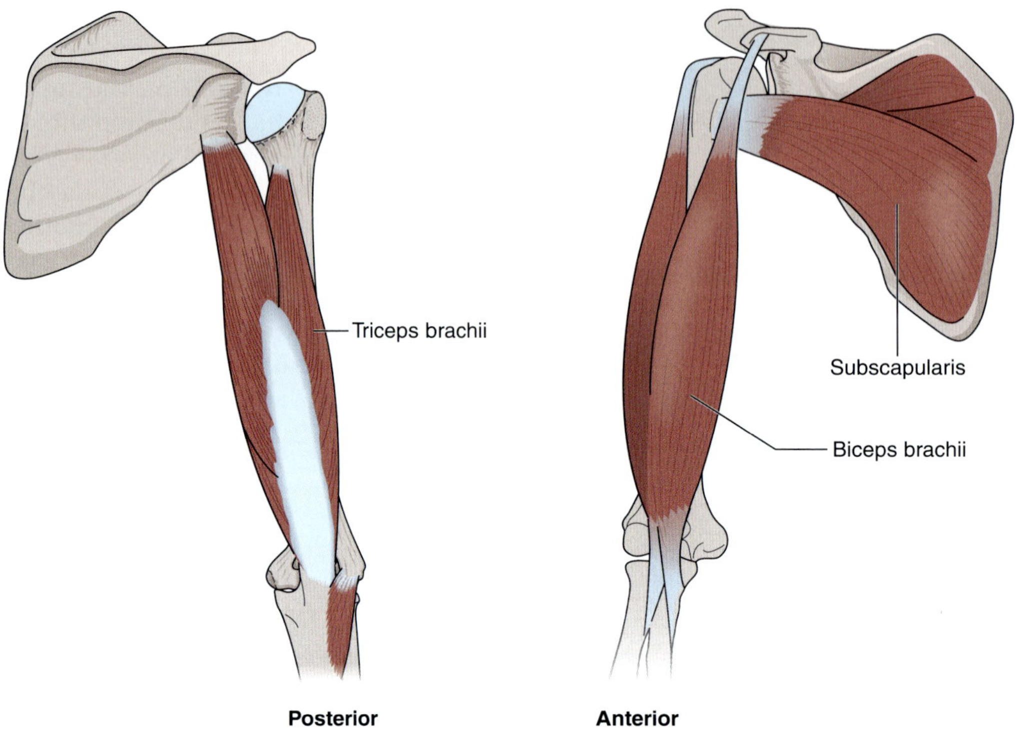

Figure 3.31 Two superficial shoulder joint muscles and the subscapularis.

Superior Muscles of the Shoulder Joint

The following muscles appear on the superior aspect of the shoulder joint.

- *Deltoid:* The deltoid is a very large muscle consisting of three parts: anterior, middle, and posterior (see figure 3.24). It covers the shoulder joint, so it is often referred to as the *shoulder cap muscle*. The anterior (clavicular) fibers originate from the lateral portion of the anterior aspect of the clavicle, the middle (acromial) fibers originate from the lateral edge of the acromion process of the scapula, and the posterior (scapular) fibers originate from the inferior edge of the spine of the scapula. All three portions combine to insert on the deltoid tubercle on the lateral surface of the middle of the humerus. Contraction of the entire deltoid muscle results in abduction of the shoulder joint; contraction of the posterior portion alone results in adduction, extension, and external rotation; and contraction of the anterior fibers alone results in adduction, flexion, and internal rotation. The middle fibers of the deltoid muscle are typically considered to be involved only in shoulder joint abduction. Once the arm has been abducted to the horizontal level from the anatomical position, all three portions of the muscle are considered abductors of the joint.

Hands On

Have your partner abduct their arm, preferably with some form of resistance, such as holding a book or a weight, as you locate all three portions of the deltoid muscle (figure 3.32).

- *Supraspinatus:* Located beneath the deltoid muscle, the supraspinatus muscle (figure 3.33) originates on the supraspinous fossa of the scapula and inserts on the proximal facet of the greater tubercle of the humerus. The muscle abducts the shoulder joint. Although it contracts throughout the full range of abduction, it is considered the primary initiator of abduction until approximately 30° of abduction, when the deltoid muscle takes over as the major abductor. The supraspinatus muscle is also one of the muscles of the rotator cuff.

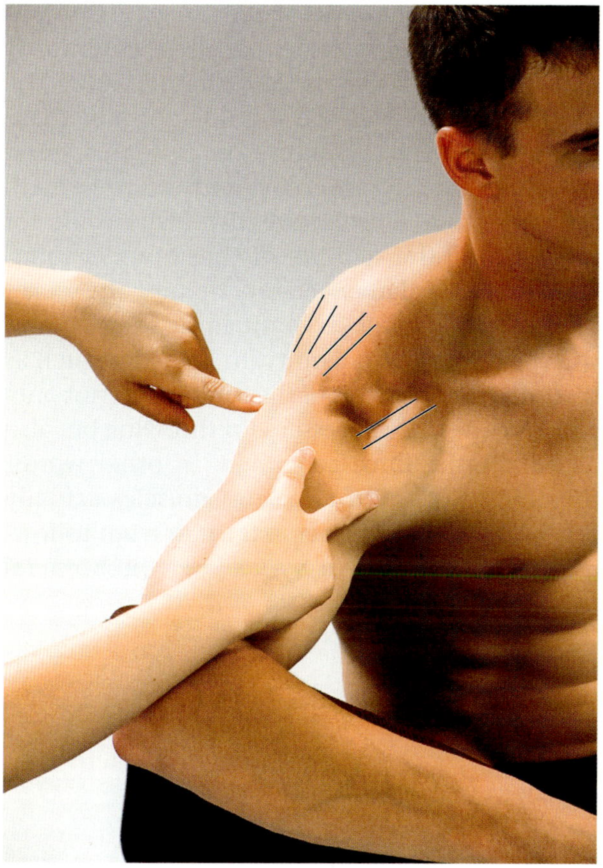

Figure 3.32 Locating the deltoid.

Posterior Muscles of the Shoulder Joint

The following muscles are found on the posterior aspect of the shoulder joint.

- *Infraspinatus:* The infraspinatus muscle (see figure 3.33) gets its name from the anatomical structure where it originates: the infraspinous fossa beneath the inferior surface of the spine of the scapula. The infraspinatus muscle inserts on the middle facet of the greater tubercle of the humerus. Contraction of the infraspinatus muscle produces external rotation and extension of the shoulder joint. The infraspinatus muscle is also part of the rotator cuff.

Hands On

Place your partner's shoulder joint in abduction, external rotation, and extension, and then locate the infraspinatus muscle (figure 3.34).

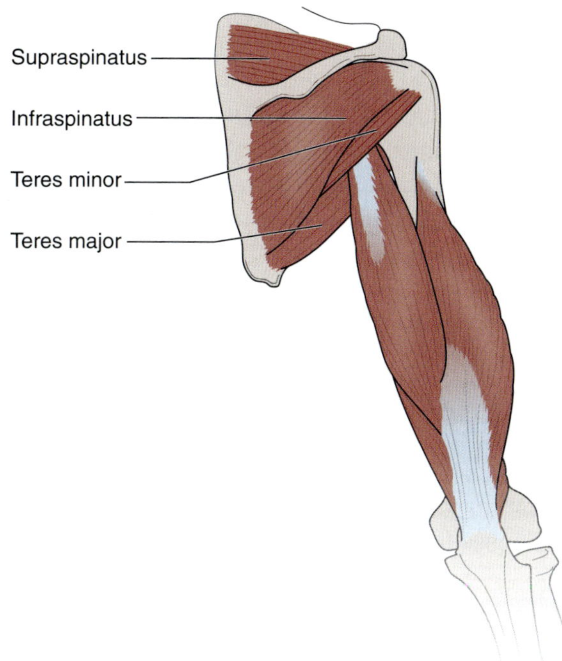

Figure 3.33 Posterior view of the supraspinatus, infraspinatus, teres minor, and teres major muscles.

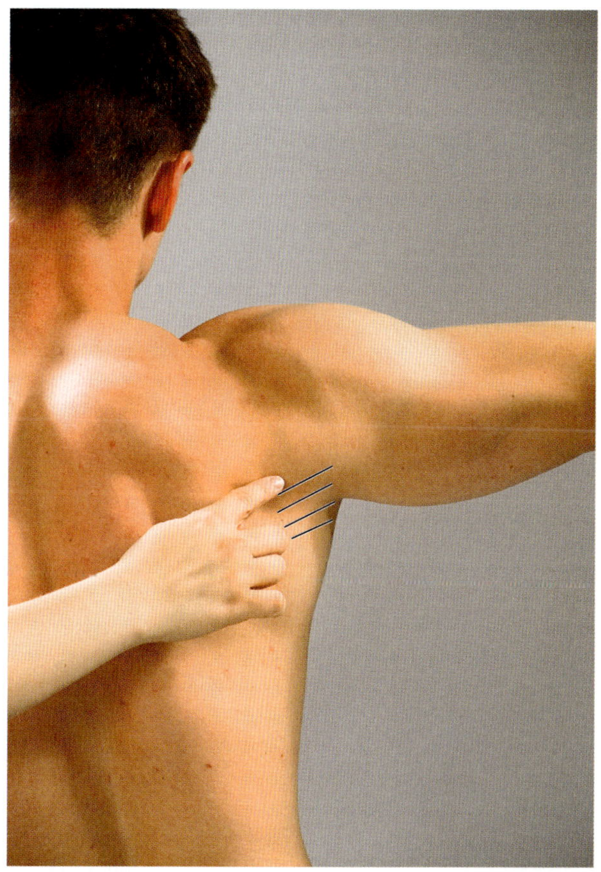

Figure 3.34 Locating the infraspinatus and teres minor.

• *Teres minor:* The teres minor muscle is often considered together with the infraspinatus muscle because they share the same function. The teres minor (see figure 3.33) originates on the upper and middle portions of the lateral border of the scapula and inserts on the distal facet of the greater tubercle of the humerus. Contraction of the teres minor muscle, like the infraspinatus muscle, produces external rotation and extension of the shoulder joint. This muscle is also one of the shoulder joint muscles of the rotator cuff.

Hands On

Place your partner's shoulder joint in abduction, external rotation, and extension, and then locate the teres minor muscle (see figure 3.34).

Rotator Cuff

Four of the muscles of the shoulder joint insert on a musculotendinous structure running between the facets located on the lesser and greater tubercles of the humerus. This structure is commonly referred to as the *rotator cuff* (figure 3.35). The motions produced at the shoulder joint by these four muscles (subscapularis, supraspinatus, infraspinatus, and teres minor) have been presented, but these four muscles are also responsible for maintaining the stability of the shoulder joint, which is particularly necessary because the socket of this ball-and-socket joint is so shallow and thus provides little stability. In an action such as throwing, the muscles of the rotator cuff not only generate the force necessary for throwing but also decelerate the force generated. In other words, the actions of the rotator cuff muscles actually prevent the entire upper extremity from following the object thrown by keeping the humeral head in the glenoid fossa.

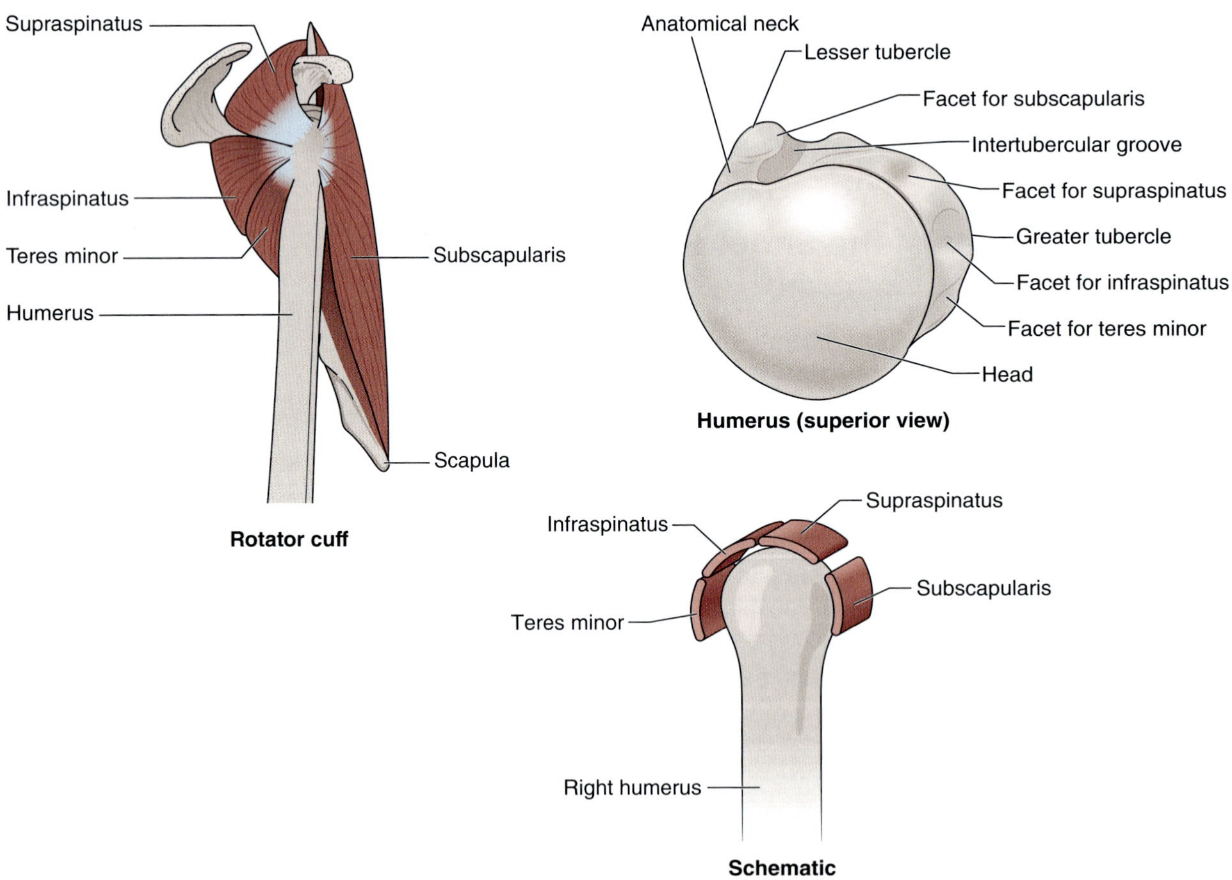

Figure 3.35 Humeral attachments for the four muscles of the rotator cuff.

FOCUS ON

Rotator Cuff

Understanding the action of the rotator cuff makes it easier to understand why some people involved in repetitive throwing activities (e.g., pitchers and quarterbacks) develop rotator cuff problems. The muscles not only generate the force needed for a throwing activity by contracting concentrically (shortening), but they also apply a braking action through an eccentric (lengthening) contraction to prevent the upper extremity from leaving the body by dissipating the force they generated. These muscles are under tremendous stress in throwing activities. Initially, they are stretched; then they must generate force; and, finally, they must dissipate the force to prevent the humeral head from leaving the glenoid of the scapula.

POP QUIZ

1. During the windup phase of throwing, what muscles of the rotator cuff are working?
2. During the middle phase of throwing, which muscles of the rotator cuff are working?
3. Upon the completion of the throwing motion, what muscles of the rotator cuff are working to prevent the humerus from leaving the shoulder joint?

Inferior Muscles

The following muscles cross the shoulder joint inferiorly (underneath).

- *Latissimus dorsi:* The latissimus dorsi muscle (see figure 3.24), a large muscle of the back, originates on the spinous processes of the lower six thoracic and all five lumbar vertebrae, the posterior aspect of the ilium (see chapter 8), the lower three ribs, and the inferior angle of the scapula; passes beneath the axilla (armpit); and inserts on the edge of the intertubercular groove on the anterior aspect of the humerus. Contraction of the latissimus dorsi muscle produces internal rotation, extension, and adduction of the shoulder joint.

Hands On

Place your partner's arm in the position of external rotation and abduction and locate the latissimus dorsi muscle (figure 3.36).

FOCUS ON

Latissimi Dorsi

In the front crawl stroke in swimming, you can see that moving the arm through the water requires strength in muscles that internally rotate, extend, and adduct the shoulder joint: the latissimi dorsi. The lat machine in the weight room is used to increase strength in what some call "the swimming muscle."

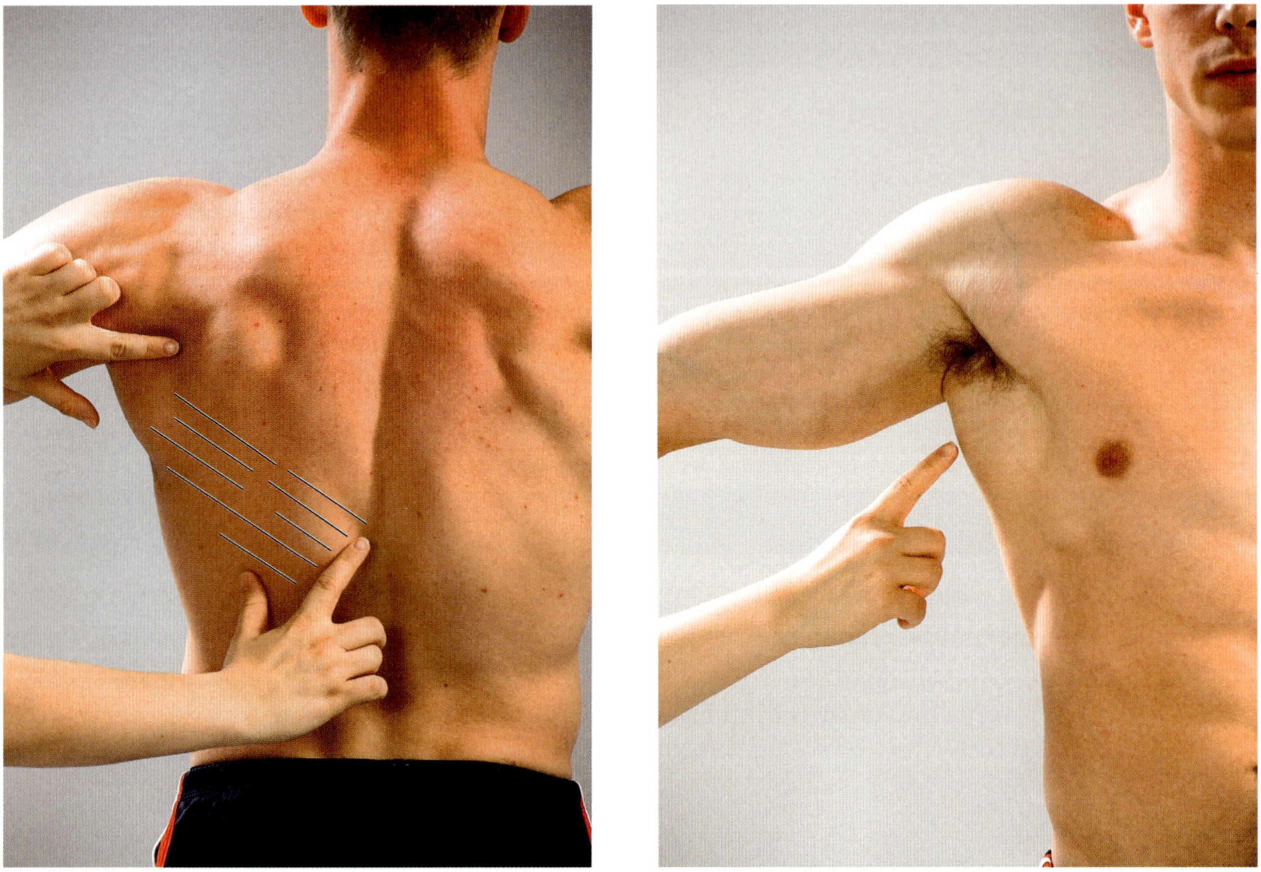

Figure 3.36 Locating the latissimus dorsi.

• *Teres major:* The teres major muscle (see figure 3.33) originates on the lower portion of the lateral border of the scapula and its inferior angle, crosses beneath the axilla (armpit), and inserts on the area just inferior to the lesser tubercle of the humerus. Contraction of the teres major produces the same action as the latissimus dorsi: internal rotation, extension, and adduction of the shoulder joint. Because the action of these two muscles is identical, the teres major is often called the latissimus dorsi's "little helper."

Hands On

Placing your partner's arm in the position of abduction, observe the posterior aspect of the shoulder joint, and locate the teres major muscle (figure 3.37).

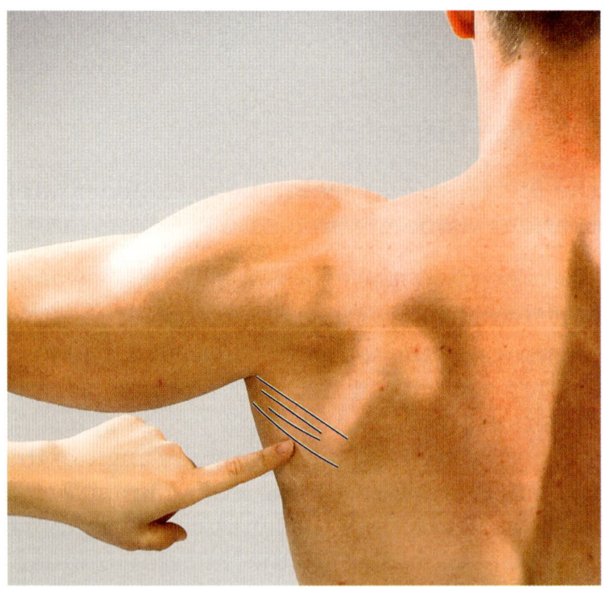

Figure 3.37 Locating the teres major.

• *Triceps brachii:* Although the triceps brachii is more often associated with elbow joint action, one of the three tendinous heads of the triceps brachii does cross the shoulder joint and assist with shoulder joint movement (see figure 3.31). Of the lateral-head, long-head, and medial-head tendons of origin of the triceps brachii, the long head originates on the infraglenoid tubercle of the glenoid lip of the scapula and joins the lateral and medial heads to insert, on a common tendon, on the olecranon process of the ulna, one of the two bones of the forearm. Contraction of the long head of the triceps brachii assists with shoulder joint extension and adduction.

Hands On

On yourself or a partner, locate all three portions of the triceps brachii muscle (figure 3.38).

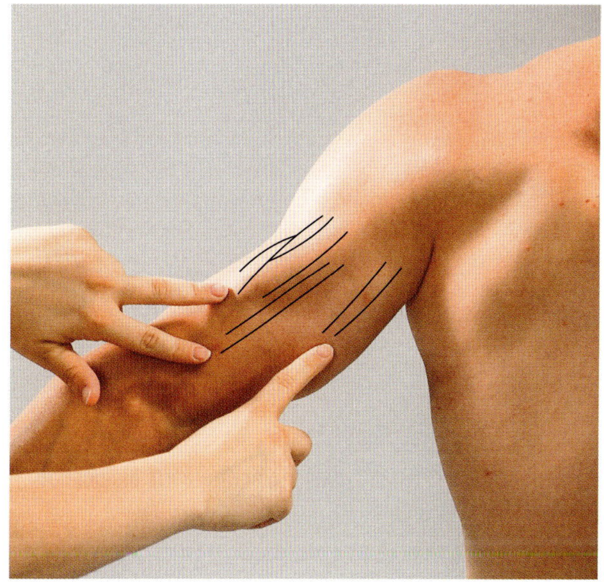

Figure 3.38 Locating the three sections of the triceps brachii.

FOCUS ON

Glenohumeral Joint

A dislocation of any joint usually results in the severe spraining of the ligaments and straining of the muscles that cross the joint. Following are three of the more common dislocations of the GH joint.

Excessive movement in any joint may result in stress being placed on the ligamentous structures tying the bones of the joint together. As noted earlier regarding the AC joint of the shoulder girdle, a sprain of a ligament is the partial or complete tearing of a ligament. When the ligaments of a joint are disrupted to the extent that the bones of the joint actually displace, it is known as a *dislocation*. Dislocations of the GH joint are not uncommon. The three most common forms of GH dislocations are anterior (subcoracoid), posterior (subspinous), and downward or inferior (subglenoid) dislocation (figure 3.39).

Anterior dislocation is the most common form of shoulder dislocation, typically resulting from excessive abduction and external rotation of the shoulder joint. The humeral head displaces anteriorly from the glenoid of the scapula and rests anterior to the glenoid just beneath the coracoid process of the scapula (thus the name subcoracoid dislocation). A second form of shoulder joint dislocation is posterior (subspinous) dislocation, which can result from excessive internal rotation and adduction of the shoulder joint. The humeral head displaces posteriorly from the glenoid of the scapula and rests posterior to the glenoid just beneath the spine of the scapula (thus the name subspinous dislocation). The third form of shoulder joint dislocation is known as a *downward* or *inferior* (subglenoid) *dislocation*, which may result from excessive shoulder joint abduction with the humerus abutting the acromion process of the scapula and the head of the humerus being forced downward or inferior (subglenoid) beneath the lower edge of the glenoid of the scapula.

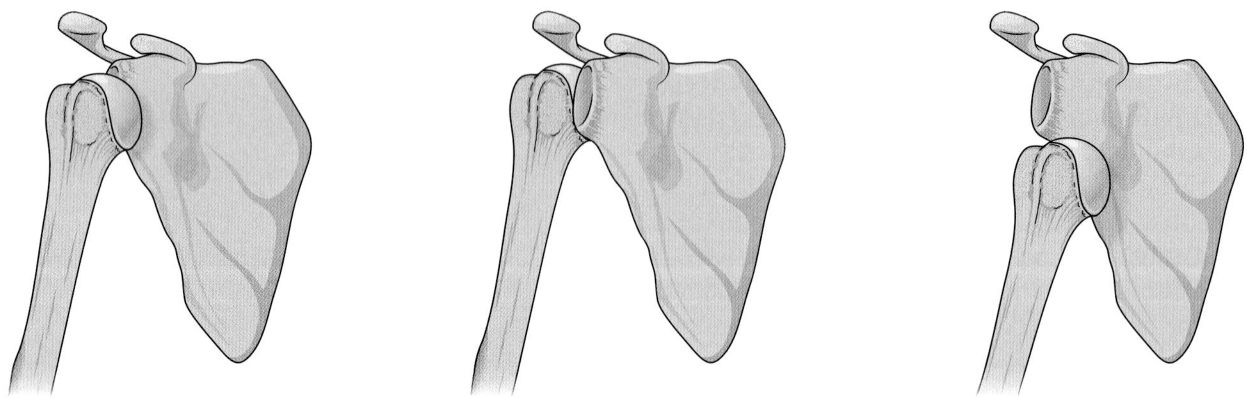

a b c

Figure 3.39 Three types of glenohumeral dislocation: *(a)* anterior (subcoracoid), *(b)* posterior (subspinous), and *(c)* downward (subglenoid).

Combined Actions of the Shoulder Girdle and Shoulder Joint

Each of the shoulder joint actions (flexion, extension, abduction, adduction, and internal and external rotation) possesses certain degrees of movement. When any particular movement of the shoulder joint reaches its end point, to move farther in that direction, the glenoid of the scapula must change its position to accommodate additional movement of the humerus. Movement of the shoulder girdle facilitates a greater range of motion in all the fundamental movements of the shoulder joint by changing the position of the glenoid, which is accomplished by movement of the scapula on the thorax (along the ribs) through motion at the AC and SC joints. The prime example of this relationship is referred to as the *scapulohumeral rhythm*. Although initial abduction of the shoulder joint is attributed to GH joint action only (approximately 120° of abduction), the combination of GH joint abduction and rotation of the scapula results in approximately 180° of shoulder joint abduction. The generally accepted ratio of motion between the GH joint and the scapula (scapulothoracic movement) is that for every 2° of GH joint abduction, the scapula rotates 1°.

When the GH joint is elevated approximately 30° anterior to the frontal plane (flexion) or at 30° of adduction in the horizontal plane, the movement is known as *scaption* and occurs in the scapular plane. In the scapular plane, GH elevation occurs in line with the scapula's position on the thorax. It is the functional position for the rotator cuff muscles and allows for a straight line of pull of these muscles from their origins to their insertions, facilitating an optimal length–tension position for GH elevation.

As it does with abduction of the shoulder joint, the scapula moves to position the glenoid to accommodate all the other fundamental movements of the shoulder joint.

Visit the web resource to complete the activities for this chapter.

CHAPTER 4

The Elbow and Forearm

The elbow and forearm are composed of three bones: the humerus, the ulna, and the radius. Together, these three bones form four joints, three at the proximal end of the forearm (radiohumeral, ulnohumeral, and proximal radioulnar) and one at the distal end of the forearm (distal radioulnar).

Bones of the Elbow and Forearm

The humerus, ulna, and radius meet to form the structure commonly known as the *elbow*, and the ulna and radius form the forearm (figure 4.1). The proximal end of the humerus was discussed in chapter 3 on the shoulder joint. The distal end of the humerus provides the bony attachments for the soft tissues that span the upper arm and the forearm to form the elbow joint. At the distal end of the humerus, the shaft widens out to form two bony prominences: the lateral and medial epicondyles.

Hands On

Find these structures on your own or your partner's elbow (figure 4.2). The epicondyles are easily palpated and visible on most people.

The parts of the humerus between the shaft and these epicondyles, where the bone actually widens, are known as the *lateral* and *medial supracondylar ridges*. There are three fossae (depressions) on the distal humerus that provide areas for other bony structures to move into during elbow joint actions. On the anterior surface, between the two epicondyles, is the coronoid fossa. This is where the coronoid process of the ulna is positioned during elbow flexion. Lateral to the coronoid fossa is the radial fossa. The head of the radius moves into this fossa during elbow flexion. Just distal to the lateral epicondyle is a smooth, round surface known as the *capitulum*. This surface is where the head of the radius rotates during forearm movement. At the very distal end of the humerus is a spool-like structure

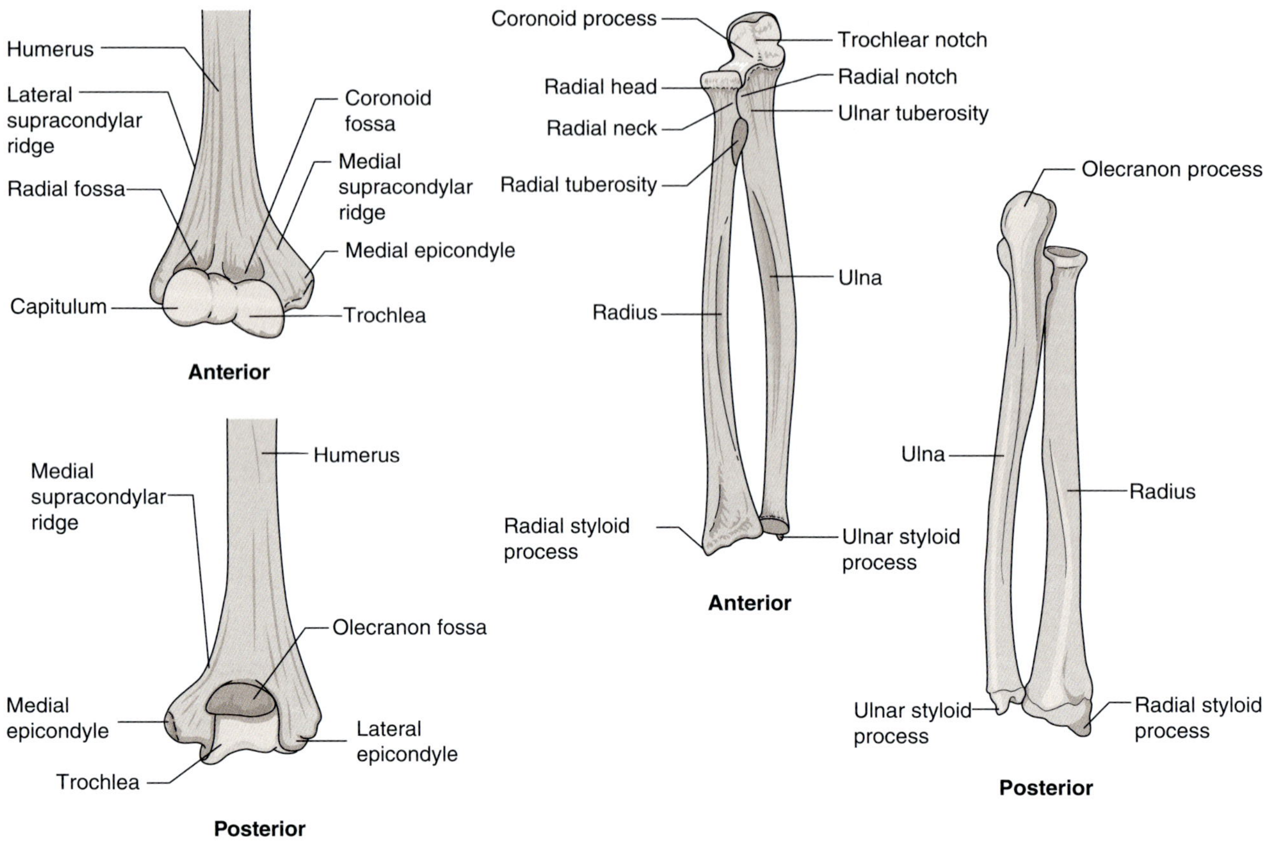

Figure 4.1 Landmarks of the elbow and forearm bones.

known as the *trochlea*. This is the structure on which the olecranon process of the ulna attaches.

A posterior view of the distal end of the humerus also reveals some structures previously discussed: the supracondylar ridges, the epicondyles, and the trochlea. In addition, the olecranon fossa, which is formed between the two epicondyles, appears. This is the depression into which the olecranon process of the ulna moves when the elbow joint is moved into extension.

Under normal conditions, when the elbow joint is in full extension, you should be able to observe that the lateral epicondyle, the olecranon process, and the medial epicondyle form a straight line. If the normal elbow joint is flexed to 90°, these three structures should form an isosceles triangle with the olecranon process distal to the epicondyles (see figure 4.2).

The two bones of the forearm are known as the *radius* and the *ulna*. From the anatomical position, the radius is on the lateral aspect and the ulna on the medial aspect of the forearm. In the

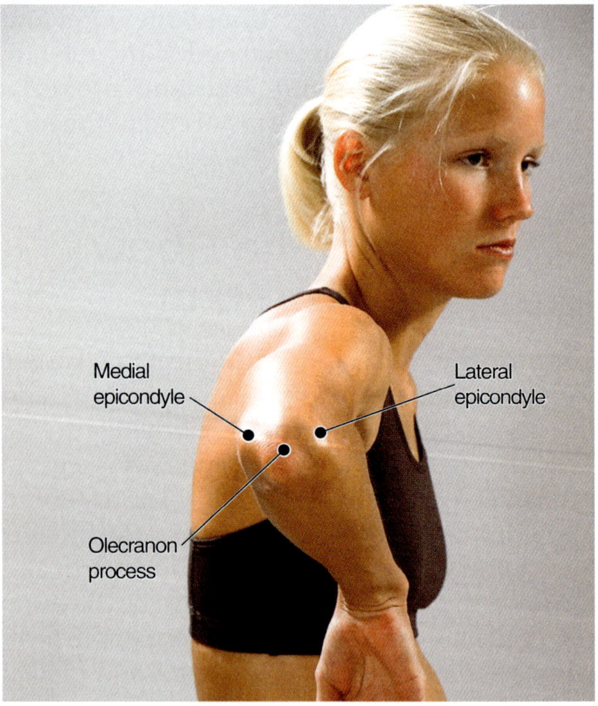

Figure 4.2 Locating the medial and lateral epicondyles.

elbow joint, the ulna has the prominent role of articulating with the humerus, whereas the radius plays the more prominent role of articulating with the bones of the wrist. The proximal end of the radius consists of the radial head, the radial neck, and a large radial tuberosity distal to the neck on the medial aspect of the upper portion of the shaft of the bone.

Hands On

You can easily palpate the head of the radius at the lateral aspect of your elbow (figure 4.3).

At the distal end of the shaft of the radius, on the lateral aspect, is a large prominence known as the *radial styloid process* (also known as the *lateral styloid process*).

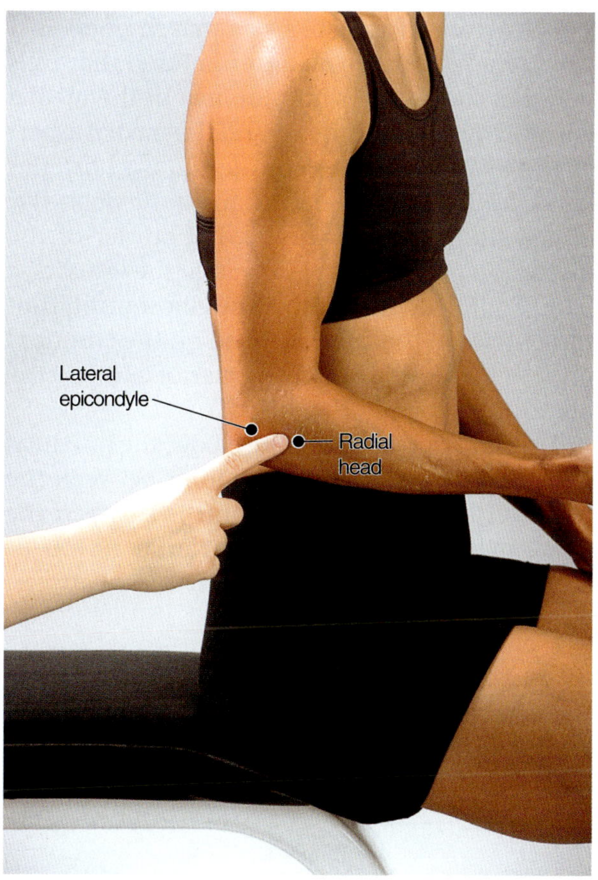

Figure 4.3 Locating the radial head.

Hands On

This structure can easily be palpated in the area where the hand and forearm come together (wrist) on the thumb side (figure 4.4).

On the medial aspect of the distal end of the radius is the ulnar notch. On the distal surface of the radius are two distinct facets where bones of the wrist articulate (figure 4.5).

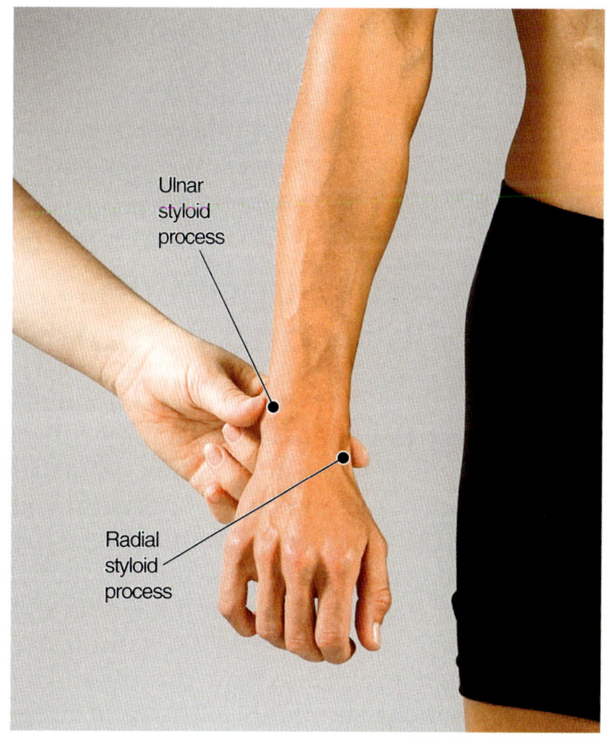

Figure 4.4 Locating the ulnar and radial styloid processes.

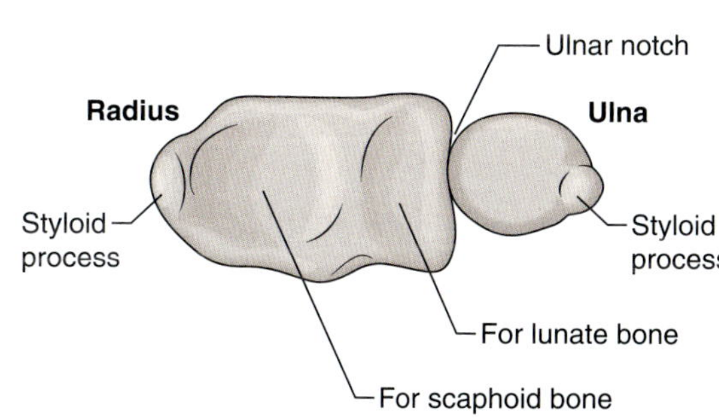

Figure 4.5 Facets on the distal surface of the radius where bones of the wrist articulate.

The medial forearm bone, the ulna, has a very large prominence at the proximal end known as the *olecranon process* (see figure 4.1). This structure has a cuplike surface, known as the *trochlear notch*, that rotates about the trochlea of the humerus to form the articulation between the humerus and the ulna, commonly known as the *elbow joint*. At the very anterior portion of the trochlear notch is a smaller prominence known as the *coronoid process*. Just lateral to the coronoid process is the radial notch of the ulna. The head of the radius articulates with the ulna (proximal radioulnar joint) at the radial notch of the ulna.

At the distal end of the ulna on the medial aspect is an easily palpated prominence known as the *ulnar styloid process* (also called the *medial styloid process*; see figures 4.1 and 4.4).

Joints and Ligaments of the Elbow and Forearm

With three bones (humerus, ulna, and radius) coming together to form the elbow joint (figure 4.6), there are actually three joints in the anatomical area between the upper arm and the forearm: the "true" elbow joint between the humerus and the ulna (the ulnohumeral joint), the radiohumeral joint between the radius and the humerus, and the proximal radioulnar joint between the radius and the ulna. The capsular ligament surrounds all three of these articulations. It is divided into an anterior and a posterior part. The anterior part extends from the anterior surface of the humerus just proximal to the coronoid fossa to the anterior surface of the coronoid process and the annular ligament. Laterally, the capsular ligament fuses with the collateral ligaments. Posteriorly, the capsular ligament attaches to the tendon of insertion of the triceps brachii muscle, the edge of the olecranon, the lateral epicondyle, and the posterior surface of the humerus in the area of the trochlea and the capitulum. Distally, the capsular ligament attaches to the lateral and superior edges of the olecranon process, the posterior aspect of the annular ligament, and posterior to the radial notch of the ulna.

The ligaments that fuse with the anterior portion of the capsular ligament are known as the *radial (lateral) collateral ligament* and the *ulnar (medial) collateral ligament*. The radial collateral ligament runs between the inferior border of the lateral epicondyle of the humerus to the annular ligament and the radial notch of the ulna.

The ulnar collateral ligament has three distinct parts (see figure 4.6): an anterior band that runs between the anterior inferior area of the medial epicondyle of the humerus and the medial aspect of the coronoid process of the ulna; a posterior band running between the medial epicondyle of the humerus and the medial border of the olecranon process of the ulna; and the transverse band, which does not cross the elbow joint, running between the anterior band on the coronoid process of the ulna and the posterior band on the olecranon process of the ulna.

The remaining joint of the elbow, the proximal radioulnar joint, is actually a joint between the bones of the forearm. The joint is between the head of the radius and the radial notch of the ulna. The annular ligament runs between

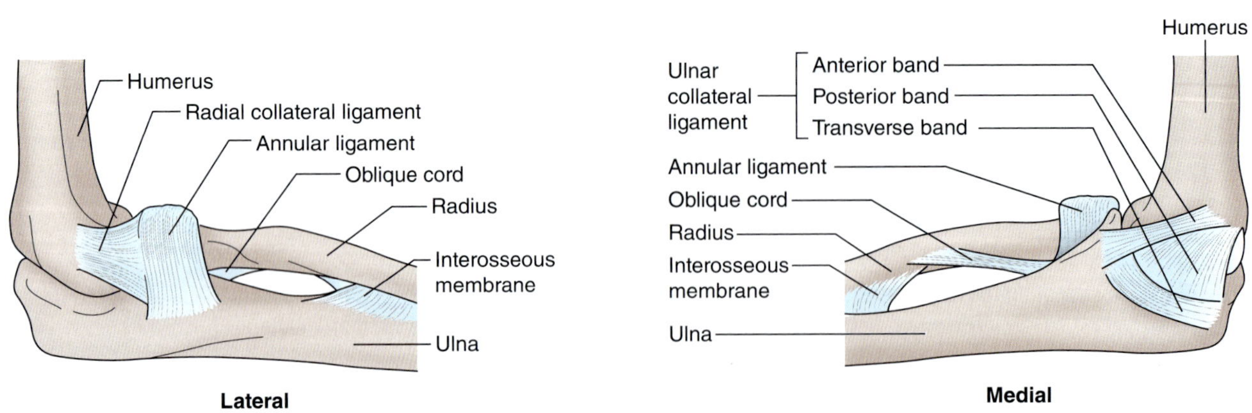

Figure 4.6 Major ligaments of the elbow.

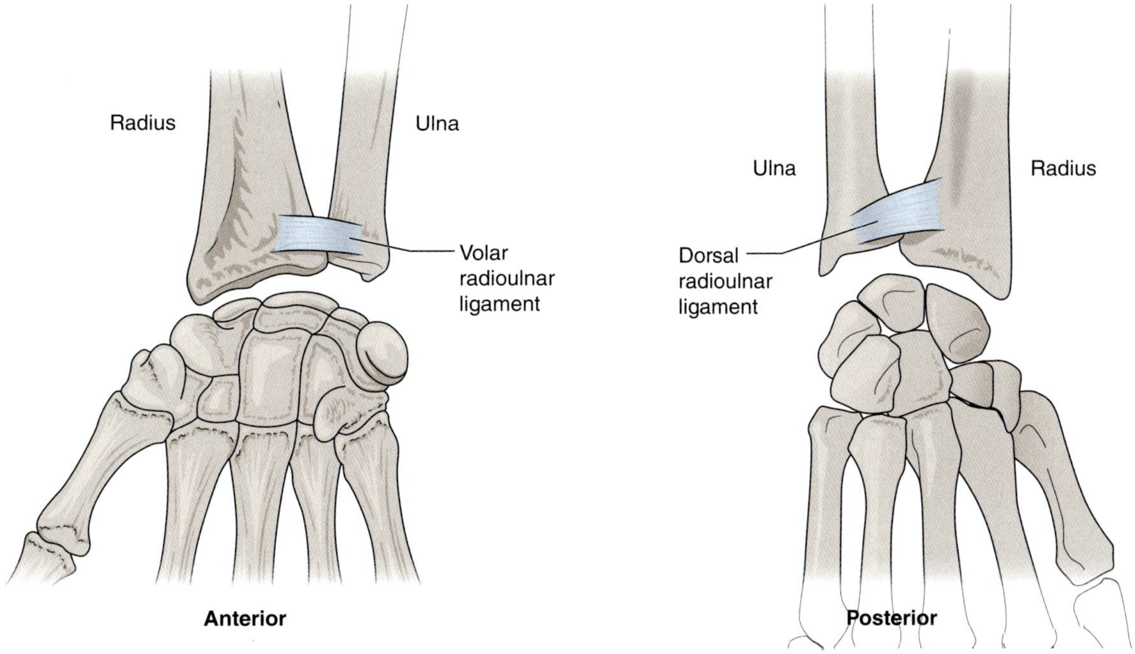

Figure 4.7 The dorsal and volar radioulnar ligaments.

the anterior and posterior edges of the radial notch on the ulna and forms a ring completely around the head of the radius.

Running between the shafts of the ulna and radius is a ligamentous band of connective tissue known as the *interosseous ligament* (also referred to as the *interosseus membrane*), which helps distribute pressure between the ulna and radius when force is applied and also serves as a place of attachment for several forearm muscles. At the proximal end of the interosseous ligament (membrane) is a ligament known as the *oblique cord*, which prevents separation between the ulna and radius.

At the distal end of the ulna and radius, just proximal to the wrist, is the distal radioulnar joint. In the concave ulnar notch of the radius, the round head of the ulna rotates, forming a pivot joint. The ligaments of this joint are the dorsal and volar radioulnar ligaments (figure 4.7).

Fundamental Movements and Muscles of the Elbow and Forearm

Remember that the starting position for describing all fundamental movements of any joint is the anatomical position. The elbow joint (articulation between the trochlea of the humerus and the olecranon process of the ulna) is a uniaxial joint capable of flexion and extension in the sagittal plane about a frontal horizontal axis. Five major muscles produce the motions of flexion and extension of the elbow joint. They are the brachialis (flexion), the brachioradialis (flexion), the biceps brachii (flexion), the triceps brachii (extension), and the anconeus (extension). The brachialis, brachioradialis, and biceps brachii muscles are anterior to the elbow joint, and the triceps brachii and anconeus muscles are posterior to the joint. Four muscles are responsible for the movements of supination and pronation of the forearm (figure 4.8). The biceps brachii (supination) has already been mentioned in its other role at the elbow joint (flexion). The other three muscles involved with forearm motion are the supinator (supination), the pronator quadratus (pronation), and the pronator teres (pronation).

Anterior Muscles of the Elbow

The anterior muscles of the elbow are three in number: the brachialis, the brachioradialis, and the biceps brachii. Two of these muscles (the brachialis and brachioradialis) are involved exclusively in one movement of the elbow joint (flexion), whereas the third (the biceps brachii)

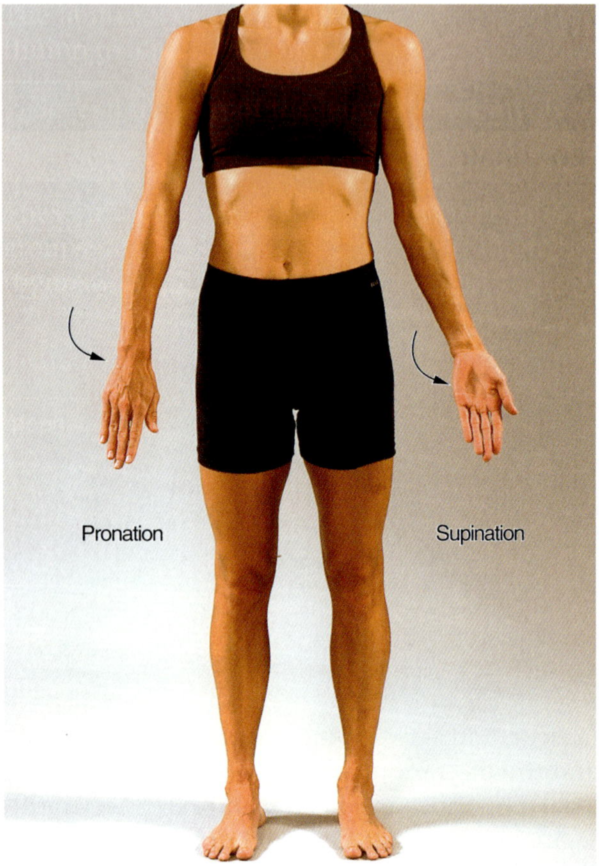

Figure 4.8 Pronation and supination of the radioulnar joints.

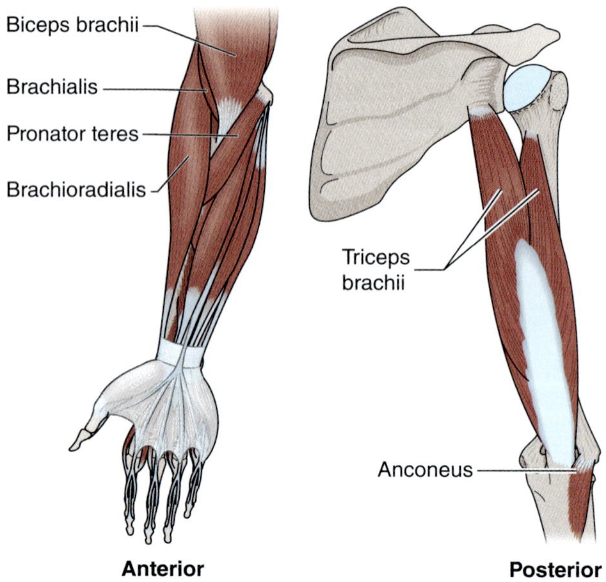

Figure 4.9 Superficial muscles of the elbow and forearm.

is involved in elbow joint movement and movement of the forearm.

• *Brachialis:* The brachialis muscle originates on the middle of the anterior shaft of the humerus and inserts on the coronoid process of the ulna. Because of its origin and insertion, its only function is flexion of the elbow joint. It is located beneath the biceps brachii muscle (figures 4.9 and 4.10).

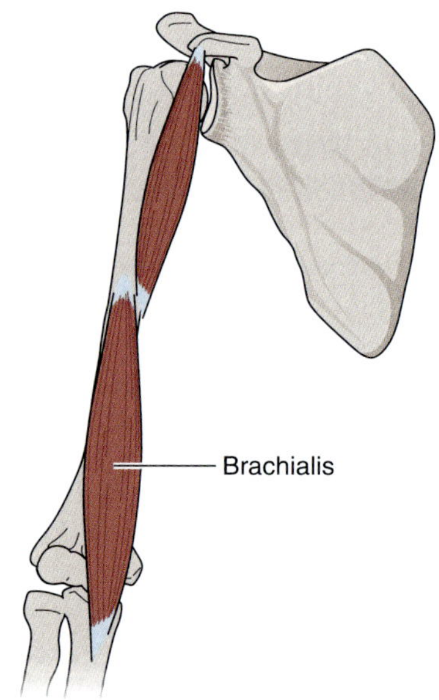

Figure 4.10 Anterior view of the brachialis.

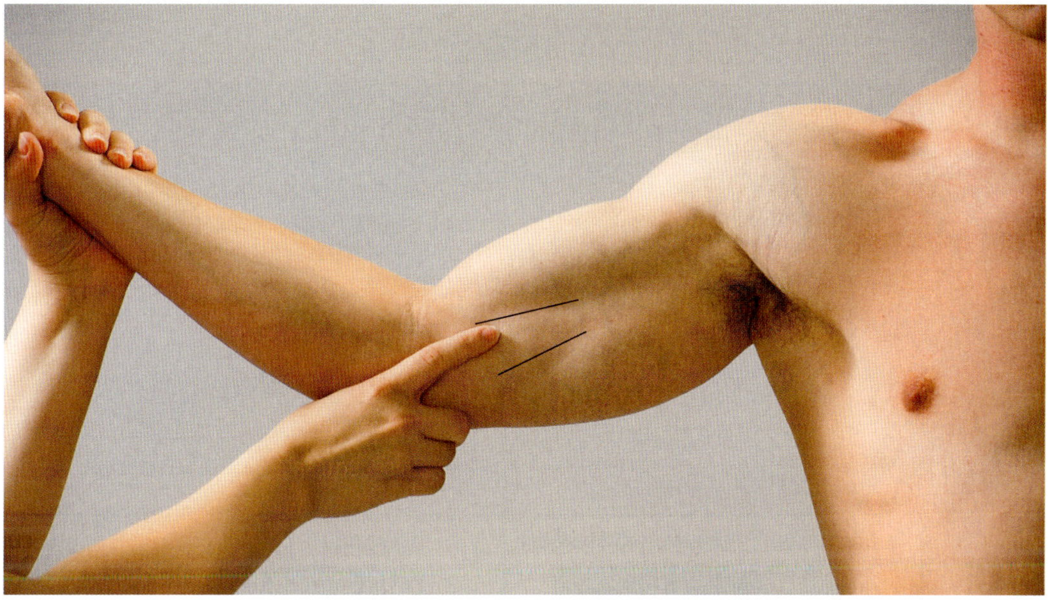

Figure 4.11 Locating the brachialis.

Hands On

Have your partner abduct and externally rotate their shoulder joint so you can locate the brachialis muscle (figure 4.11).

- *Brachioradialis:* The brachioradialis muscle originates on the lateral epicondyle of the humerus and inserts on the radial styloid process. The muscle crosses the anterior aspect of the elbow joint and, therefore, is a flexor of the elbow joint (see figure 4.9).

Hands On

Have your partner flex their elbow joint against a slight resistance applied by the opposite hand as you locate the brachioradialis muscle (figure 4.12).

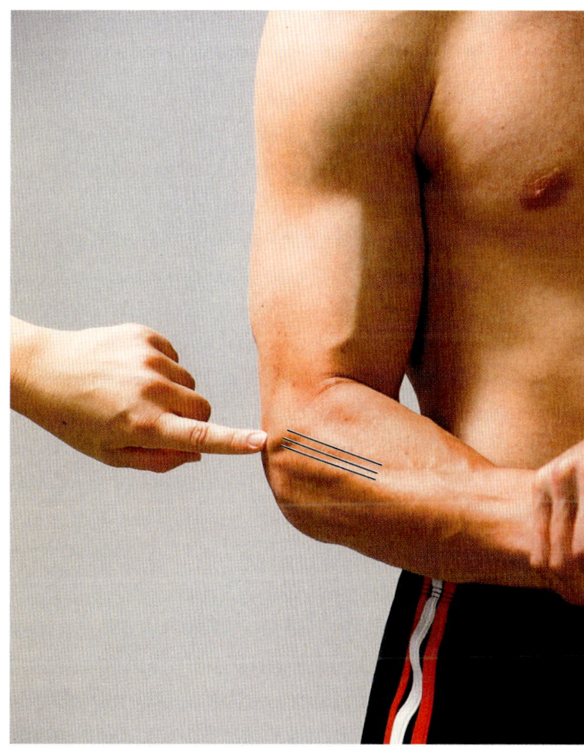

Figure 4.12 Locating the brachioradialis.

- *Biceps brachii:* The biceps brachii is commonly considered a flexor of the elbow (see figure 4.9), but both the long-head and short-head tendons of the biceps brachii also cross the shoulder joint and contribute to shoulder motion

Figure 4.13 Demonstrating the biceps brachii.

(figure 4.13). The long head originates on the supraglenoid tubercle on the superior edge of the glenoid of the scapula, and the short head originates on the coracoid process of the scapula (and is conjoined with the coracobrachialis tendon of origin). Both heads combine into the belly of the muscle, which inserts on the tuberosity of the radius. Actions produced by contraction of this muscle are flexion at the elbow joint and supination at the forearm.

Posterior Muscles of the Elbow

The anconeus and triceps brachii are posterior muscles of the elbow.

• *Triceps brachii:* The triceps brachii is most often associated with elbow joint extension (see figure 4.9), but one of its three tendinous heads crosses the shoulder joint and assists with shoulder joint movement. The long head originates on the infraglenoid tubercle of the glenoid lip of the scapula and joins the lateral and medial heads to insert, on a common tendon, on the olecranon process of the ulna. Contraction of the triceps brachii extends the elbow joint.

Hands On

To palpate the three heads of the triceps brachii (the lateral head, long head, and medial head), have a partner extend their elbow against a resistance while also extending the shoulder joint as you locate all three heads (figure 4.14).

• *Anconeus:* The anconeus originates on the lateral epicondyle of the humerus and inserts on the olecranon process of the ulna. This muscle assists the triceps brachii in elbow extension (see figures 4.9 and 4.14). A common error is to assume that the anconeus is also involved in the movement of the forearm; on careful inspection, this muscle's origin and insertion points allow only one action when the muscle contracts: elbow extension.

FOCUS ON

Throwing Action

The pronator teres muscle and a group of muscles that originate in the common flexor tendon, presented in chapter 5 on the wrist and hand, come under great stress in the overhead throwing motion, particularly at the end of the windup stage of throwing. This pronator–flexor muscle group is also stressed at the release stage of throwing a curveball in baseball pitching. Inflammation of the pronator–flexor group at its origin in the area of the medial epicondyle of the humerus is often referred to as *pitcher's elbow* or *Little League elbow*.

A further complication from this throwing action is the spraining and possible complete disruption of the medial or ulnar collateral ligament of the elbow joint. In 1974, Dr. Frank Jobe, a Los Angeles orthopedic surgeon, reconstructed the torn ulnar collateral ligament of Major League Baseball pitcher Tommy John. Dr. Jobe used a section of the tendon of insertion of the palmaris longus, an anterior muscle of the forearm, and wove it through drill holes in both the humerus and the ulna, making the tendon a substitute for the disrupted ligament. Today, the sports world refers to this reconstruction technique as "the Tommy John surgery."

An understanding of proper throwing mechanics is essential to prevent elbow and shoulder joint problems. Studies in kinesiology, biomechanics, athletic training, and sports medicine provide that understanding.

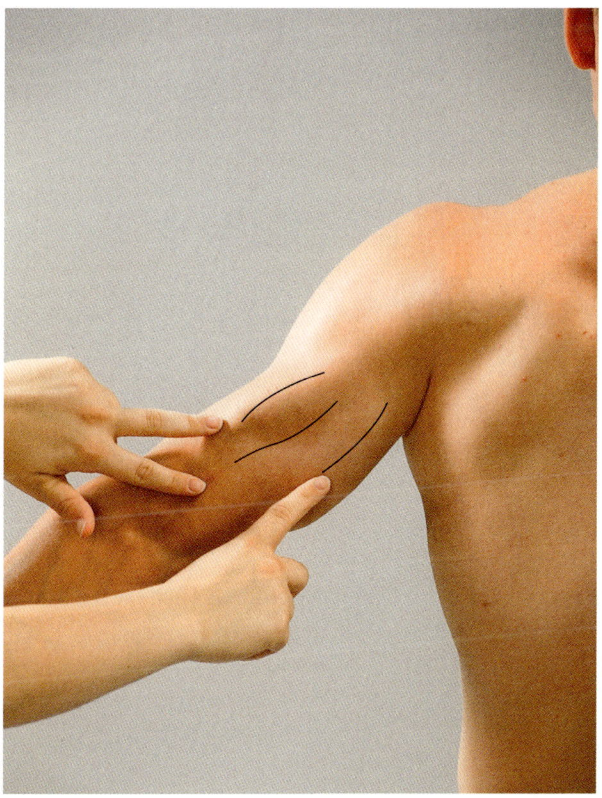

Figure 4.14 Locating the three sections of the triceps brachii.

Hands On

Locate the anconeus muscle on the posterior lateral aspect of your or your partner's elbow (figure 4.15).

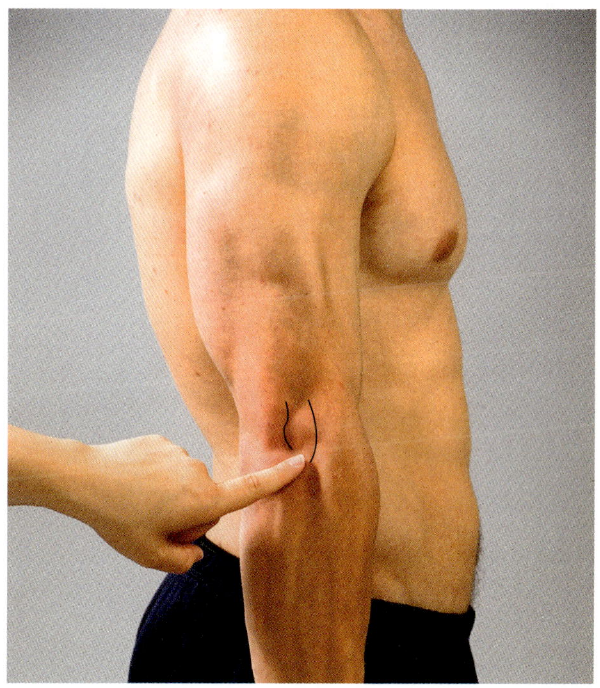

Figure 4.15 Locating the anconeus.

Muscles of the Forearm

Muscles of the forearm include two pronator and two supinator muscles. All are major producers of forearm movement. Only the biceps brachii has another function, that of elbow flexion, as previously mentioned.

- *Pronator teres:* Originating on the coronoid process of the ulna and inserting on the lateral surface of the radius, the pronator teres, as indicated by its name, is responsible for pronation of the forearm (see figure 4.9).

Hands On

Manually resist your partner's attempt to flex their elbow and pronate their wrist while you look for the pronator teres muscle (figure 4.16).

Figure 4.16 Locating the pronator teres muscle.

- *Pronator quadratus:* This muscle originates on the radius and inserts on the ulna just proximal to the wrist (figure 4.17). Its name reflects its function and shape. Because the ulna is the stable bone in the distal radioulnar articulation, when the pronator quadratus muscle contracts, the radius is pulled over the ulna, and forearm pronation takes place.
- *Supinator:* The supinator originates on the ulna and inserts on the radius at the proximal ends of the bones on the posterior aspect (figure 4.17). The name of this muscle indicates its function: forearm supination.
- *Biceps brachii:* Because of the position of the biceps brachii attachment to the radial tuberosity, when the forearm is in pronation, contraction of the biceps brachii causes the radius to rotate externally (laterally), causing the forearm to supinate (see figures 3.31 and 4.9).

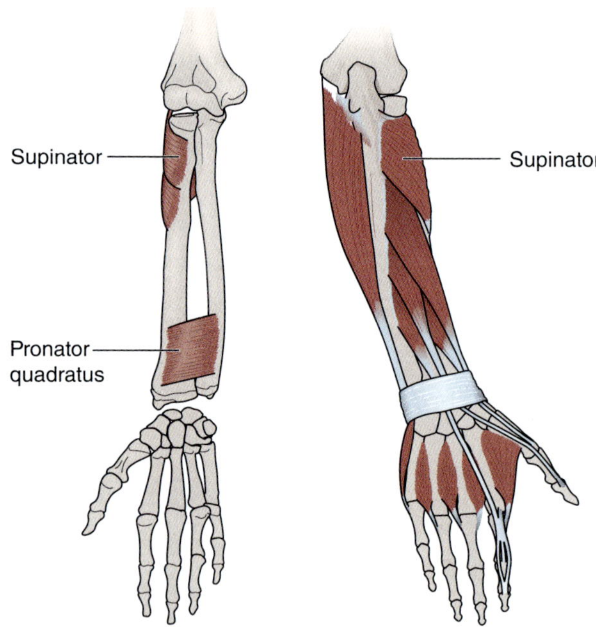

Figure 4.17 The pronator quadratus and the supinator muscles.

POP QUIZ

Look carefully at this pitcher's elbow at the end of the windup. What ligament is under stress (repeatedly during every pitch)? Hint: This ligament is often subject to reconstructive surgery, commonly called "the Tommy John surgery."

POP QUIZ

Observe the player's elbow at the top end of the service at the point of contact between the racket and the ball. At this point, which ligament of the elbow joint is placed under stress?

Observe this player hitting a backhand, and identify what muscles and bony attachments are under stress. Hint: Inflammation of this area is often referred to as tennis elbow.

Observe this player hitting a two-hand backhand shot. While the force and amount of topspin may be less than that of the one-hand backhand, stress on what muscles and bony attachments will be reduced?

FOCUS ON

Tennis Elbow

Tennis elbow is the bane of tennis players. In tennis, when you strike the ball using a backhand stroke, particularly if you attempt to put topspin on the ball as you strike it, the forearm moves from pronation to supination, the wrist moves from flexion to extension, and the supinator muscle, along with the muscles that originate in the common extensor tendon (presented in chapter 5 on the wrist and hand), contracts. If the tennis ball is hit off center (outside the so-called sweet spot), a torque (turning force) is applied to the racket that opposes the force applied by the supinator–extensor muscle group. This torque can result in tennis elbow, an inflammation around the supinator–extensor muscle group's origin on the lateral epicondyle of the humerus. An understanding of proper stroke technique is essential to prevent tennis elbow from developing. Sometimes, whether in person or on television, carefully observe the forearms of the top professional tennis players. You will be amazed at the difference in forearm size between the serving arm and the nonracket arm.

As a challenge, think about how the two-hand backhand stroke might help prevent tennis elbow. Hint: The biomechanical action of the forearm is different between the one-hand and two-hand backstrokes. How is the stress reduced?

Visit the web resource to complete the activities for this chapter.

CHAPTER 5

The Wrist and Hand

The wrist and hand are complicated structures with multiple bones (figure 5.1), ligaments, joints, and muscles. Because of the fine movements performed by the hand and thumb, these areas are very complex and require more time and effort to learn about than the areas previously presented. Although the thumb is often considered one of the hand's five fingers, its movements are unique, so it is discussed separately. Humans have prehensile hands (i.e., capable of grasping). The capacity to grasp is a direct result of the ability of the thumb to perform opposition, which is discussed under the movements of the thumb. The structure of the thumb joint and its muscles of the thenar eminence (and, to a lesser extent, the hypothenar eminence of the little finger) contribute to the ability of primates (the mammalian order that includes monkeys, apes, and human beings) to grasp things in their hands. These structures are presented in detail later in this chapter.

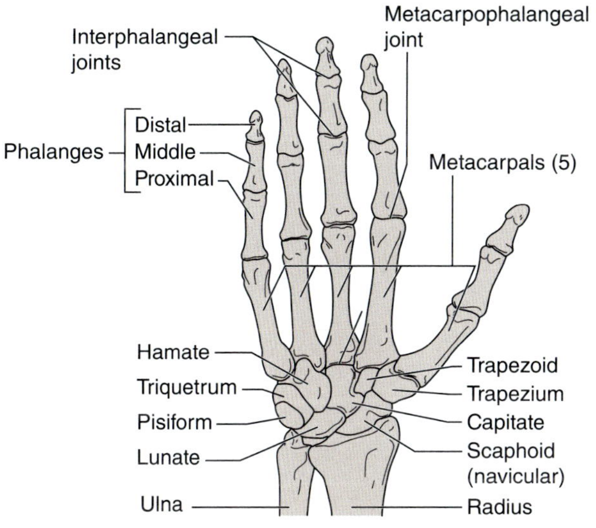

Figure 5.1 Bones of the wrist and hand.

Bones of the Wrist and Hand

The wrist contains eight bones, roughly aligned in two rows, known as the *carpal bones*. The proximal row of carpal bones articulates with the forearm (radius and ulna), and the distal row of carpal bones articulates with the long bones of the hand (the metacarpals). The proximal row of carpal bones, from lateral to medial, is identified as the scaphoid (also known as the *navicular*), the lunate, the triquetrum, and the pisiform. The pisiform and the scaphoid bones in the proximal row are easy to palpate.

Hands On

Place your index finger on the spots indicated in figures 5.2 and 5.3, and apply pressure downward to feel these bones.

The scaphoid (navicular) is a peanut-shaped bone and the most frequently fractured bone of the wrist, typically when forced into the distal end of the radius from a fall on an extended wrist. The lunate, because of its smooth, dome-shaped proximal end, is the most frequently dislocated bone of the wrist, typically when it is forced into the distal end of the radius from a fall on a flexed wrist. The bones in the distal row of carpals articulate with the bones of the hand. They are, from lateral to medial, the trapezium (also known as the *greater multangular* or *multangulum major*), the trapezoid (also known as the *lesser multangular* or *multangulum minor*), the capitate, and the hamate.

Hands On

Downward pressure on the spot illustrated in figure 5.2 will bring you in contact with the hamate bone in the distal row of carpal bones.

There are five bones of the hand known as the *metacarpal bones*. The metacarpal bone of the thumb is usually referred to as the *first metacarpal*. The second metacarpal is that of the index finger, the third metacarpal is that of the middle finger, the fourth metacarpal is that of the ring finger, and the fifth metacarpal bone is that of the little finger. Distal to each of the metacarpal bones of the hand are the phalanges of the fingers. The thumb has two phalanges known as the *proximal phalanx* and the *distal phalanx*. The other fingers each consist of three phalanges: a proximal phalanx, a middle phalanx, and a distal phalanx.

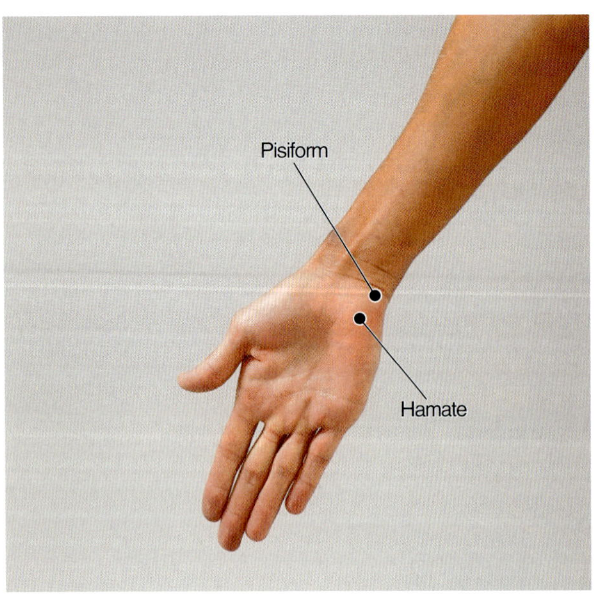

Figure 5.2 Locating the pisiform and the hamate.

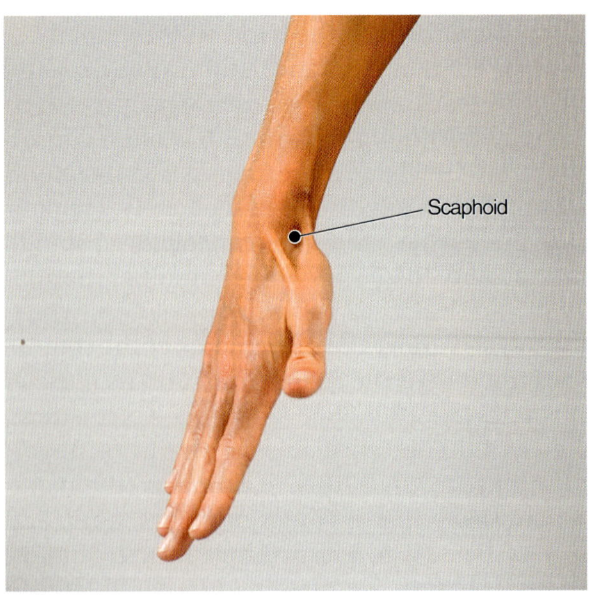

Figure 5.3 Locating the scaphoid.

Joints and Ligaments of the Wrist and Hand

With the large number of bones composing the wrist (ulna, radius, eight carpals, and five metacarpals), it makes sense that several joints make up the structure known as the *wrist* (figure 5.4). There are joints between the forearm bones and the proximal row of the carpals (radiocarpals), joints between the proximal and distal rows of carpals (midcarpals), and joints between the distal row of carpals and the five metacarpal bones of the hand (carpometacarpals). In addition to these wrist joints, there are joints between the carpal bones within each row (intercarpals).

What is commonly referred to as the wrist joint is the articulation between the distal end of the radius and primarily two bones of the proximal row of carpal bones: the scaphoid (navicular) and the lunate. The movement between these bones produces a gliding type of action as they roll or slide over each other. The radiocarpal joints are classified as condyloid joints because of their movements.

There are five main ligaments of the wrist (figure 5.5). A capsular ligament runs between the distal ends of the ulna and radius to the

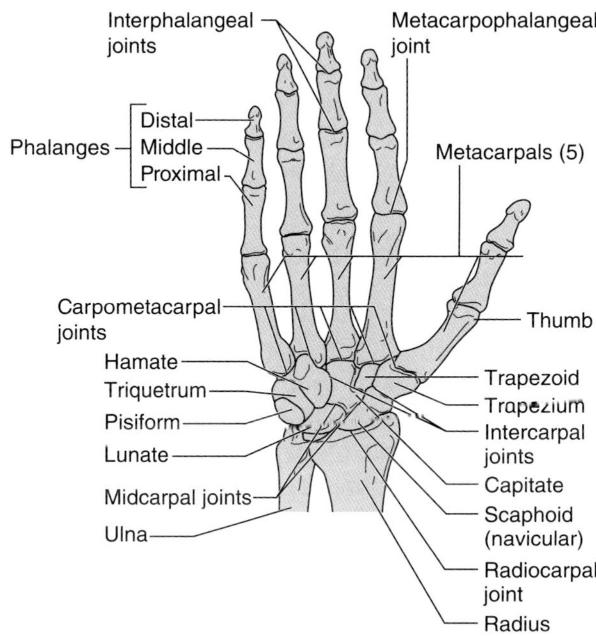

Figure 5.4 The wrist and intercarpal joints.

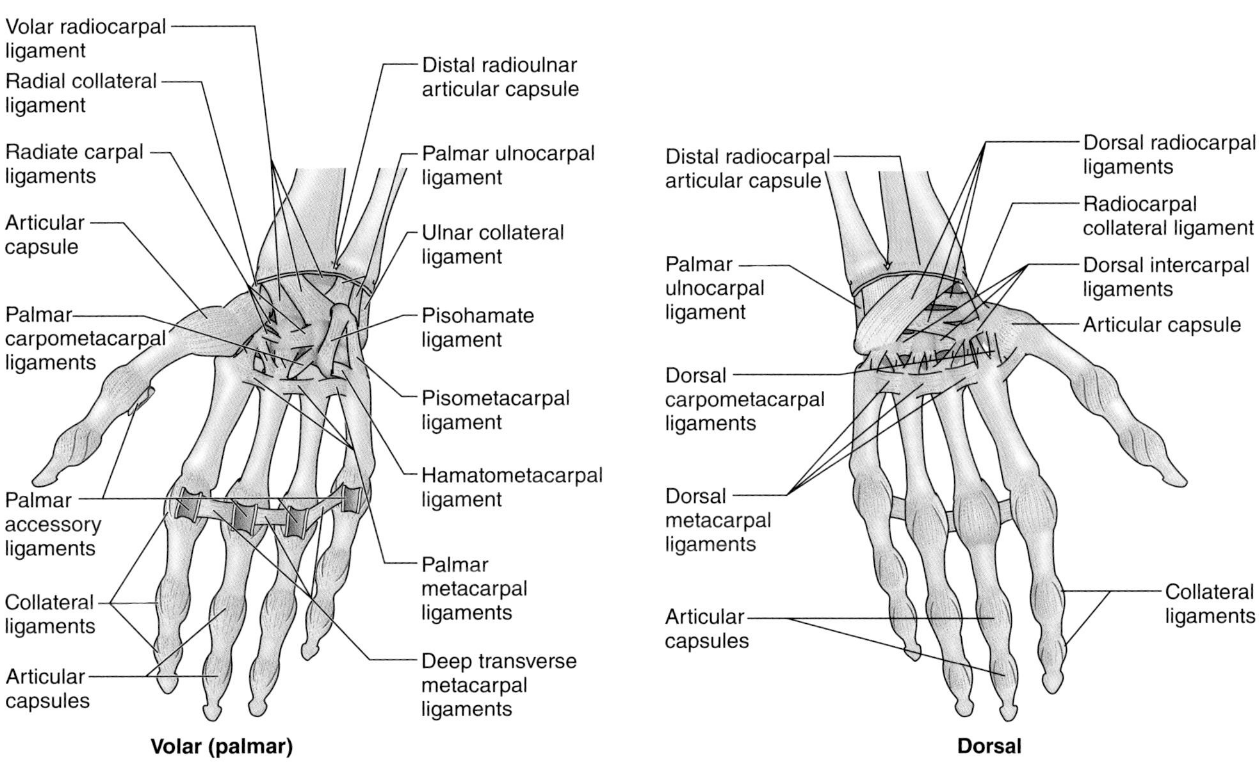

Figure 5.5 The capsular ligament between the distal ends of the ulna and radius; the volar radiocarpal ligament; the dorsal radiocarpal ligament; the radial and ulnar collateral ligaments of the wrist; and the capsular and collateral ligaments of the interphalangeal joints.

proximal row of carpal bones. There is a volar (palmar) radiocarpal ligament and a dorsal radiocarpal ligament. The volar radiocarpal ligament is found between the anterior surface of the radius and its styloid process and the proximal row of carpal bones (figure 5.5). The dorsal radiocarpal ligament is found between the distal end of the radius and the proximal row of carpal bones (figure 5.5). The two additional ligaments of the wrist are collateral ligaments: the radial (lateral) and ulnar (medial) collateral ligaments (figure 5.5). The radial collateral ligament of the wrist runs between the styloid process of the radius and the scaphoid carpal bone. The ulnar collateral ligament runs between the styloid process of the ulna and the medial portions of the pisiform and triquetrum bones.

The intercarpal joints between the carpal bones of the wrist are connected by three forms of intercarpal ligaments: those that connect the four carpal bones in the proximal row, those that connect the four carpal bones in the distal row, and those that connect the carpals of the proximal row to those of the distal row. The intercarpal ligaments can be further divided into volar, dorsal (see figure 5.5), interosseous, radial and ulnar collateral, pisohamate, and pisometacarpal ligaments. All of these ligaments are referred to in this text as simply the *intercarpal ligaments*. The intercarpal joints move in a gliding motion.

The last group of joints considered to be part of the wrist are the carpometacarpal joints, of which there are five: four between the four carpal bones in the distal carpal row and the bases of the four metacarpal bones of the hand, and one between the trapezium and the base of the first (thumb) metacarpal bone. The motion of these joints is gliding. The ligaments of the four carpometacarpal joints of the hand are the dorsal, volar, interosseous, and capsular carpometacarpal ligaments. The dorsal and volar (palmar) carpometacarpal ligaments (see figure 5.5) are found between the dorsal and volar surfaces of the distal row of carpal bones and the bases of the metacarpal bones. The interosseous ligaments are found between the hamate and capitate bones and the bases of the third and fourth metacarpal bones. The capsular ligaments are located between the distal row of carpals and the bases of the four metacarpals of the hand. The first (thumb) carpometacarpal joint is unique compared with the other four carpometacarpal joints; its ligamentous structure consists of a loose capsular carpometacarpal ligament that is found between the trapezium and the base of the first (thumb) metacarpal bone. The carpometacarpal joint of the thumb is referred to as a *saddle joint* because of its shape.

Two additional ligamentous structures of the wrist are the flexor (volar) and extensor (dorsal) retinacula (figure 5.6), which are bands of connective tissue over the volar surface (flexor retinaculum) and the dorsal surface (extensor retinaculum) of the wrist. The flexor retinaculum forms a bridge over the carpal bones to form the carpal tunnel through which the flexor muscle tendons of the wrist and hand pass. There is a much smaller space between the extensor retinaculum and the carpal bones through which pass the extensor tendons of the wrist and hand.

The joints of the hand and fingers include five metacarpophalangeal (MCP) joints, which are articulations between the five long metacarpal bones of the hand and the five proximal phalanges of the fingers. Also, the four fingers, each having three phalanges, have proximal interphalangeal (PIP) joints and distal interphalangeal (DIP) joints. The thumb, having only two phalanges, has only one interphalangeal (IP) joint. All five MCP, four PIP, and four DIP joints of the fingers and the IP joint of the thumb have capsular ligaments and ulnar and radial collateral ligaments (see figure 5.5).

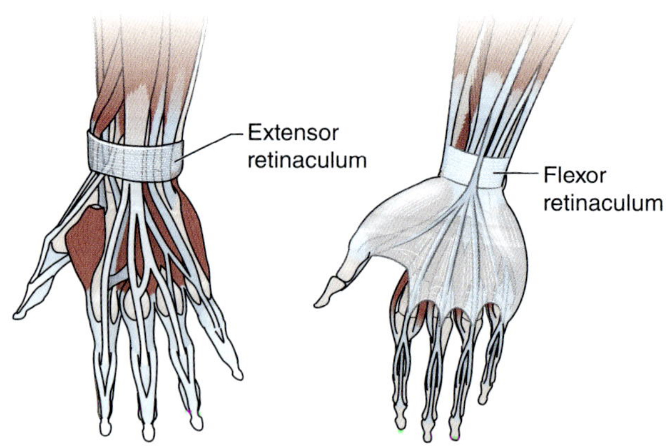

Figure 5.6 The flexor and extensor retinacula.

POP QUIZ

In the photo, the gymnast has the wrists taped to reinforce what structures?

a. flexor retinaculum
b. carpal tunnel
c. extensor retinaculum
d. proximal carpal row

FOCUS ON

Carpal Tunnel Syndrome

In addition to the various tendons, blood vessels and nerves pass through the carpal tunnel. Anything causing inflammation of these tissues in the tunnel, such as direct trauma or overuse of muscles that have tendons passing through the tunnel, can cause swelling within the tunnel. Pressure on the tendons, blood vessels, or nerves can result in pain and diminished functioning of any of these tissues. This condition is often referred to as *carpal tunnel syndrome*.

Repetitive motions that stress the wrist may lead to carpal tunnel syndrome. A common cause of carpal tunnel syndrome is typing at a computer keyboard. In an effort to alleviate this stress, design engineers and biomechanists use an understanding of anatomy to produce new and safer keyboards. The field of study that attempts to improve biomechanical working conditions is known as *ergonomics*.

Fundamental Movements of the Wrist and Hand

Remember that the primary movement across all the numerous joints of the wrist is defined as *gliding*. A combination of these gliding joint actions results in the wrist having four fundamental movements (figures 5.7 and 5.8): flexion (which takes place primarily in the radiocarpal joints), extension (which takes place primarily in the midcarpal joints), ulnar deviation (adduction), and radial deviation (abduction). Although movement at the wrist that results in the hand moving toward the midline of the body might be considered adduction, it is known as *ulnar deviation*. Likewise, moving the hand away from the midline of the body might be considered abduction, but it is known as *radial deviation*. Keep in mind that the wrist is a biaxial joint (i.e., it can move in two planes about two axes) and is capable of circumduction. Circumduction was defined in chapter 2 as a combination of fundamental movements of a biaxial or triaxial joint. The MCP, PIP, DIP, and IP joints of the fingers and thumb are capable of flexion and extension. The MCP joints are also capable of abduction and

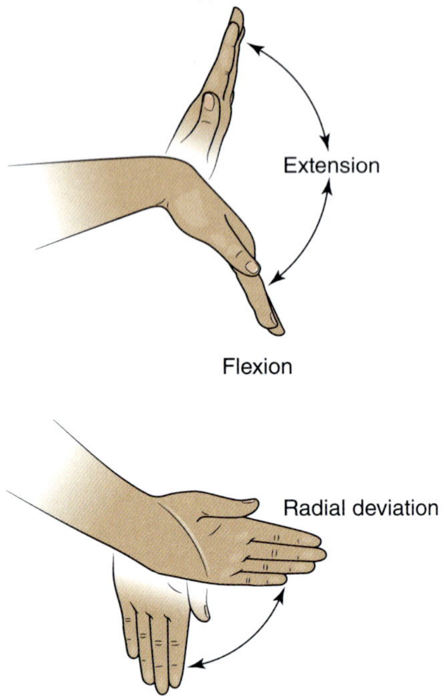

Figure 5.7 Movements of the wrist.

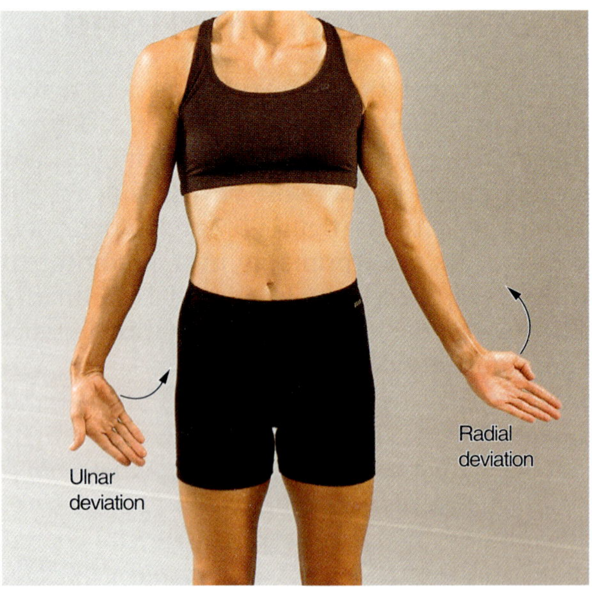

Figure 5.8 Radial and ulnar deviation.

adduction. The thumb is capable of additional movements that are presented later when the muscles of the thumb are discussed.

Extrinsic Muscles of the Wrist and Hand

Several muscles that are responsible for movement in the wrist actually originate above the elbow joint on either the medial or lateral epicondyle of the humerus. Although these muscles cross the elbow joint, they are not typically considered muscles that create movement in the elbow joint. They are considered muscles of the wrist and hand. Muscles that originate externally to the hand (on the humerus, ulna, or radius) and insert within the hand are referred to as *extrinsic muscles of the hand*.

Anterior Muscles

There are five major muscles of the hand and wrist that appear on the anterior (volar) surface of the forearm. Four of these muscles (figure 5.9) originate on the medial epicondyle of the humerus on a structure known as the *common flexor tendon*: the flexor carpi radialis, the flexor carpi ulnaris, the flexor digitorum superficialis, and the palmaris longus. The fifth muscle, not part of the common flexor tendon, is the flexor digitorum profundus.

- *Flexor carpi radialis:* Part of the group of muscles originating from the medial epicondyle of the humerus, off the structure known as the *common flexor tendon*, this muscle inserts on the bases of the second and third metacarpal bones of the hand (see figure 5.9). Contraction of the flexor carpi radialis muscle produces flexion and radial deviation of the wrist.

Hands On

Make a fist and flex your wrist, find the tendon of the flexor carpi radialis, and trace it upward with your finger until you feel the belly of the muscle (figure 5.10).

- *Flexor carpi ulnaris:* Originating on the common flexor tendon, the flexor carpi ulnaris inserts on the pisiform and hamate carpal bones of the wrist and the base of the fifth metacarpal bone of the hand (see figure 5.9). This muscle produces flexion and ulnar deviation of the wrist.

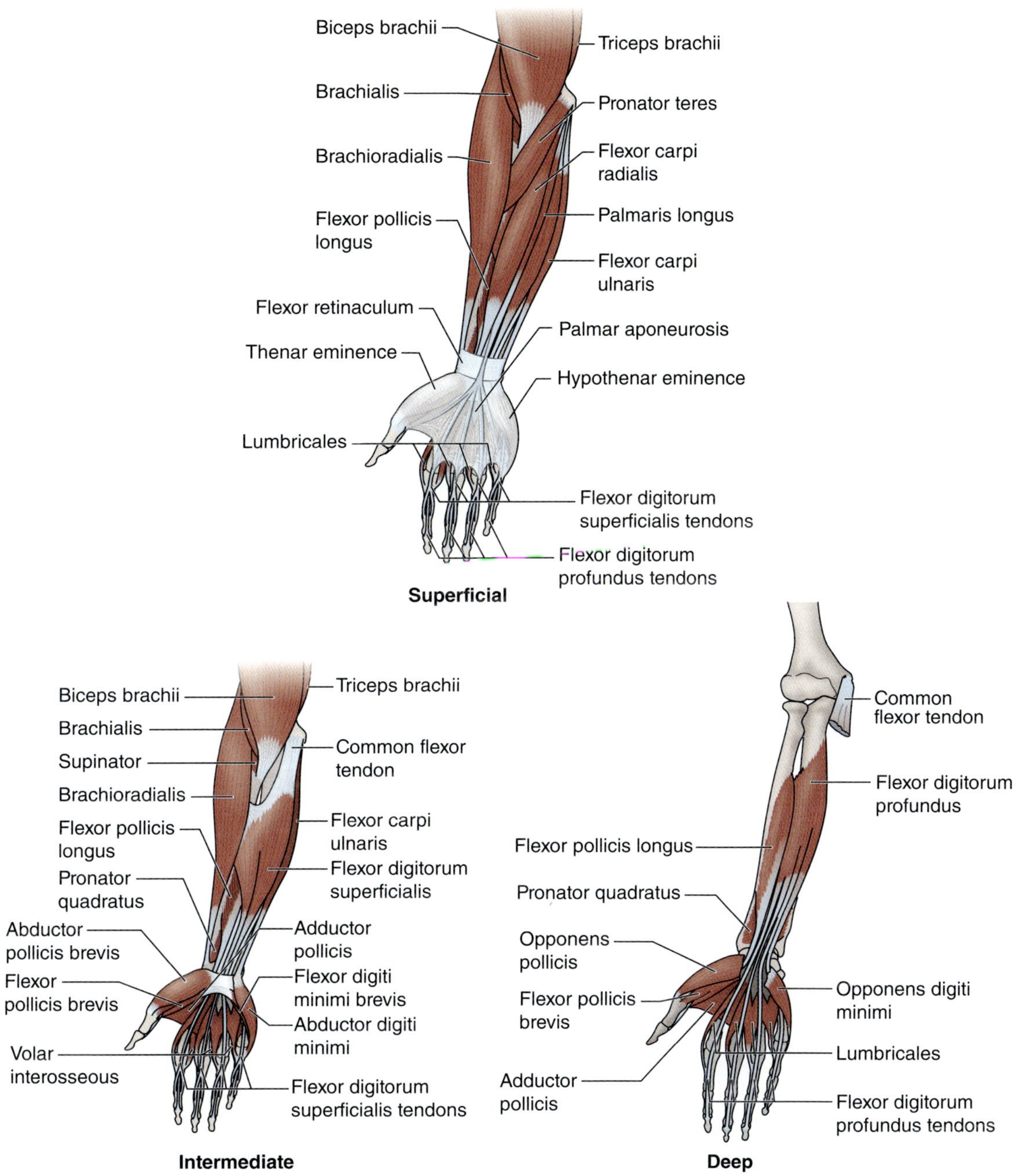

Figure 5.9 Anterior view of the superficial, intermediate, and deep muscles of the wrist and hand.

Hands On

Flex your wrist against a resistance and note the tendon of the flexor carpi ulnaris (figure 5.11).

- *Flexor digitorum superficialis:* This muscle originates on the common flexor tendon; its tendon of insertion splits into four separate tendons that split again and insert on each side of the bases of the middle phalanges of the four fingers (see figure 5.9). This muscle flexes the wrist, the MCP joint, and the PIP joint of the four fingers.

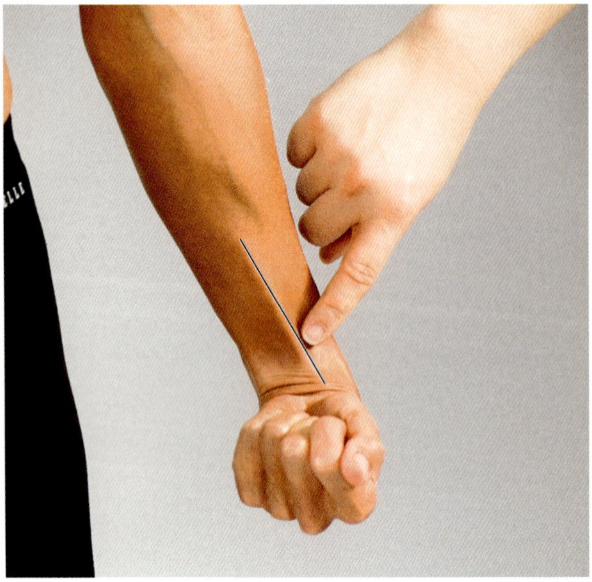

Figure 5.10 Locating the flexor carpi radialis.

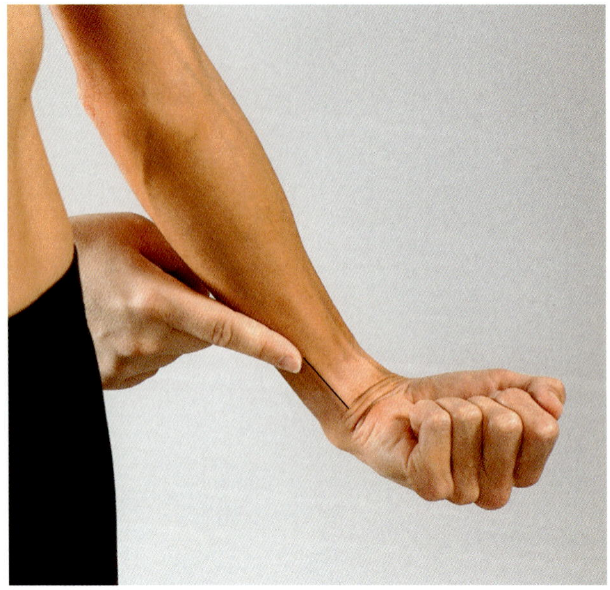

Figure 5.11 Locating the flexor carpi ulnaris.

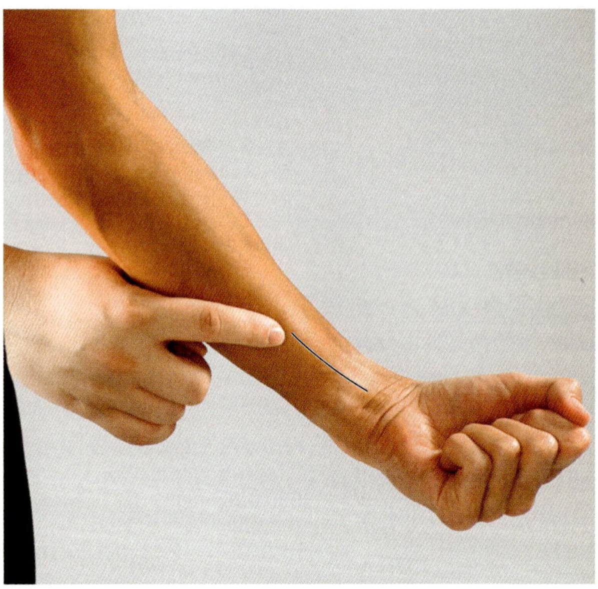

Figure 5.12 Locating the flexor digitorum superficialis.

Hands On

Apply pressure to the area of the forearm containing the tendon of the flexor digitorum superficialis, and flex and extend your fingers. You should feel or see movement. This movement is produced by the flexor digitorum superficialis (figure 5.12).

- *Palmaris longus:* This is the fourth and last muscle that has its origin on the common flexor tendon, and it inserts on the palmar aponeurosis (see figure 5.9). This muscle's very long tendon passes above the flexor retinaculum, unlike the tendons of the other common flexor tendon muscles, which pass beneath the retinaculum and through the carpal tunnel. Contraction of this muscle results in flexion of the wrist and tightening of the palmar fascia. The palmaris longus is absent in 20% of people, but its absence is not a problem because its function is to assist the other flexor muscles and not to act as a prime flexor. The tendon of this muscle is often used to reinforce elbow ligaments that need surgical reconstruction.

Hands On

Press your fingers and thumb together, flex your wrist, and feel for movement in the lower third of the anterior aspect of your forearm (figure 5.13). The palmaris longus muscle creates this movement.

- *Flexor digitorum profundus:* This muscle is not part of the common flexor tendon group. It originates on the proximal portion of the volar surface of the ulna and divides into four tendons that pass through the carpal tunnel and split to

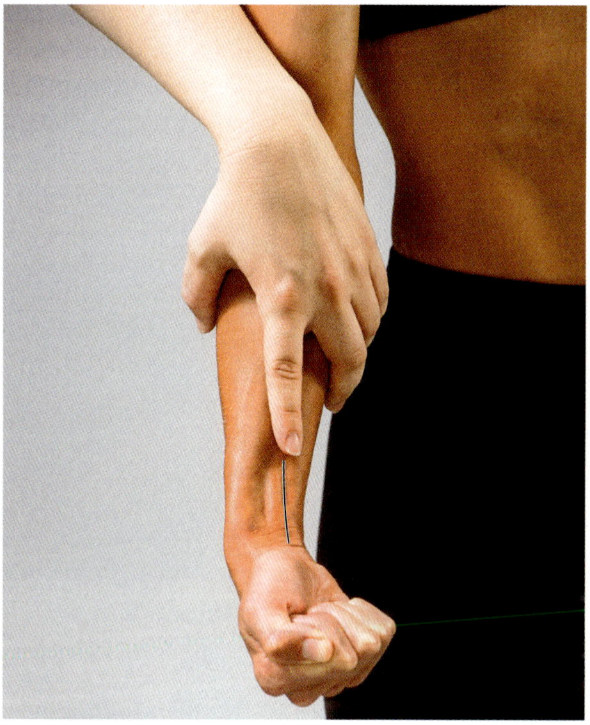

Figure 5.13 Locating the palmaris longus.

insert on either side of the bases of the distal phalanges of the four fingers (see figure 5.9). This muscle is involved in flexion of the wrist, the four MCP joints, the four PIP joints, and the four DIP joints of the fingers.

Posterior Muscles

Six major muscles of the hand and wrist appear on the posterior (dorsal) surface of the forearm (figure 5.14). Four of these muscles originate on the lateral epicondyle of the humerus on a structure known as the *common extensor tendon*: the extensor carpi radialis brevis, the extensor carpi ulnaris, the extensor digitorum (communis), and the extensor digiti minimi (proprius). The fifth and sixth muscles, not part of the common extensor tendon, are the extensor carpi radialis longus and the extensor indicis.

- *Extensor carpi radialis brevis:* This muscle (see figure 5.14) is one of the four muscles originating from the common extensor tendon on the lateral epicondyle of the humerus and inserts on

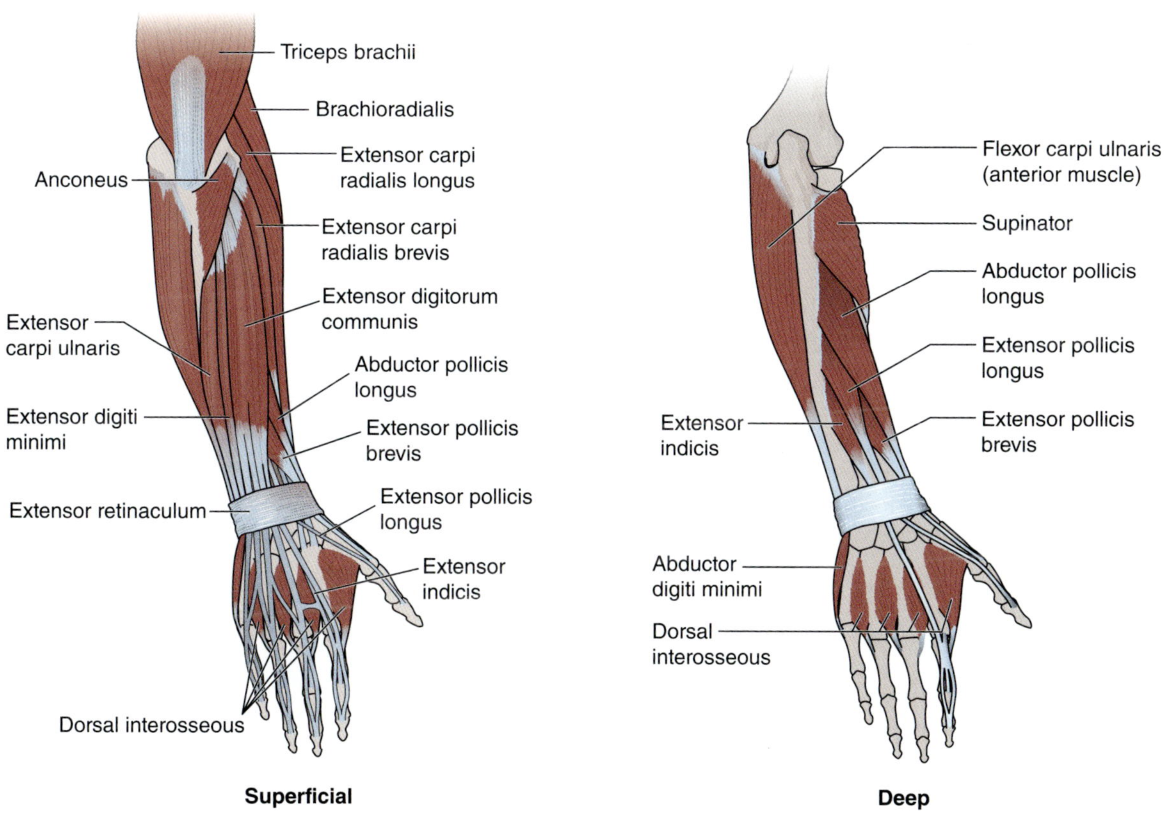

Figure 5.14 Posterior view of the superficial and deep muscles of the wrist and hand.

the dorsal aspect of the third metacarpal bone. The action produced by this muscle is extension and radial deviation of the wrist.

Hands On

Against resistance, extend your wrist. On the radial side of the dorsal aspect of the lower third of the forearm, you should see and feel two tendons: the extensor carpi radialis longus and the extensor carpi radialis brevis (figure 5.15).

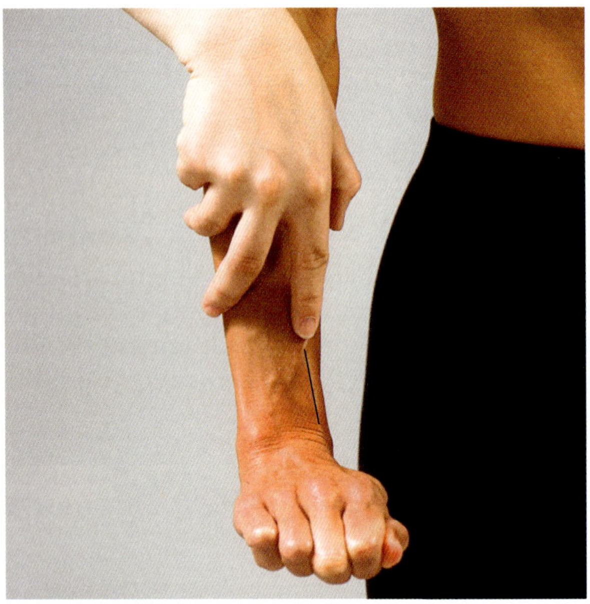

Figure 5.15 Locating the extensor carpi radialis longus and brevis.

- *Extensor carpi ulnaris:* Another of the four muscles originating on the common extensor tendon, this muscle (see figure 5.14) inserts on the dorsal aspect of the fifth metacarpal bone and extends and ulnarly deviates the wrist.

Hands On

Extend your wrist while it is in ulnar deviation, and you will feel the extensor carpi ulnaris muscle (figure 5.16).

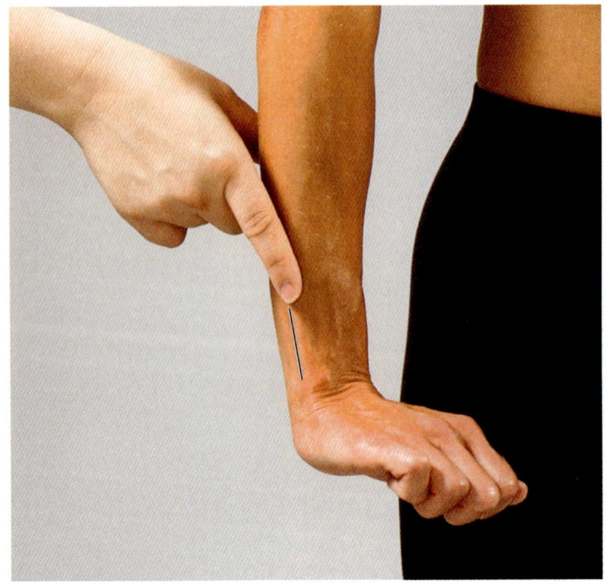

Figure 5.16 Locating the extensor carpi ulnaris.

- *Extensor digitorum (communis):* The third muscle originating from the common extensor tendon, this muscle is typically identified as the extensor digitorum, although some texts add the additional term communis to the name to distinguish it from similarly named muscles in the foot. The muscle divides into four tendons and inserts on the bases of the distal phalanges of the four fingers (see figure 5.14). The muscle extends the wrist, the MCP joints, the PIP joints, and the DIP joints of all four fingers.

Hands On

Extend your wrist and flex and extend your fingers. Observe the movement of the four tendons of the extensor digitorum on the dorsal aspect of your hand (figure 5.17).

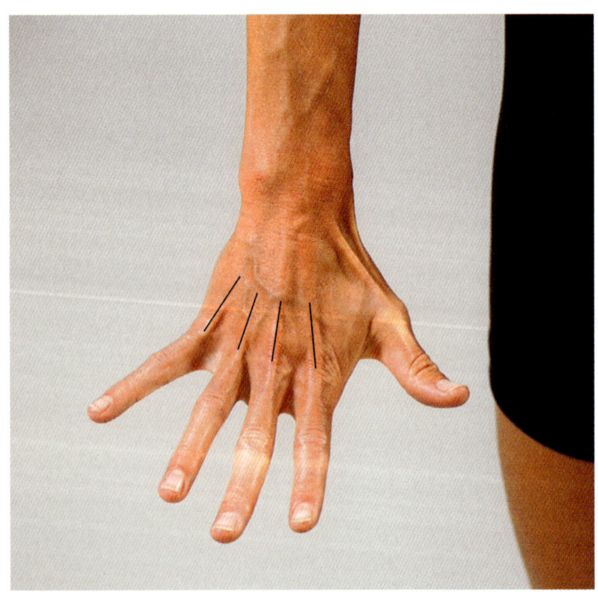

Figure 5.17 Viewing the extensor digitorum communis.

• *Extensor digiti minimi (proprius):* The fourth muscle originating from the common extensor tendon sometimes includes the term proprius. Most texts simply call the muscle the extensor digiti minimi, indicating that the muscle extends the little finger (digiti minimi). The muscle inserts on the base of the proximal phalanx of the fifth finger and extends the wrist and the MCP joint of the fifth finger (see figure 5.14).

Hands On

Extend your little finger and feel for movement by placing pressure over the fifth MCP joint. This movement is created by the extensor digiti minimi (figure 5.18).

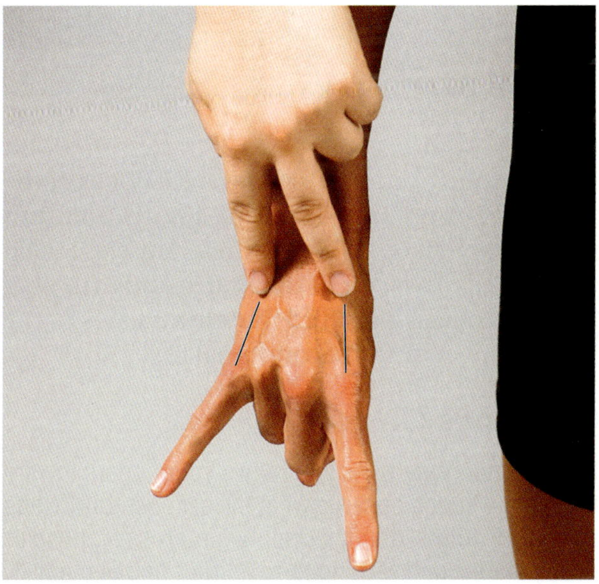

Figure 5.18 Locating the extensor indicis and extensor digiti minimi.

• *Extensor carpi radialis longus:* This is not a muscle of the common extensor tendon. It originates on the lateral supracondylar ridge of the humerus and inserts on the lateral side of the base of the second metacarpal bone (see figure 5.14). Its actions include extension and radial deviation of the wrist.

Hands On

Against resistance, extend your wrist. On the radial side of the dorsal aspect of the lower third of the forearm, you should see and feel two tendons: the extensor carpi radialis longus and the extensor carpi radialis brevis (see figure 5.15).

• *Extensor indicis:* This muscle originates on the dorsal surface of the distal portion of the ulna and inserts on the base of the proximal phalanx of the index finger (see figure 5.14). Its primary function is to extend the MCP joint of the index finger and secondarily to assist in extension of the wrist.

Hands On

Place pressure on your second MCP joint, and feel for movement as you flex and extend your index finger. The movement is created by the tendons of the extensor digitorum and, just medial to it, the extensor indicis (see figure 5.18).

Intrinsic Muscles of the Hand

Whereas the preceding extrinsic muscles of the hand originate outside the hand and insert within the hand, the following muscles are the intrinsic muscles of the hand, originating and inserting totally within the hand. Three of these muscles (see figure 5.9) make up an anatomical structure known as the *hypothenar eminence* (the abductor digiti minimi, the flexor digiti minimi brevis, and the opponens digiti minimi). *Hypo* (from Greek origin, meaning "less than") indicates that these are a group of muscles involved with the movement of the fifth (little) finger.

• *Abductor digiti minimi:* The first of three muscles making up the hypothenar eminence, this muscle originates on the pisiform bone of the wrist and inserts on the proximal phalanx of the little finger. It both abducts and assists with flexion of the fifth MCP joint.

• *Flexor digiti minimi brevis:* The second muscle of the hypothenar eminence originates on the hamate bone of the wrist and inserts on the proximal phalanx of the fifth finger. It assists in flexion of the fifth MCP joint.

• *Opponens digiti minimi:* This is the third and last muscle making up the hypothenar eminence. It originates on the hamate bone of the wrist and inserts on the fifth metacarpal and proximal phalanx bone of the fifth finger and assists with flexion and adduction of the fifth MCP joint.

The ability to bring the little finger toward the thumb to allow grasping is known as *opposition.*

Several muscles of the hand are not involved with the movement of the fifth finger but are involved in the movement of other finger joints. These muscles appear as three distinct muscle groups: the dorsal and volar (palmar) interossei (figure 5.19) and the lumbricals.

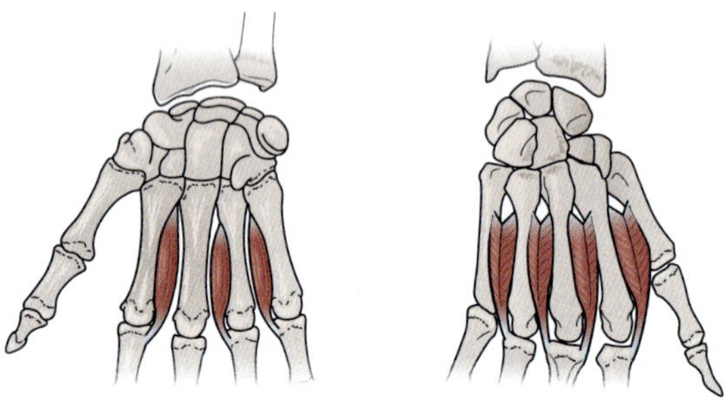

Figure 5.19 The volar (palmar) and dorsal interossei muscles of the hand.

- *Dorsal interossei:* There are four dorsal interossei (between the bones) muscles that originate on the four metacarpal bones of the hand (index, middle, ring, and little fingers) and insert on the second (index), third (middle), and fourth (ring) proximal phalanges. Two of these muscles attach to the lateral aspect of the second and third phalanges, whereas the other two attach to the medial aspect of the third and fourth phalanges. The first interosseus can be easily located on the dorsal side of the hand (figure 5.20).

Hands On

Spread your fingers and observe what the second, third, and fourth MCP joints do. The dorsal interossei muscles cause the second MCP joint to abduct (move laterally from the midline of the hand) and the fourth MCP joint to adduct (move medially from the midline of the hand), whereas the third MCP joint remains stationary. Why? Depending on the angle of the MCP joints, these muscles also assist with flexion and extension of the joints.

- *Palmar interossei:* There are three palmar (volar) interossei muscles that originate on the second, fourth, and fifth metacarpal bones of the hand and insert on the second, fourth, and fifth proximal phalanges. Two of these muscles attach to the medial aspects of the fourth and fifth phalanges, whereas the third muscle attaches to the lateral aspect of the second phalanx.

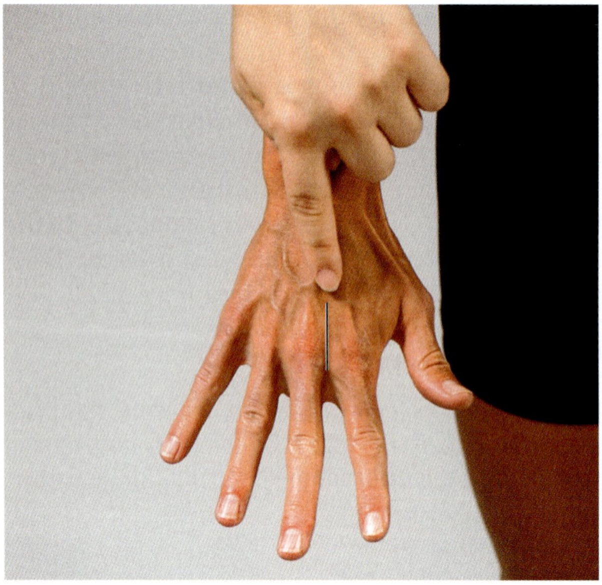

Figure 5.20 Locating the first dorsal interosseus.

Hands On

Spread your fingers. As you return your fingers toward the middle finger, observe what the second, fourth, and fifth MCP joints do. The palmar interossei muscles cause the second MCP joint to adduct (move medially toward the midline of the hand) and the fourth and fifth MCP joints to abduct (move laterally to the midline of the hand), whereas the third MCP joint remains stationary. Why? Depending on the angle of the MCP joints, these muscles also assist with flexion and extension of the joints.

- *Lumbricals:* These four muscles are found deep within the hand; they originate on the tendons of the flexor digitorum profundus and insert on the tendons of the extensor digitorum communis in the area of the proximal phalanges (see figure 5.9). This group of muscles assists with the flexion of the MCP joints and extension of the PIP and DIP joints (figure 5.21).

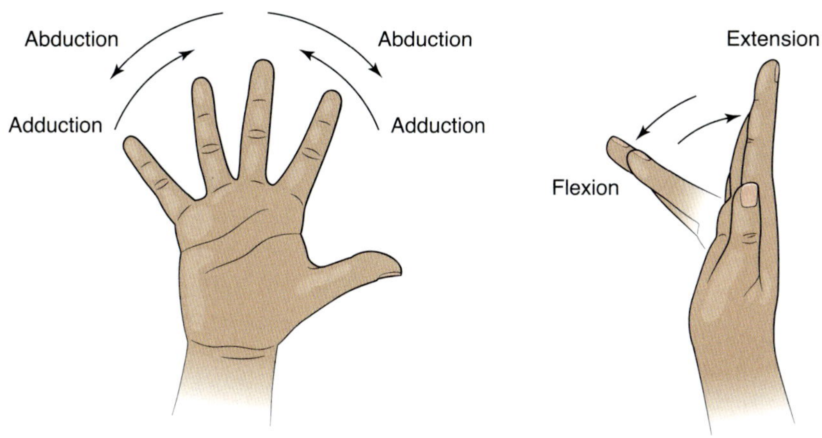

Figure 5.21 Movements of the metacarpophalangeal joints.

Muscles of the Thumb

Muscles involved with the movement of the thumb are divided into extrinsic (originating outside of and inserting within the hand) and intrinsic (originating and inserting within the hand) groups.

Extrinsic Muscles

These muscles all originate extrinsically, proximal to the thumb, and insert within the thumb. Note that all names of the muscles contain the word *pollicis* (from the Latin word *pollex*, for thumb).

- *Extensor pollicis longus:* This muscle originates on the ulna, inserts on the distal phalanx of the thumb, and extends both the IP and MCP joints of the thumb (see figure 5.14). This muscle also assists with radial deviation of the wrist and supination of the forearm.
- *Extensor pollicis brevis:* This muscle originates on the radius and inserts on the proximal phalanx of the thumb (see figure 5.14). It extends the MCP joint and assists with radial deviation of the wrist.

The tendons of the extensor pollicis brevis and the abductor pollicis longus, as they cross the wrist, form a depression (fossa) directly over the scaphoid carpal bone. This depression is commonly known as the *anatomical snuffbox* (figure 5.22). This term dates to the period in history when tobacco in the form of snuff was used by placing it on this area and then inhaling it through the nose.

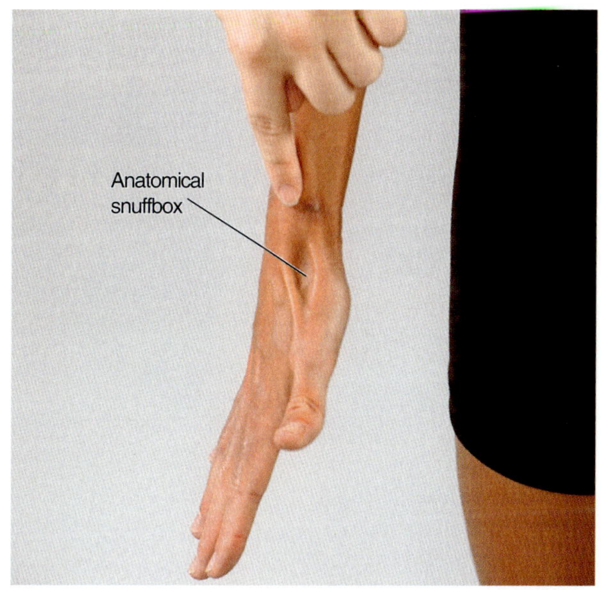

Figure 5.22 Locating the extensor pollicis longus and brevis and the anatomical snuffbox.

- *Abductor pollicis longus:* This muscle originates on the ulna and inserts on the base of the first metacarpal bone (see figure 5.14). It abducts the first metacarpal bone and assists with radial deviation and flexion of the wrist.

Hands On

Abduct your thumb, and apply pressure at the base of the thumb between the two tendons (see figure 5.22). The abductor pollicis longus is difficult to palpate on some people; try examining several people until you can palpate this muscle as your subject abducts the thumb (figure 5.23).

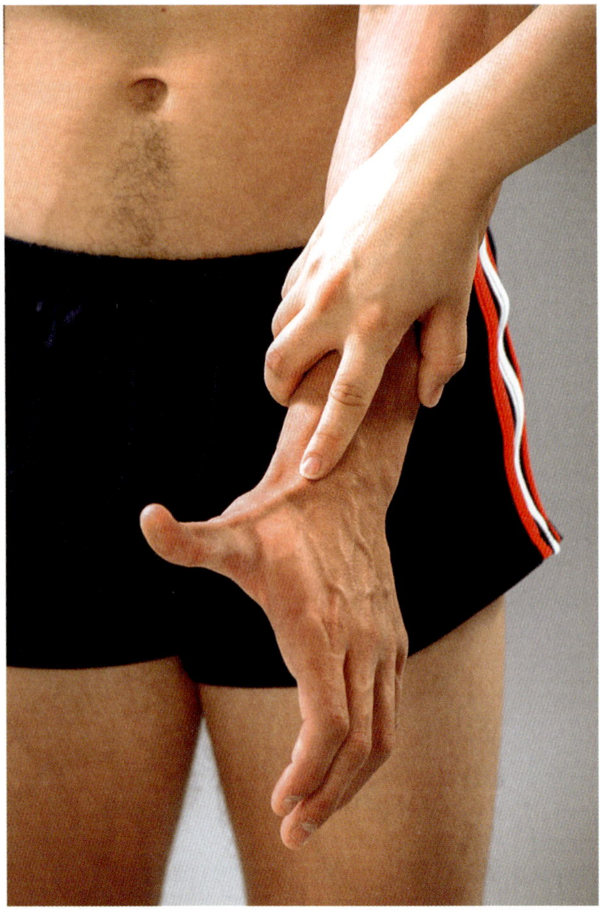

Figure 5.23 Locating the abductor pollicis longus.

- *Flexor pollicis longus:* This muscle originates on the radius and inserts on the distal phalanx of the thumb (see figure 5.9). It flexes both the IP and MCP joints of the thumb and assists with flexion of the wrist (figure 5.24).

Intrinsic Muscles

The following group of muscles both originate and insert totally within (intrinsically to) the hand. The bellies of these muscles form a thick pad of tissue just proximal to the thumb that is commonly known as the *thenar eminence* (see figure 5.9). Again, note that all four muscles have the term *pollicis* (thumb) in their names.

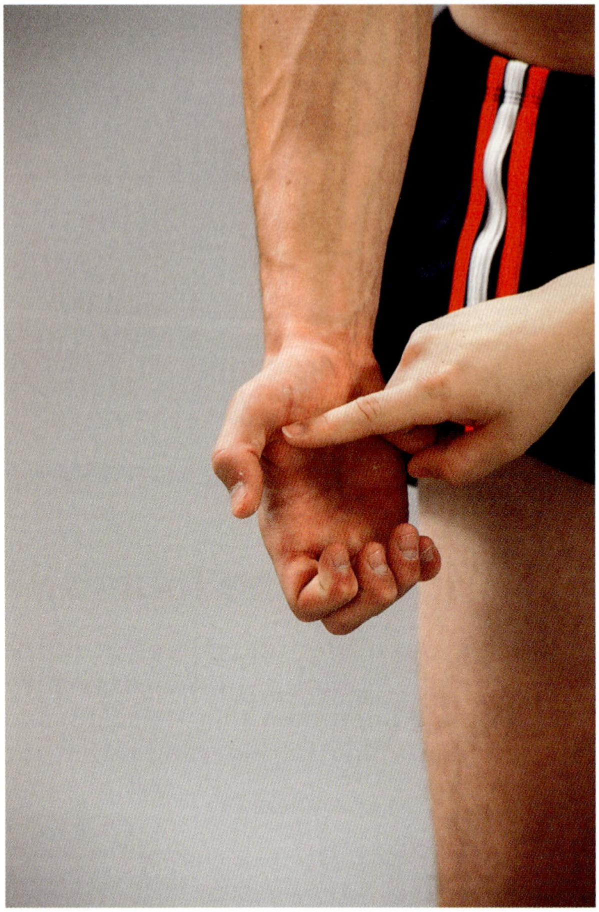

Figure 5.24 Locating the flexor pollicis longus.

- *Abductor pollicis brevis:* Originating from the trapezium and scaphoid bones and inserting on the base of the proximal phalanx of the thumb, this muscle assists with abduction of the thumb (see figure 5.9).

Hands On

Abduct your thumb and observe or feel the movement on the lateral aspect of the thenar eminence. This movement is produced by the abductor pollicis brevis.

- *Flexor pollicis brevis:* This muscle (see figure 5.9) originates on the trapezium bone and inserts on the proximal phalanx of the thumb. It assists the flexor pollicis longus in flexing the MCP joint of the thumb.

Hands On

Flex your thumb toward your second finger, and feel the flexor pollicis brevis move (figure 5.25).

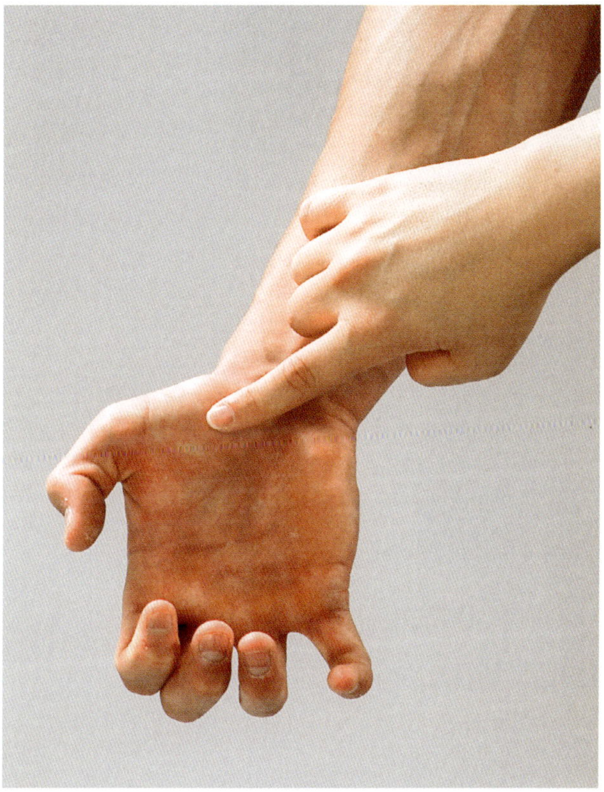

Figure 5.25 Locating the flexor pollicis brevis.

- *Opponens pollicis:* Originating on the trapezium and inserting on the first metacarpal bone, this muscle adducts (opposes) the thumb (see figure 5.9).

Hands On

Move your thumb toward the four fingers, and palpate along the lateral aspect of the thumb's MCP joint. The movement you feel is created by the opponens pollicis (figure 5.26).

- *Adductor pollicis:* This two-headed muscle has one head (oblique) originating on the capitate and second and third metacarpal bones and the other head (transverse) originating on the third metacarpal bone. The two heads combine

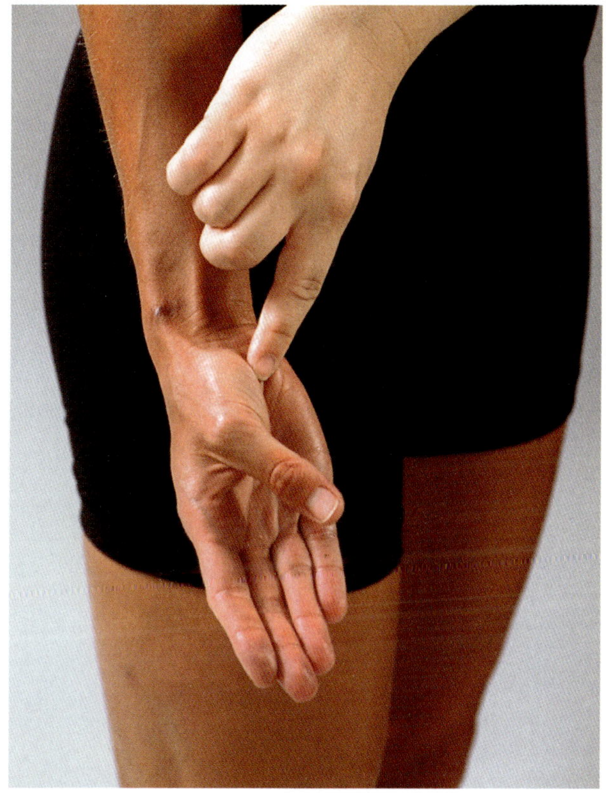

Figure 5.26 Locating the opponens pollicis.

and insert on the base of the thumb's proximal phalanx (see figure 5.9). The action of this muscle is adduction (opposition) of the thumb.

Hands On

Apply the thumb and index finger of your opposite hand to the space between the thumb and index finger, feeling for the tendon of the adductor pollicis.

Figures 5.27 through 5.29 illustrate the movements that the thumb is capable of performing as a result of actions by its extrinsic and intrinsic muscles. The thumb is capable of flexion and extension (figure 5.27), adduction and abduction (figure 5.28), and movement in opposition (figure 5.29). Note that, in describing thumb movements in the anatomical position, there are exceptions to the usual references. Thumb flexion and extension appear to take place in the frontal plane, whereas abduction and adduction appear to take place in the sagittal plane.

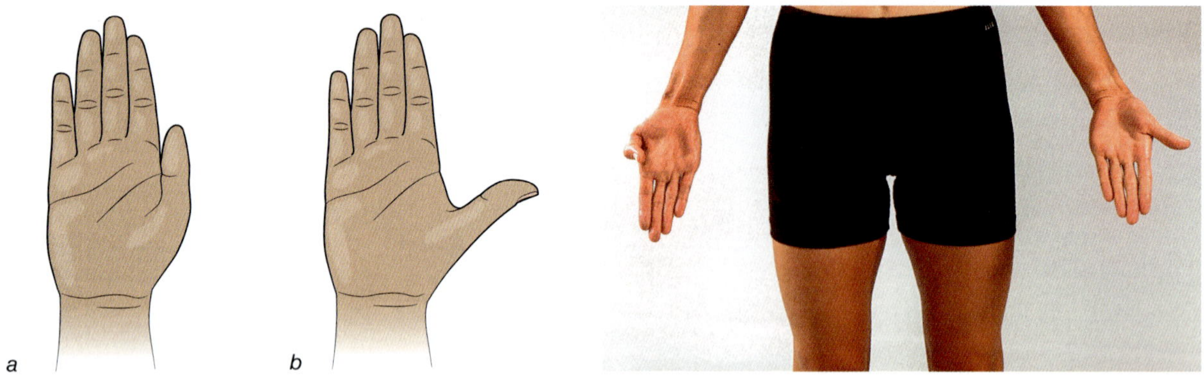

Figure 5.27 *(a)* Flexion and *(b)* extension of the thumb.

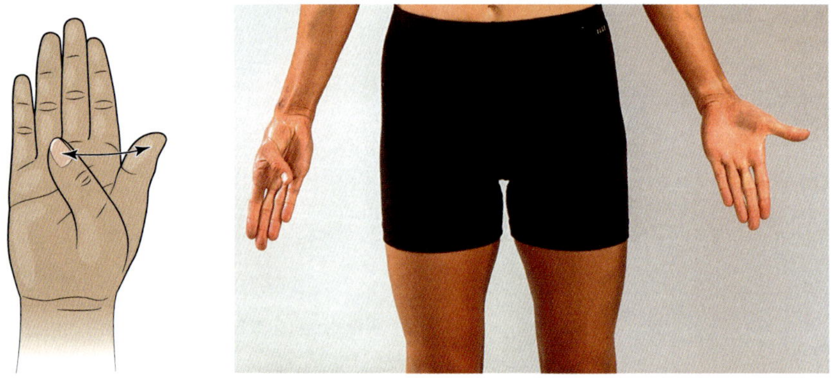

Figure 5.28 Abduction and adduction of the thumb.

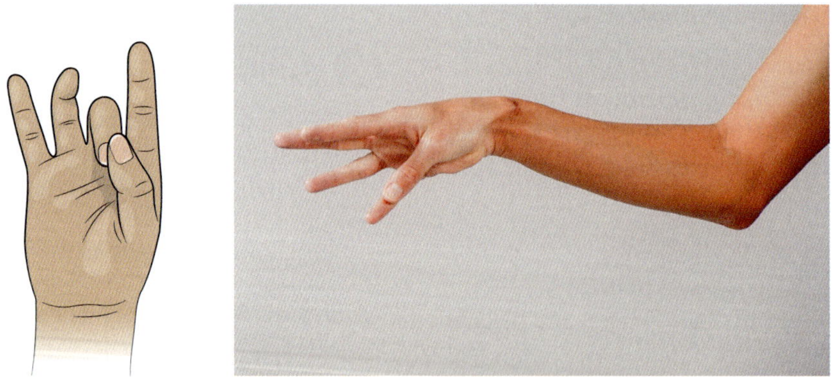

Figure 5.29 Opposition of the thumb.

Visit the web resource to complete the activities for this chapter.

CHAPTER 6

Nerves and Blood Vessels of the Upper Extremity

The structures known as nerves, arteries, and veins were presented in chapter 1. These structures are important to movement because they provide the stimulus to contract (nerves), supply nutrients via blood (arteries), and remove by-products of the muscle's efforts (veins) to move the bones. The nervous and vascular systems are very complex and have various functions. The nerves and blood vessels pertinent to the musculature of each anatomical section of this text (upper extremity; head, spinal column, and thorax, including the heart and lungs; and lower extremity) are presented after the bones, ligaments, and muscles have been discussed.

The nerves of the upper extremity originate in the spinal cord from the first cervical to the first thoracic vertebral section and are commonly classified as motor, sensory, or mixed nerves containing both motor and sensory capabilities. Motor nerves innervate muscle and, when stimulated voluntarily or involuntarily, cause muscle fibers to contract and create movement in the joints that these muscle fibers cross. These nerves are presented in the summary table at the end of the upper-extremity part of this text (see part II summary table).

The major upper-extremity arteries supply the fuel essential for muscular contraction, and the major veins carry away the waste products of muscular contraction. The major arteries supplying muscles are also listed in the summary table at the end of this part (see part II summary table).

The intent of this text is to examine the anatomy of the nervous system and the vascular system of those structures that are essential to the function of the body's musculature. Figure 6.1 shows some of the relevant nerves that are covered in this text.

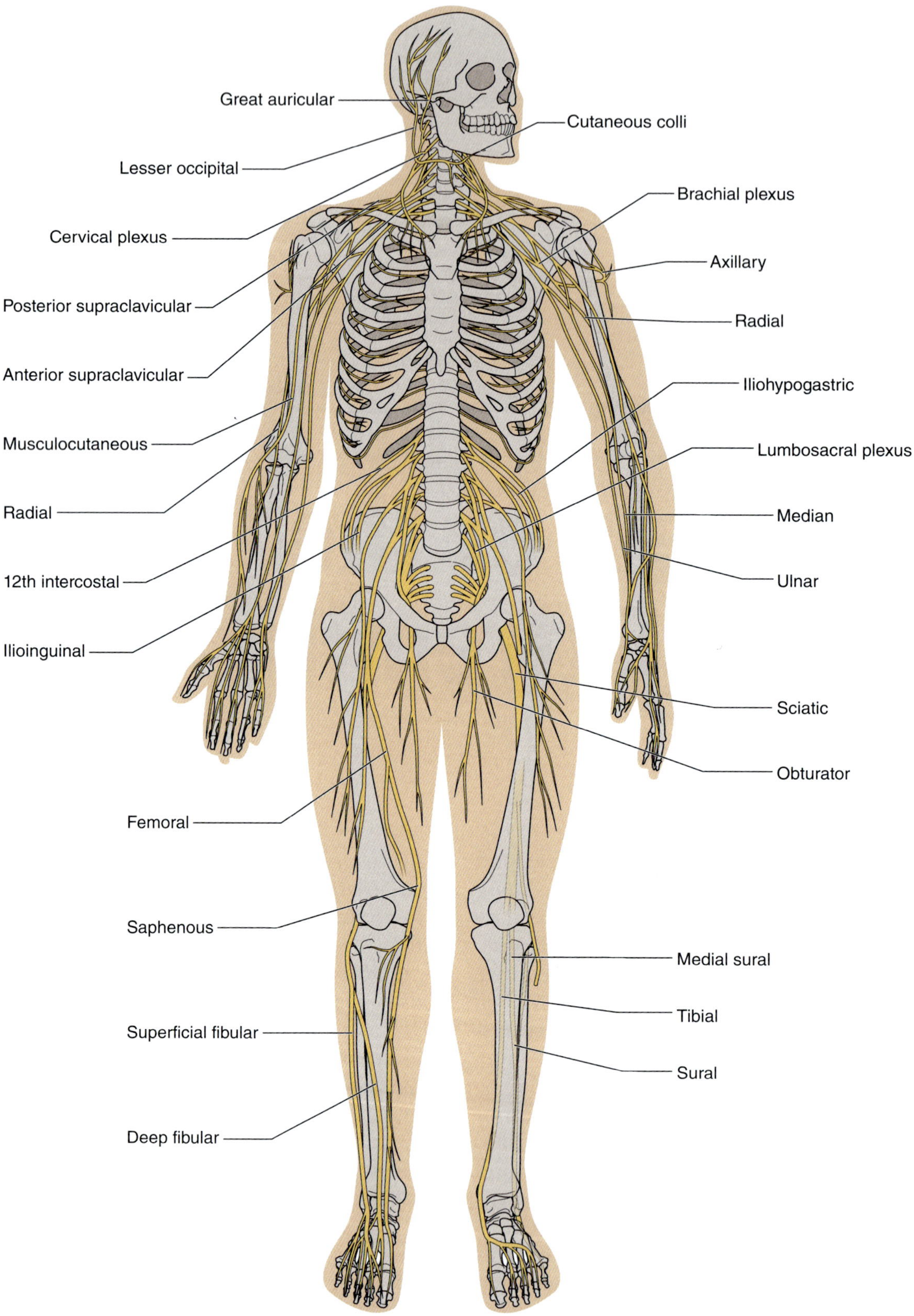

Figure 6.1 The nervous system superimposed over the skeleton.

Spinal nerves innervating the musculature form plexuses, or networks. In the upper extremity, the nerves arise from the brachial plexus (C5, C6, C7, C8, and T1 nerve roots and often the nerve roots from C4 and T2; figure 6.2). Although there are only seven cervical vertebrae, there are eight cervical spinal nerve roots. This is easily understood when you realize that the first cervical nerve root originates *above* the first cervical vertebra and the eighth cervical nerve root originates *below* the seventh cervical vertebra.

Nerves of the Brachial Plexus

Spinal nerves have both anterior and posterior rami (branches or arms). The anterior rami form the brachial plexus (figure 6.3), which is typically divided into two parts based on their position relative to the clavicle: the supraclavicular (above the clavicle) and the infraclavicular (below the clavicle) parts.

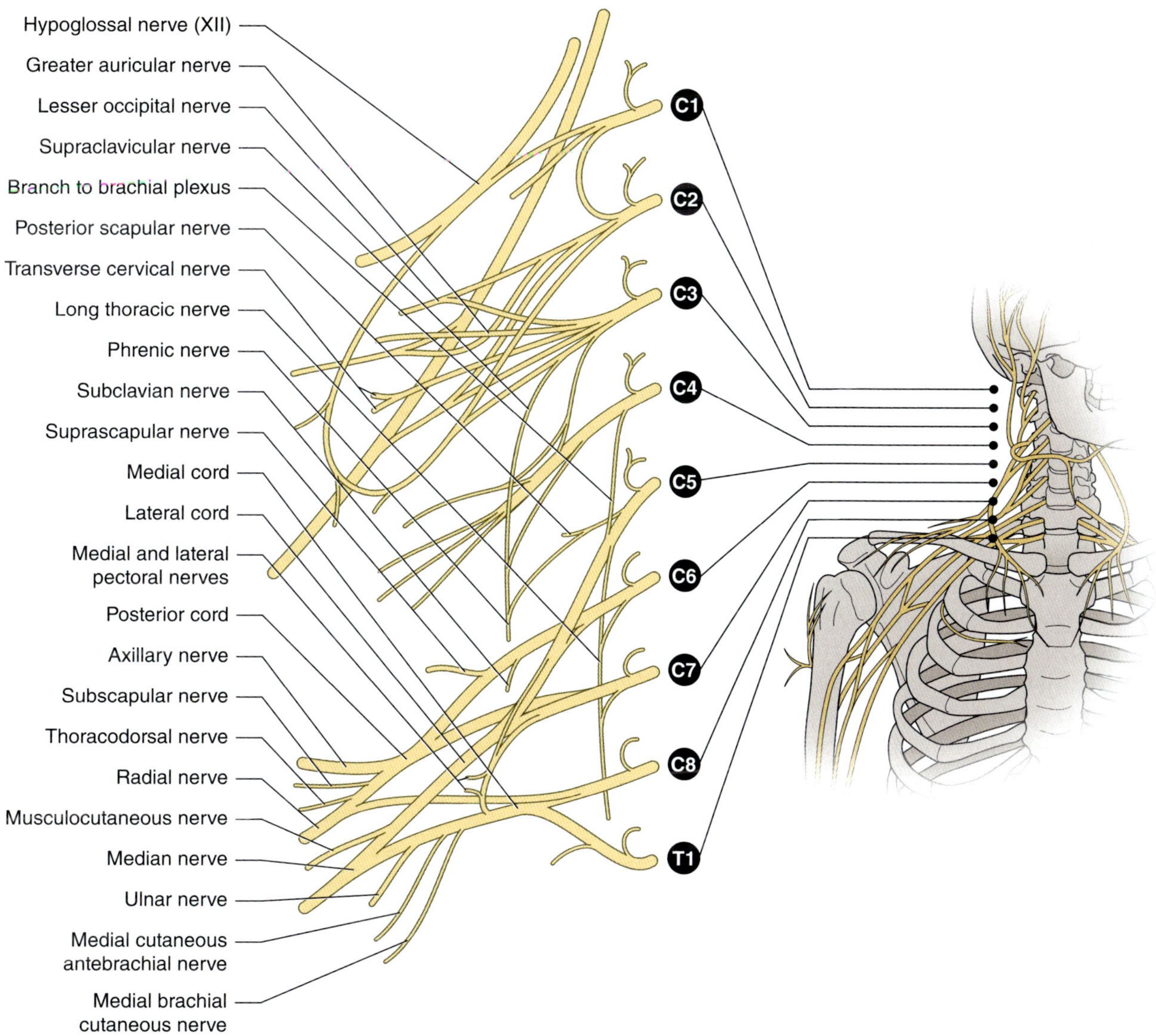

Figure 6.2 View of C1 through T1 nerve roots.

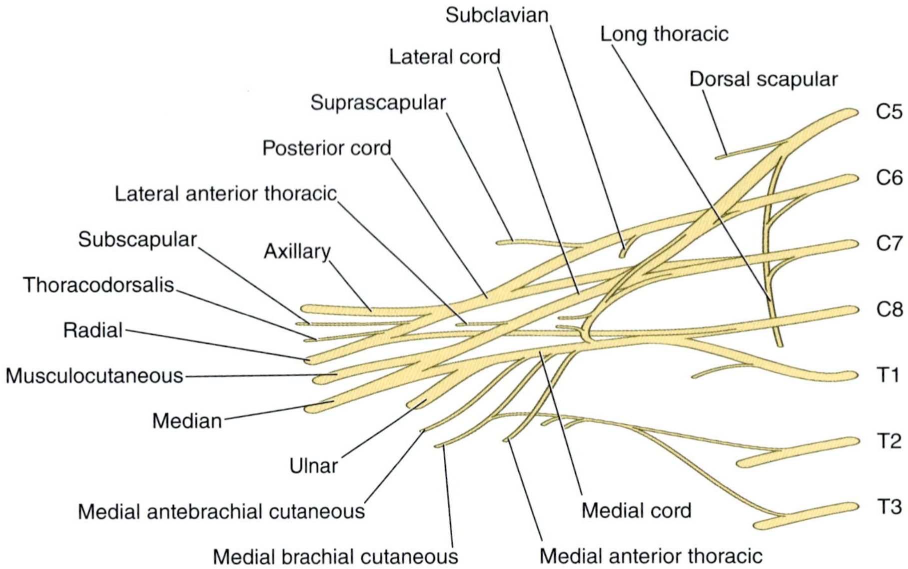

Figure 6.3 Isolated view of the brachial plexus.

Supraclavicular Nerves

The dorsal scapular nerve (C5 anterior and posterior rami) innervates the levator scapulae and rhomboid major and minor muscles (see figure 6.3). The long thoracic nerve (C5, C6, and C7 anterior and posterior rami; see figures 6.2 and 6.3) innervates the serratus anterior muscle. Not illustrated are nerves to the scalene and longus colli muscles of the cervical spine (C2-C8) and a communicating nerve (C5) to the phrenic nerve of the cervical plexus, which are discussed in chapter 10 on the spinal column. Two nerves coming from one of three groups of nerves of the brachial plexus, known as the *upper*, *middle*, and *lower trunks*, are the subclavian nerve (C4, C5, and C6), innervating the subclavian muscle, and the suprascapular nerve (C4, C5, and C6; see figures 6.2 and 6.3), innervating the supraspinatus and infraspinatus muscles of the rotator cuff.

Infraclavicular Nerves

Three large cords of nerves (lateral, medial, and posterior) are formed in the brachial plexus from divisions of the upper, middle, and lower trunks of spinal nerves (see figure 6.3). The infraclavicular nerves arising from the lateral cord are the lateral anterior thoracic nerve (C5, C6, and C7), innervating the pectoralis major muscle; the lateral aspect of the median nerve (C5, C6, and C7),

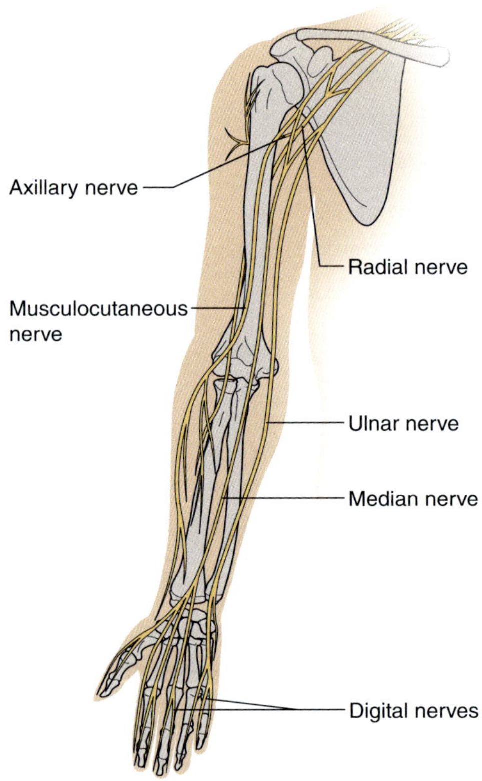

Figure 6.4 Major nerves of the upper extremity.

innervating most of the anterior forearm muscles and some in the hand; and the musculocutaneous nerve (C4, C5, and C6; figure 6.4), innervating the anterior arm muscles.

Hands On

Applying pressure between the triceps brachii and the biceps brachii on the medial aspect of your arm may produce a tingling sensation in your hand because you are compressing the median nerve (figure 6.5).

There are five major nerves of the brachial plexus that arise from the medial cord (see figure 6.3). The medial anterior thoracic nerve (C8 and T1) innervates the pectoralis major and minor muscles (see figure 6.3). Joining the previously mentioned lateral aspect is the medial aspect of the median nerve (C8 and T1), which innervates most of the anterior muscles of the forearm and some of the hand (see figures 6.4 and 6.5). The medial cutaneous nerve of the forearm (medial antebrachial cutaneous nerve; T1) is *not* a motor nerve and innervates the skin and fascia on the medial surface of the arm. The ulnar nerve (C8 and T1) innervates the flexor carpi ulnaris and the muscles of the medial aspect of the hand not innervated by the median nerve (see figure 6.4).

Hands On

The ulnar nerve is easily located between the medial epicondyle of the humerus and the olecranon process of the ulna (figure 6.6). Sudden pressure on this area often produces a tingling sensation in the forearm and is referred to as "hitting your funny bone."

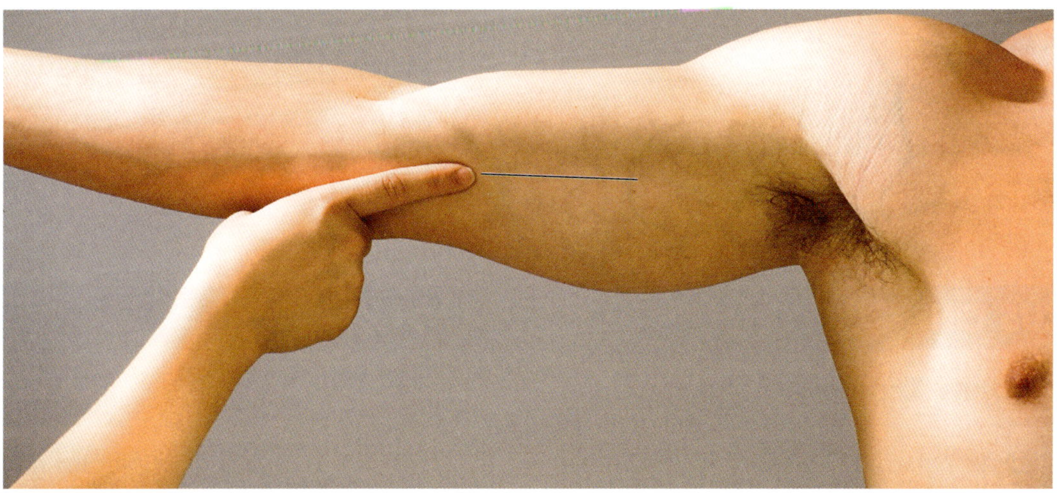

Figure 6.5 Locating the median nerve.

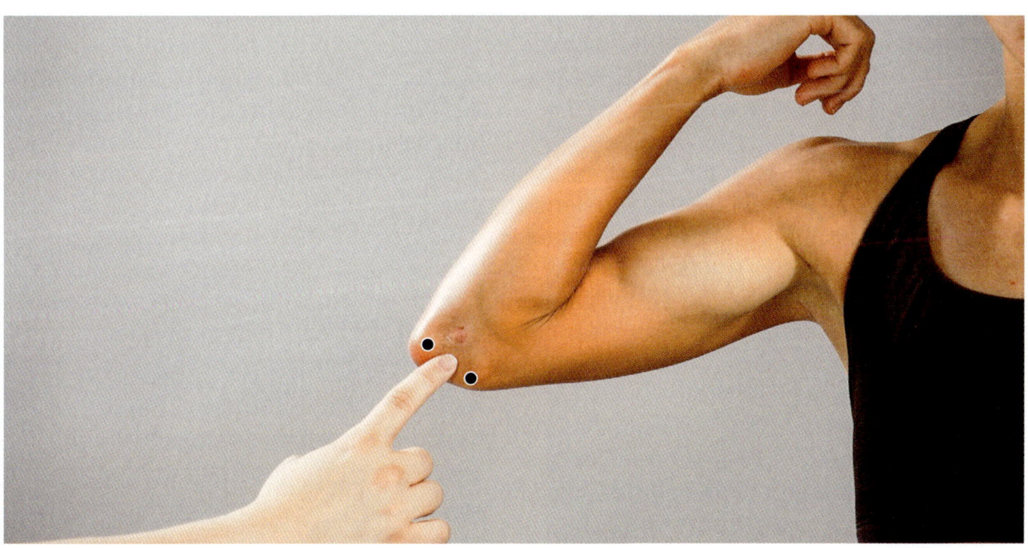

Figure 6.6 Locating the ulnar nerve.

The fifth nerve, the medial brachial cutaneous nerve, is *not* a motor nerve and innervates the skin and fascia on the medial aspect of the upper arm down to the medial epicondyle and olecranon process areas of the elbow.

Four brachial plexus nerves compose the posterior cord (see figure 6.3). The axillary nerve (C5 and C6; see figure 6.4) innervates the deltoid and teres minor muscles. The radial nerve, also known as the *musculospiral nerve* (C5, C6, C7, C8, and T1), innervates the posterior muscles of the arm and forearm (see figure 6.4). The subscapular nerves (C5 and C6), the upper and lower subscapular nerves, innervate the subscapularis and teres major muscles (see figure 6.3). The dorsal thoracic nerve, also known as the *thoracodorsalis* (C6, C7, and C8), innervates the latissimus dorsi muscle (see figure 6.3).

POP QUIZ

The use of the illustrated tackling technique can possibly result in trauma to which of the following structures?

a. cervical spine vertebra
b. cervical plexus nerves
c. cervical spine musculature
d. stinger

jpbcpa/iStockphoto/Getty Images

FOCUS ON

Vertebral Trauma

In a "pinched" nerve or a "stinger" type of trauma in the cervical region or upper extremity, the nerves of the brachial plexus are impinged or stretched. This type of trauma may occur from the performance of questionable (if not illegal in football) techniques, such as butt or face blocking and spear tackling. Butt or face blocking is an offensive player's technique that occurs when a blocker makes initial contact with an opponent by striking the opponent's chest ("in the numbers") with the blocker's helmet (face, head, forehead, or nose). This can result in the blocker incurring a compression of the cervical vertebrae, causing the impingement of brachial plexus nerves (from a direct head-on blow) or causing the blocker's cervical spine to be excessively laterally flexed. In turn, either or both of the brachial plexus nerves are stretched on one side of the cervical spine while being impinged between cervical vertebrae on the opposite side (when the blow is struck off center, causing the cervical spine to laterally flex). Spear tackling (or spearing or headhunting) is a defensive player's technique in which a ballcarrier is struck by the tackler's head as the initial point of contact with the opponent. The same or similar results that could occur from butt blocking are possible with this tackling technique. Recent football rule changes and coaching techniques have made it difficult to determine whether these events are accidental or intended. Aside from these football examples, any force that may cause compression of the cervical spine or excessive lateral floxion of the cervical spine can result in trauma, impinging or stretching the brachial plexus nerves, or even both.

For players to avoid injury to the structures of the brachial plexus, coaches need to learn and teach proper tackling techniques. In the event of injury to the brachial plexus, coaches and athletic trainers must have sufficient background in first aid to minimize further injury. An understanding of anatomy is vital for the prevention and treatment of pinched nerves.

Major Arteries of the Upper Extremity

The blood vessels of the body are divided into arteries, arterioles, capillaries, venules, and veins. The arteries (see figure 1.26), which carry blood away from the heart, and some of the larger arterioles are easily observed, whereas the smaller arterioles and capillaries are microscopic. The artery walls are lined with smooth muscle fibers that add pumping action; the veins have few or no smooth muscle fibers but have small valves that permit blood flow in only one direction. Any attempt to reverse this direction is blocked by the closing of these valves. The veins, which carry blood to the heart, and some of the larger venules are easily observed, whereas the smaller venules and capillaries are microscopic. Cells receive their nutrients and release their byproducts at the capillary level.

The major arteries of the upper extremity include the subclavian, axillary, brachial, radial, deep volar arch, and ulnar arteries (figure 6.7).

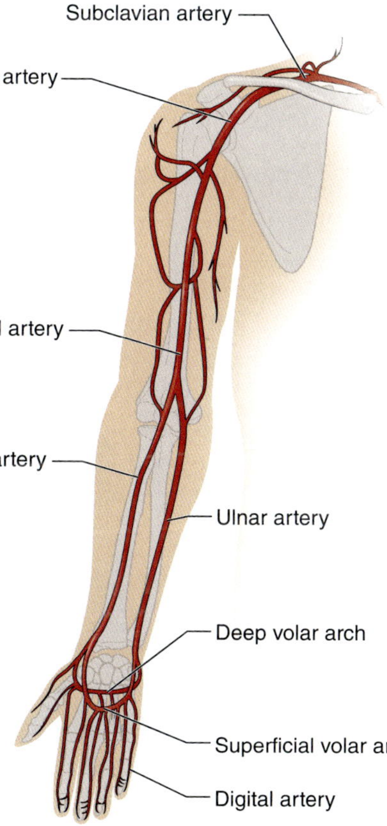

Figure 6.7 Major arteries of the upper extremity.

The subclavian artery originates near the heart, runs posterior to the clavicle, and has three major parts: the costocervical trunk, the internal mammary, and the thyrocervical trunk. Each of these three parts has branches that supply various structures of the thorax and upper extremity.

Hands On

Apply finger pressure to the middle superior aspect of your clavicle (take your pulse) to palpate the subclavian artery (figure 6.8). Note that using your fingers rather than your thumb is the preferred technique for taking a pulse. Some authorities believe that the thumb also has a pulse and could give a false impression when used to palpate for the presence of a pulse in another anatomical area.

The axillary artery is an extension of the subclavian artery; it begins at the outside border of the first rib and runs to the lower portion of the teres major muscle, where it becomes known as the *brachial artery*.

Hands On

Palpate the axillary area beneath the coracobrachialis muscle for a pulse (figure 6.9). This is a pulse from the axillary artery.

The brachial artery extends from the axilla to the elbow, where it divides into the radial and

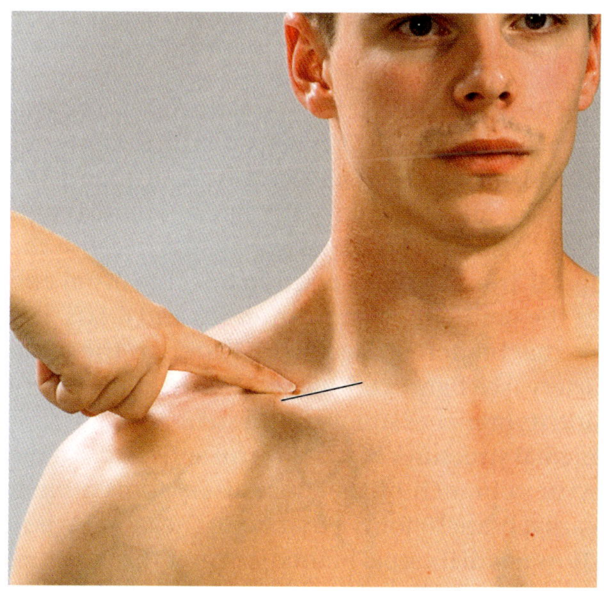

Figure 6.8 Locating the subclavian artery.

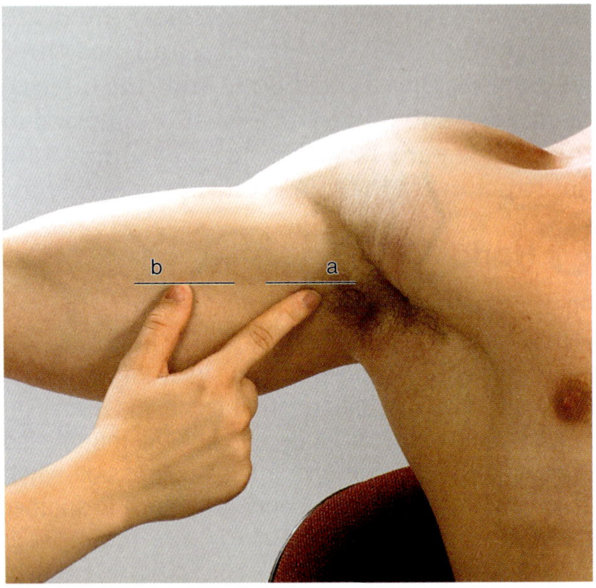

Figure 6.9 Locating the a) axillary and b) brachial arteries.

ulnar arteries. The branches of the brachial artery supply the structures of the upper arm (shoulder to elbow).

Hands On

Palpate the area on the medial aspect of the upper arm between the triceps brachii and the biceps brachii muscles (take your pulse) to locate the brachial artery (see figure 6.9). This area is often considered a *pressure point* for attempting to reduce blood flow below that site (as in applying pressure in first aid for bleeding in the forearm or hand).

The radial artery (the lateral branch of the brachial artery) and its branches supply the anterior lateral structures from the elbow, through the forearm and wrist, into the hand. This artery is often associated with taking a pulse because the radial artery is compressed against the distal end of the radius bone (see figures 6.7 and 6.10). The medial branch of the brachial artery is the ulnar artery (see figures 6.7 and 6.11). The ulnar artery and its branches supply the posterior and anterior medial structures of the forearm and hand. In the wrist, a branch of the ulnar artery (dorsal ulnar carpal branch) joins a branch of the radial artery (dorsal radial carpal branch) to form the structure known as the *dorsal carpal branch* (figure 6.12). In the palm of the hand (volar surface), the radial

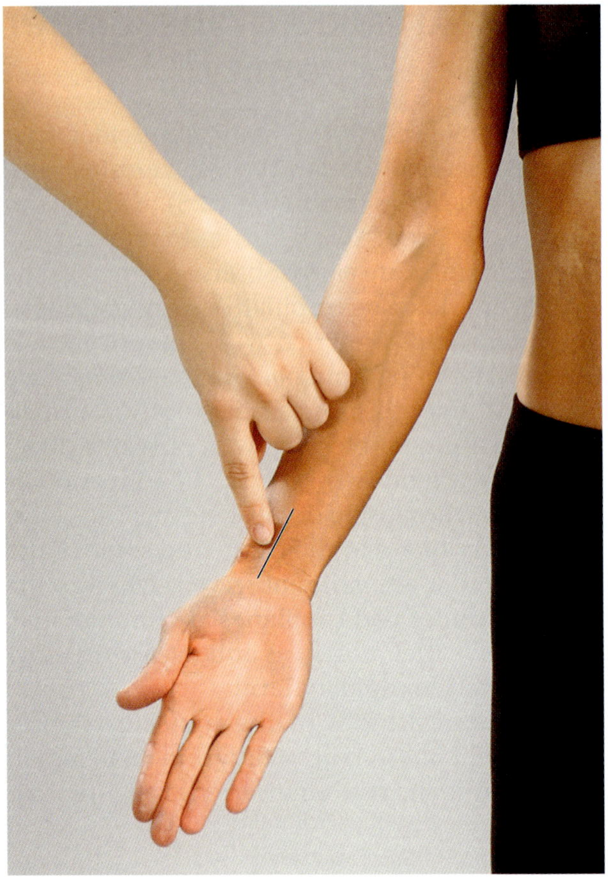

Figure 6.10 Locating the radial artery.

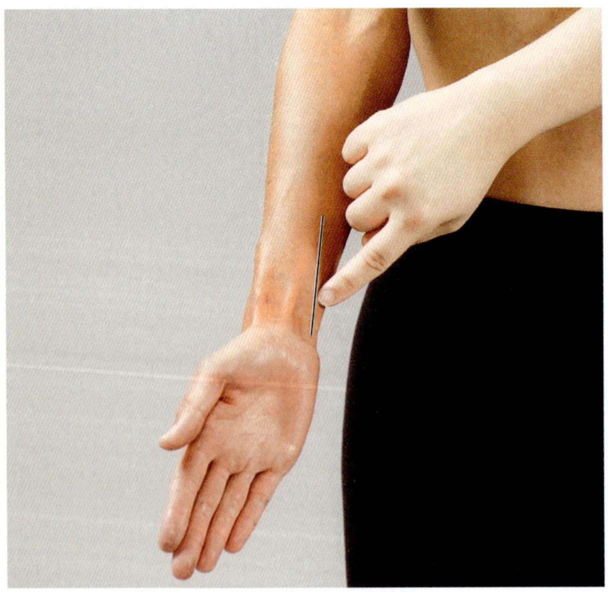

Figure 6.11 Locating the ulnar artery.

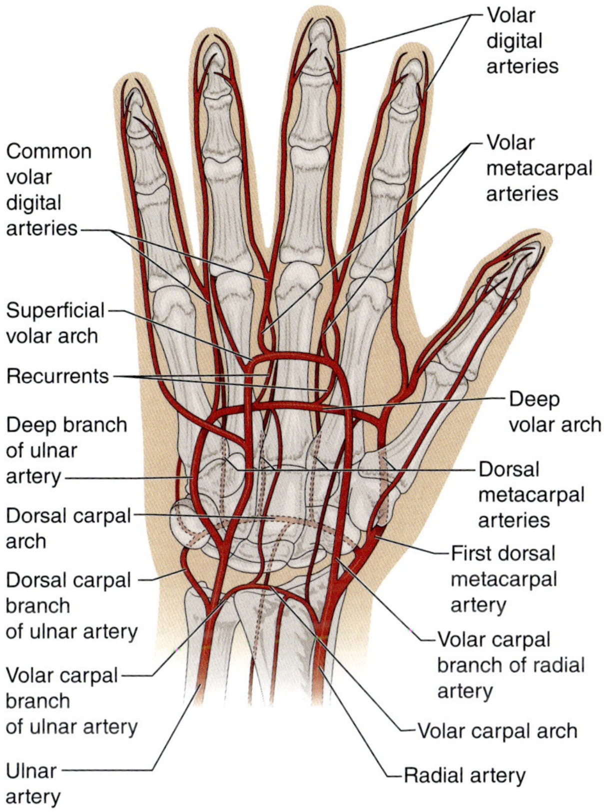

Figure 6.12 The dorsal carpal branch of the ulnar artery, the dorsal carpal arch, and the superficial and deep volar arches.

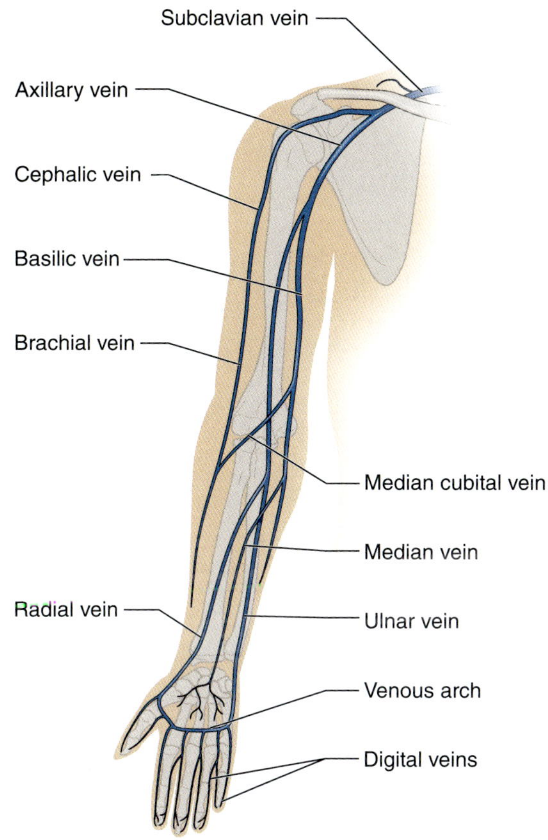

Figure 6.13 Major veins of the upper extremity.

artery joins with branches of the ulnar artery to form two structures known as the *superficial volar arch* and the *deep volar arch* (see figures 6.7 and 6.12). Branches from these volar arches, including the digital arteries on either side of the fingers, supply the structures of the fingers and thumb (see figures 6.7 and 6.12).

Major Veins of the Upper Extremity

The veins that return blood to the heart (see figure 1.27) are commonly divided into deep veins and superficial veins. With a few exceptions, the deep veins have the same names as the arteries they parallel, such as the subclavian vein and axillary vein. The superficial veins have specific names and are located near the skin. Unlike arteries, veins do not always appear exactly where one might expect, and often they may be absent altogether.

The major deep veins of the upper extremity (figure 6.13) are the subclavian and axillary veins. The axillary vein drains the blood from the upper extremity into the subclavian vein, which combines with the internal jugular vein to form the brachiocephalic vein (figure 6.14). Feeding into the axillary vein are the veins paralleling the major arteries of the upper extremity: the brachial vein, radial vein, ulnar vein, venous arch, and digital veins (see figure 6.13).

The major superficial veins of the upper extremity (see figure 6.13) include the basilic, the cephalic, and the median veins. The median cubital vein crosses the anterior aspect of the elbow joint (cubital fossa) laterally to medially, and the cephalic vein can be observed superior to the brachioradialis muscle and slightly lateral to the biceps brachii muscle (figure 6.15). The basilic vein originates on the ulnar aspect of the venous arch in the hand. At the elbow, the basilic vein receives the median cubital vein and, on reaching the lower portion of the teres major muscle,

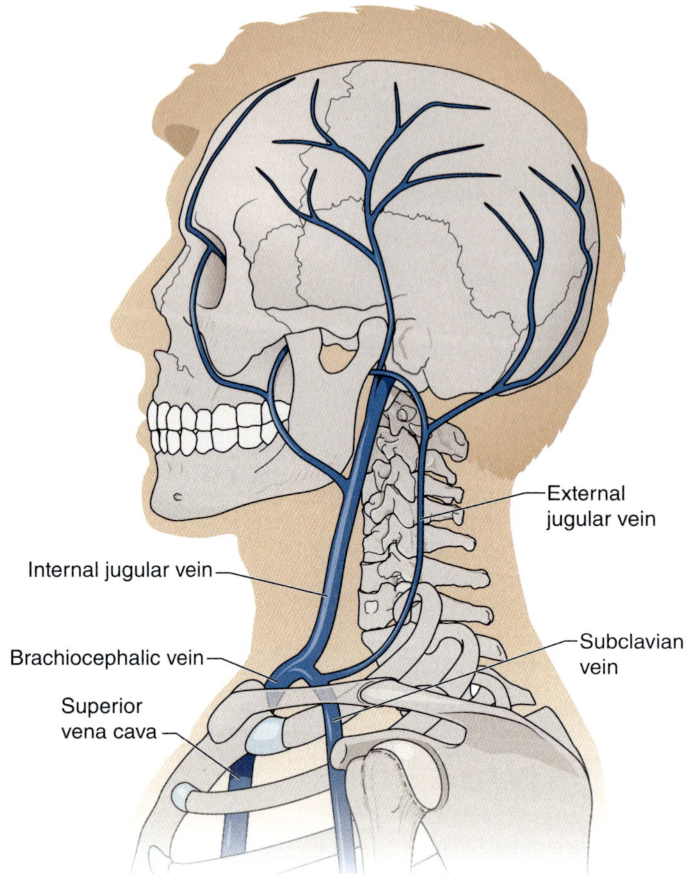

Figure 6.14 The brachiocephalic and internal jugular veins.

joins the axillary vein (figure 6.16). It drains the structures on the ulnar side of the volar and dorsal aspects of the hand, forearm, and upper arm. The cephalic vein originates on the radial side of the dorsal venous arch in the hand; at the elbow, it is connected to the basilic vein by the median cubital vein and then joins the axillary vein. Also at the elbow, the cephalic vein connects to the accessory cephalic vein (figure 6.17). The cephalic vein drains the structures from the lateral aspect of the hand, forearm, and upper arm. The median vein originates in the palmar venous vessels of the hand and connects to the basilic cubital vein. In the elbow area, the median vein is joined by the median cubital vein, which drains the medial aspect of the forearm and hand (figure 6.18). The median cubital vein is often used when blood is donated or a sample of blood is needed for testing purposes.

Here is a bit of trivia now that the blood vessels of the upper extremity have been presented.

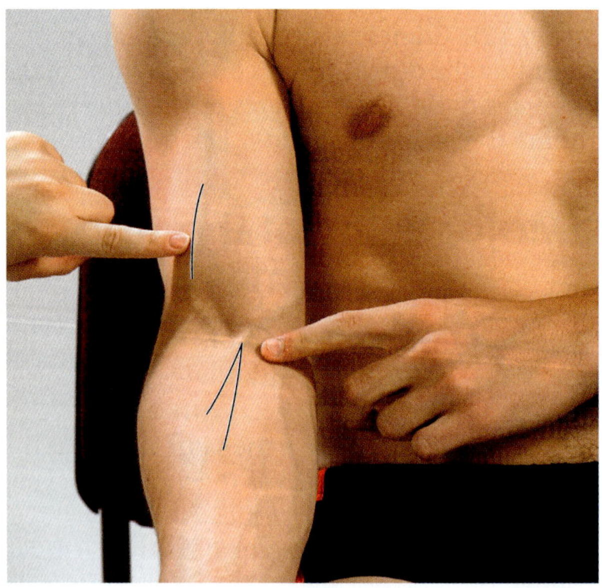

Figure 6.15 Locating the median cubital and cephalic veins.

Legend has it that the reason the wedding ring is worn on the left hand's fourth finger is because it is the only finger with a blood vessel that leads directly to the heart. Based on what was just presented, do you believe the legend? Is this blood vessel an artery or a vein?

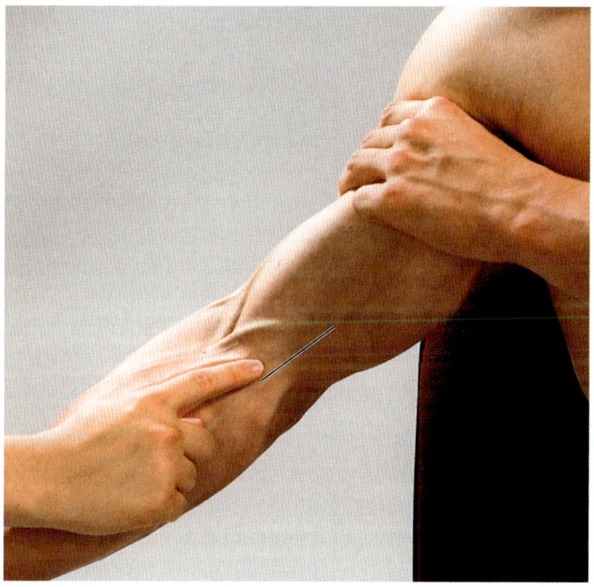

Figure 6.16 Locating the basilic vein.

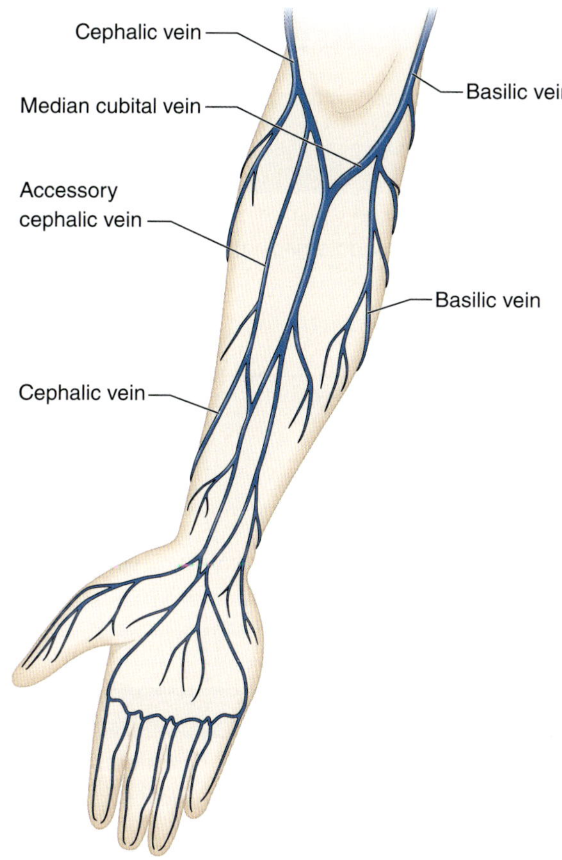

Figure 6.17 The cephalic and accessory cephalic veins.

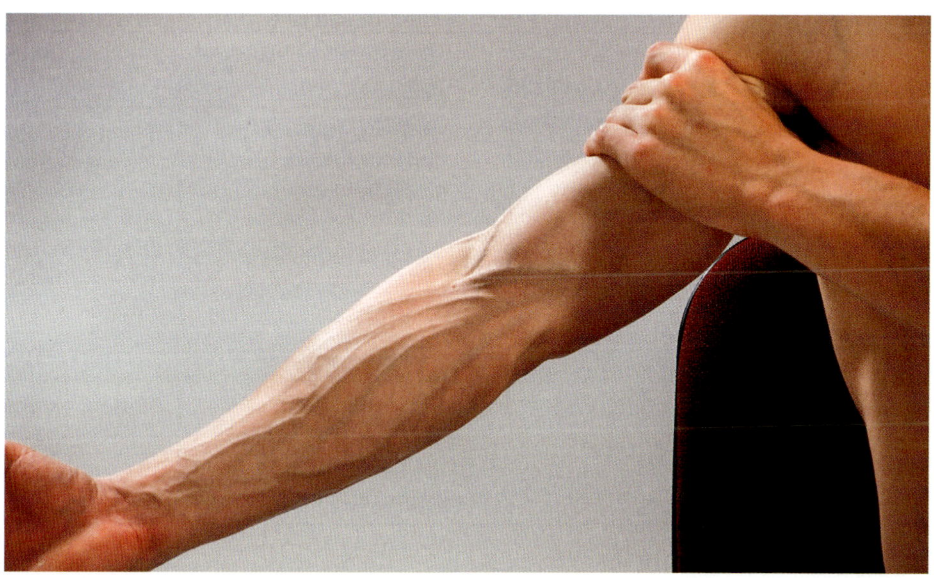

Figure 6.18 Venous network of the median vein of the forearm.

Visit the web resource to complete the activities for this chapter.

Articulations of the Upper Extremity

Joint	Type	Bones	Ligaments	Movement
SHOULDER GIRDLE				
Sternoclavicular	Arthrodial (saddle/gliding)	Sternum and clavicle	• Sternoclavicular (anterior, superior, and posterior) • Costoclavicular • Interclavicular • Articular disc	Elevation, depression, rotation, protraction, retraction
Acromioclavicular	Arthrodial (plane)	Clavicle and scapula	• Acromioclavicular • Coracoclavicular (conoid and trapezoid) • Coracoacromial (for protection of glenohumeral [GH] joint) • Articular disc	Elevation, depression, rotation, protraction, retraction, winging (posterior movement to keep scapula close to the thorax during scapular abduction), tipping (rotation to keep scapula close to the thorax during scapular elevation)
SHOULDER JOINT				
GH	Ball and socket	Humerus and scapula	• GH (superior, middle, and inferior) • Coracohumeral • Transverse humeral (keeps long-head biceps tendon in intertubercular groove) • Labrum (deepens socket of the glenoid fossa) • Capsule	Flexion, extension, hyperextension, abduction, adduction, internal rotation, external rotation, and circumduction
Scapulothoracic	False joint	Scapula and thorax	No ligaments (fascia of the serratus anterior muscle and fascia from the thorax)	Elevation, depression, upward rotation and lateral tilt (abduction, protraction), and downward rotation and medial tilt (adduction, retraction) of the scapula on the thorax
ELBOW				
Humeroulnar	Hinge	Humerus and ulna	• Capsule • Medial (ulnar) collateral (three bands: anterior, posterior, and transverse)	Flexion, extension
Radiohumeral	Irregular (gliding)	Humerus and radius	• Lateral (radial) collateral • Annular	Flexion, extension, rotation (pronation, supination)
Radioulnar (proximal)	Pivot	Radius and ulna	Annular	Pronation, supination

Joint	Type	Bones	Ligaments	Movement
FOREARM				
Radioulnar (middle)	Syndesmosis	Radius and ulna	• Oblique cord • Interosseous membrane	Slight (if any)
Radioulnar (distal)	Pivot	Radius and ulna (and articular disc)	• Capsule • Dorsal radioulnar • Volar (palmar) radioulnar	Pronation, supination
WRIST				
Radiocarpal	Condyloid (gliding)	Radius, scaphoid, lunate, triquetrum	• Capsule • Radial collateral • Dorsal radioulnar • Volar (palmar) radioulnar	Flexion, extension, abduction (radial deviation), adduction (ulnar deviation)
Ulnocarpal	Irregular (gliding)	Ulna, pisiform, triquetrum	• Capsule • Ulnar collateral • Articular disc	Gliding
WRIST: INTERCARPALS				
Proximal row	Arthrodial (gliding)	Radius, scaphoid, lunate, triquetrum	• Capsule • Flexor (volar) retinaculum • Extensor (dorsal) retinaculum • Radial collateral • Ulnar collateral • Volar (palmar) intercarpal • Dorsal intercarpal	Gliding
Middle	Hinge (gliding)	Proximal and distal rows of carpals	• Capsule • Volar (palmar) intercarpal • Dorsal intercarpal • Interosseous intercarpal • Radial collateral • Ulnar collateral • Pisohamate • Pisometacarpal	Gliding
Distal row	Irregular (gliding)	Trapezius, trapezoid, capitate, hamate	• Capsule • Volar (palmar) intercarpal • Dorsal intercarpal • Interosseous intercarpal	Gliding

(continued)

Articulations of the Upper Extremity *(continued)*

Joint	Type	Bones	Ligaments	Movement
HAND AND FINGERS				
Carpometacarpal	Modified saddle	Trapezium, trapezoid, capitate, hamate, metacarpals (4)	• Capsule • Volar carpometacarpal • Dorsal carpometacarpal • Interosseous carpometacarpal • Volar accessory	Flexion, extension, abduction, adduction, slight opposition of fifth finger
Metacarpophalangeal (MCP)	Condyloid	Metacarpals (4) and proximal phalanges (4)	• Transverse metacarpal • Capsule • Radial collateral • Ulnar collateral • Volar accessory	Flexion, extension, abduction, adduction
Proximal interphalangeal (PIP)	Hinge	Proximal phalanges (4) and middle phalanges (4)	• Capsule • Radial collateral • Ulnar collateral • Volar accessory	Flexion, extension
Distal interphalangeal (DIP)	Hinge	Middle phalanges (4) and distal phalanges (4)	• Capsule • Radial collateral • Ulnar collateral • Volar accessory	Flexion, extension
THUMB				
Carpometacarpal	Saddle	Trapezium and first metacarpal	Capsule	Flexion, extension, abduction, adduction, opposition
MCP	Hinge	First metacarpal and proximal phalanx	• Capsule • Radial collateral • Ulnar collateral • Dorsal accessory	Flexion, extension
Interphalangeal (IP)	Hinge	Proximal phalanx and distal phalanx	• Capsule • Radial collateral • Ulnar collateral • Volar accessory	Flexion, extension

Muscles, Nerves, and Blood Supply of the Upper Extremity

Muscle	Origin	Insertion	Action	Nerve	Blood supply
SHOULDER GIRDLE, ANTERIOR					
Pectoralis minor	Ribs (3-5)	Coracoid process of scapula	Downward rotation and depression of scapula	Medial pectoral	Lateral thoracic and thoracoacromial branches of axillary
Serratus anterior	Upper nine ribs	Medial border of scapula	Upward rotation and protraction of scapula	Long thoracic	Lateral thoracic and subscapular branches of axillary, transverse cervical branch of thyrocervical branch of subclavian
Subclavius	First rib	Subclavian groove of clavicle	Depress the clavicle	Medial anterior thoracic	Clavicular branch of thoracoacromial branch of axillary thoracic
SHOULDER GIRDLE, POSTERIOR					
Levator scapulae	Transverse processes of first four cervical vertebrae	Medial border of scapula	Elevation and downward rotation of scapula	Dorsal scapular	Dorsal scapular
Rhomboid major	Spinous processes of vertebrae T2-T5	Medial border of scapula	Retraction, elevation, and downward rotation of scapula	Dorsal scapular	Transverse cervical branch of thyrocervical branch of subclavian *or* dorsal scapular
Rhomboid minor	Spinous processes of vertebrae C7-T1	Medial border of scapula	Retraction, elevation, and downward rotation of the scapula	Dorsal scapular	Transverse cervical branch of thyrocervical branch of subclavian *or* dorsal scapular
Trapezius	External occipital protuberance, ligamentum nuchae, and C7-T12 spinous processes	Spine of scapula, acromion process, lateral 1/3 of posterior clavicle	• Superior fibers: elevation and upward of the scapula, extension and lateral flexion of the neck • Middle fibers: retraction of scapula • Inferior fibers: depression of scapula	C2, C3, C4, and spinal accessory (11th cranial)	Transverse cervical branch of thyrocervical branch of subclavian

(continued)

Muscles, Nerves, and Blood Supply of the Upper Extremity *(continued)*

Muscle	Origin	Insertion	Action	Nerve	Blood supply
SHOULDER JOINT, ANTERIOR					
Pectoralis major	Ribs (2-6), sternum, medial 1/2 of clavicle	Lateral lip of the intertubercular groove of the humerus	Flexion, adduction, horizontal adduction, and internal rotation of GH joint	Lateral and medial pectoral	Lateral thoracic and thoracoacromial branches of axillary
Coracobrachialis	Coracoid process of scapula	Middle 1/3 of medial surface of humerus	Flexion and adduction of GH joint	Musculocutaneous	Brachial
Biceps brachii (long head)	Supraglenoid tubercle	Radial tuberosity	Flexion of GH joint, flexion of elbow, supination of forearm	Musculocutaneous	Brachial
Biceps brachii (short head)	Coracoid process of scapula	Radial tuberosity	Flexion of GH joint, flexion of elbow, supination of forearm	Musculocutaneous	Brachial
Subscapularis	Subscapular fossa	Lesser tubercle of humerus	Internal rotation of GH joint, GH joint stability	Upper and lower subscapular nerves	Subscapular
SHOULDER JOINT, SUPERIOR					
Deltoid (clavicular)	Lateral 1/3 of anterior border of clavicle	Deltoid tuberosity	Flexion, horizontal, adduction, and internal rotation of GH joint	Axillary	Posterior humeral circumflex, thoracoacromial
Deltoid (acromial)	Lateral border of acromion process	Deltoid tuberosity	Abduction of GH joint		
Deltoid (scapular)	Inferior lip of spine of scapula	Deltoid tuberosity	Extension, horizontal abduction, and external rotation of GH joint		
Supraspinatus	Supraspinous fossa of scapula	Proximal facet of greater tubercle of humerus	Abduction of GH joint, GH joint stability	Suprascapular	Suprascapular
SHOULDER JOINT, POSTERIOR					
Infraspinatus	Infraspinatus fossa of scapula	Middle facet of greater tubercle of humerus	External rotation, abduction, and horizontal abduction of GH joint, GH joint stability	Suprascapular	Suprascapular and circumflex scapular
Teres minor	Upper 2/3 of lateral border of scapula	Distal facet of greater tubercle of humerus	External rotation and horizontal abduction of GH joint, GH joint stability	Axillary	Circumflex scapular

Muscle	Origin	Insertion	Action	Nerve	Blood supply
SHOULDER JOINT, INFERIOR					
Latissimus dorsi	Spinous processes of lower six thoracic and all lumbar vertebrae, iliac crest, lower three ribs, inferior angle of scapula	Intertubercular groove	Internal rotation, extension, and adduction of GH joint	Thoracodorsal	Thoracodorsal
Teres major	Inferior 1/3 of lateral border and inferior angle of scapula	Beneath the lesser tubercle on anterior humerus	Internal rotation, extension, and adduction of GH joint	Lower subscapular	Circumflex scapular
Triceps brachii (long head)	Infraglenoid tubercle of scapula	Olecranon process of ulna	Extension and adduction of GH joint, extension of elbow	Radial	Posterior humeral circumflex, profundus branch of brachial
ELBOW JOINT AND FOREARM, ANTERIOR					
Brachialis	Distal 1/3 of anterior surface of humerus	Coronoid process and tuberosity of ulna	Flexion of elbow	Musculocutaneous	Brachial
Brachioradialis	Proximal aspect of lateral supracondylar ridge of humerus	Lateral aspect of styloid process of radius	Flexion of elbow	Radial	Radial
Biceps brachii (long head)	Supraglenoid tubercle	Radial tuberosity	Flexion of GH joint, flexion of elbow, supination of forearm	Musculocutaneous	Brachial
Biceps brachii (short head)	Coracoid process of scapula	Radial tuberosity	Flexion of GH joint, flexion of elbow, supination of forearm	Musculocutaneous	Brachial
Pronator teres	Common flexor tendon and coronoid process of ulna	Middle of lateral aspect of radius	Pronation of forearm	Median	Radial branch of brachial
Pronator quadratus	Distal part of volar aspect of ulna	Distal part of volar aspect of radius	Pronation of forearm	Median	Ulnar and radial

(continued)

Muscles, Nerves, and Blood Supply of the Upper Extremity *(continued)*

Muscle	Origin	Insertion	Action	Nerve	Blood supply
ELBOW JOINT AND FOREARM, POSTERIOR					
Triceps brachii (lateral head)	Proximal 1/3 of posterolateral aspect of humerus	Olecranon process of ulna	Extension of elbow	Radial	Posterior humeral circumflex, profundus branch of brachial
Triceps brachii (long head)	Infraglenoid tubercle of scapula	Olecranon process of ulna	Extension of elbow, extension and adduction of GH joint	Radial	Posterior humeral circumflex, profundus branch of brachial
Triceps brachii (medial head)	Distal 2/3 of posteromedial aspect of humerus	Olecranon process of ulna	Extension of elbow	Radial	Posterior humeral circumflex, profundus branch of brachial
Anconeus	Posterior aspect of lateral epicondyle of humerus	Radial aspect of olecranon process and dorsal surface of proximal 1/4 of ulna	Extension of elbow	Radial	Profundus branch of brachial and dorsal interosseous
Supinator	Lateral epicondyle of humerus and supinator fossa of ulna	Proximal 1/3 of lateral and volar aspect of radius	Supination of forearm	Radial	Dorsal interosseous and radial recurrent
WRIST AND HAND, ANTERIOR WRIST AND EXTRINSIC HAND					
Flexor carpi radialis	Common flexor tendon	Bases of second and third metacarpals	Flexion and radial deviation of wrist	Median	Radial
Flexor carpi ulnaris	Common flexor tendon	Pisiform, hamate, and base of fifth metacarpal	Flexion and ulnar deviation of wrist	Ulnar	Ulnar branch of brachial
Flexor digitorum superficialis	Common flexor tendon, coronoid process, and proximal 2/3 of radius	Split tendons on each side of base of middle phalanges of digits (2-5)	Flexion of PIP, MCP, and wrist joints	Median	Ulnar branch of brachial
Palmaris longus	Common flexor tendon	Palmar aponeurosis	Flexion of wrist	Median	Radial branch of brachial
Flexor digitorum profundus	Proximal 3/4 of coronoid process of ulna and interosseous membrane	Four tendons—passing split superficialis tendons—on bases of distal phalanges of digits (2-5)	Flexion of DIP, PIP, MCP, and wrist joints	Median and ulnar	Ulnar and volar interosseous

Muscle	Origin	Insertion	Action	Nerve	Blood supply
WRIST AND HAND, POSTERIOR WRIST AND EXTRINSIC HAND					
Extensor carpi radialis longus	Lateral supracondylar ridge of humerus	Base of second metacarpal	Extension and radial deviation of wrist	Radial	Radial
Extensor carpi radialis brevis	Common extensor tendon	Base of third metacarpal	Extension and radial deviation of wrist	Radial	Radial
Extensor carpi ulnaris	Common extensor tendon	Base of fifth metacarpal	Extension and ulnar deviation of wrist	Radial	Posterior interosseous
Extensor digitorum communis	Common extensor tendon	Bases of distal phalanges of digits (2-5)	Extension of DIP, PIP, and MCP joints of digits (2-5); extension of wrist	Radial	Posterior interosseous
Extensor digiti minimi (proprius)	Common extensor tendon	Base of proximal phalanx of fifth digit	Extension of fifth MCP joint and wrist	Radial	Posterior interosseous
Extensor indicis	Dorsal aspect of distal ulna and interosseous membrane	Base of proximal phalanx of index finger	Extension and radial deviation of second MCP joint, extension of wrist	Radial	Posterior interosseous
WRIST AND HAND, INTRINSIC HAND					
Abductor digiti minimi	Pisiform bone	Base of proximal phalanx of fifth digit	Abduction and flexion of fifth MCP joint	Ulnar	Ulnar
Flexor digiti minimi brevis	Hamate bone	Base of proximal phalanx of fifth digit	Flexion of fifth MCP joint	Ulnar	Ulnar
Opponens digiti minimi	Hamate bone	Medial surface of fifth metacarpal	Flexion and adduction of little finger	Ulnar	Ulnar
Dorsal interossei (4)	Adjacent sides of all four metacarpals	Bases of proximal phalanges (2-4) and extensor digitorum communis tendon	Abduction of second and fourth MCP joints	Ulnar	Metacarpal branches of radial

(continued)

Muscles, Nerves, and Blood Supply of the Upper Extremity *(continued)*

Muscle	Origin	Insertion	Action	Nerve	Blood supply
WRIST AND HAND, INTRINSIC HAND					
Palmar interossei (3)	Ulnar side of second metacarpal and radial side of fourth and fifth metacarpals	Radial side of fourth and fifth phalanges and ulnar side of second proximal phalanx	Adduction and flexion of second, fourth, and fifth MCP joints	Ulnar	Metacarpal branches of deep volar arch
Lumbricals	Flexor digitorum profundus tendons	Extensor digitorum communis tendons	Flexion of four MCP joints, extension of four PIP and DIP joints	Median and ulnar	Volar metacarpal branch of deep volar
THUMB, EXTRINSIC					
Extensor pollicis longus	Middle 1/3 of ulna and interosseous membrane	Base of distal phalanx of thumb	Extension of IP and MCP joints of thumb	Radial	Posterior interosseous
Extensor pollicis brevis	Middle dorsal aspect of radius and interosseous membrane	Base of proximal phalanx of thumb	Extension of MCP joint of thumb, radial deviation of wrist	Radial	Posterior interosseous
Abductor pollicis longus	Middle dorsolateral aspect of ulna and interosseous membrane	Base of first metacarpal	Abduction of first metacarpal	Radial	Posterior interosseous
Flexor pollicis longus	Middle 1/2 of volar surface of radius	Base of distal phalanx of thumb	Flexion of IP and MCP joints of thumb	Median	Anterior interosseous
Abductor pollicis brevis	Trapezius and scaphoid bones	Base of proximal phalanx of thumb	Abduction of MCP joint of thumb	Median	Radial
Flexor pollicis brevis	Trapezium	Base of proximal phalanx of thumb	Flexion of MCP joint of thumb	Median	Radial
Opponens pollicis	Trapezium	Lateral surface of first metacarpal	Opposition of MCP joint of thumb	Median	Radial
Adductor pollicis	Oblique head: capitate and second and third metacarpal bones; transverse head: third metacarpal	Volar surface of base of proximal phalanx of thumb	Adduction of MCP joint of thumb	Ulnar	Deep volar arch

PART III

The Head, Spinal Column, Thorax, and Pelvis

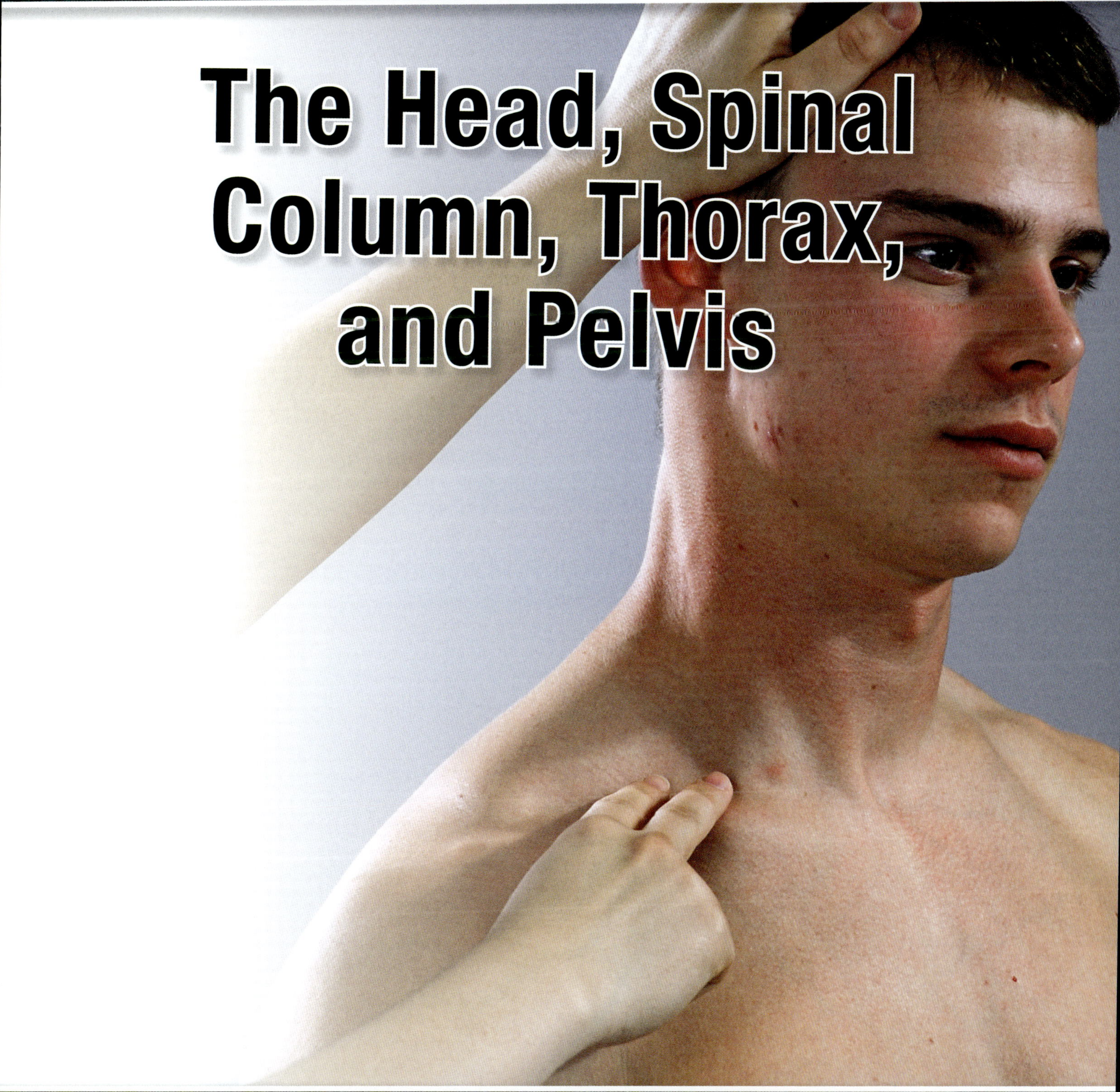

CHAPTER 7

The Head

Although the structures of the skull are not usually associated with a great degree of movement in the human body (the only movable joint in the head is the temporomandibular joint, or TMJ), some of the structures within the head are infinitely involved with all movement. When studying human anatomy, particularly the structures involved in movement, it is also appropriate to take a brief look at the extensive number of bones, ligaments, muscles, and blood vessels of the head, as well as the contents within the skull related to the nervous system. This chapter concentrates on the bones, ligaments, and muscles of the skull, while chapter 10 presents the blood vessels and nerves of the head and spinal column.

Bones of the Head

The head (or skull) consists of 28 bones divided into the 8 bones of the skull (cranium), 14 of the face (facial bones), and 6 within the ears. (A 29th bone, the hyoid, is considered a head bone by some, but, since it has no bony attachment to any of the other bones, it is not considered a skull bone in this text.) The bones of the skull consist of one ethmoid bone, one frontal bone, two temporal bones, two parietal bones, one sphenoid bone (figure 7.1), and one occipital bone (figure 7.2). The facial bones include two lacrimal bones, one mandible bone, two maxilla bones, two nasal bones, one vomer bone, two zygomatic bones, and two palatine bones (figure 7.3). Additionally, there are two nasal concha bones, with superior, middle, and inferior portions, that form the lateral wall of the nasal cavity.

The bony markings of the skull are listed here. Most of them are named based on their purpose, and many of them are very small and typically of greater concern in advanced studies of human anatomy. However, this listing is presented to illustrate the functions of common bony markings of the skull and facilitate the use of the part III summary table, where bony markings are noted as the origins and insertions of the 84 muscles of the head.

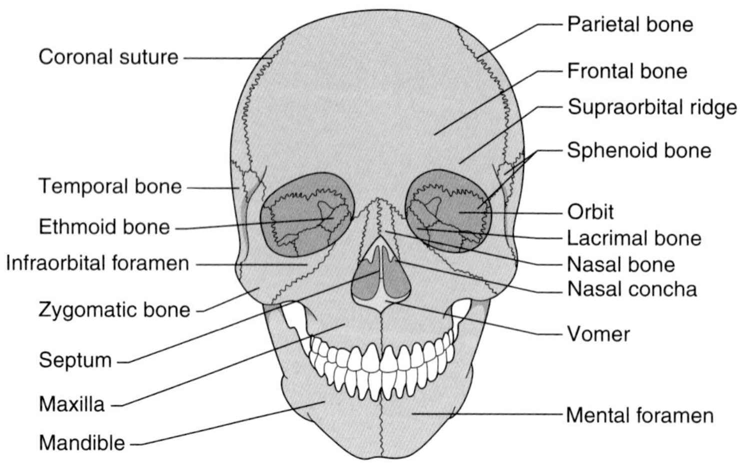

Figure 7.1 Anterior structures of the skull.

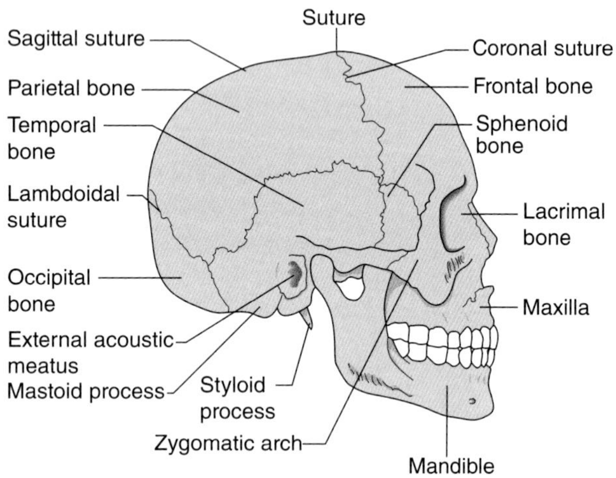

Figure 7.2 Lateral structures of the skull.

Alveolus: a socket that holds a tooth

Anterior cranial fossa: area that supports the brain's frontal lobe

Body of the mandible: source of attachments for the buccinator, digastric, triangularis, mentalis, platysma, mylohyoid, quadratus labii inferior, geniohyoid, and genioglossus muscles

Canine fossa: source of muscular attachments for the caninus

Carotid canal: temporal bone structure for the internal carotid artery

Cerebellar fossa: surface for the cerebellum

Condyloid fossa: area for a vein from the transverse sinus

Coronoid process: source of the temporalis muscle attachment

Cribriform: plate area for the olfactory nerve between the nasal cavity and the olfactory bulb (which is responsible for the sense of smell)

Crista galli: source of attachment for the falx cerebri

External acoustic meatus: the canal that conducts sound through the temporal bone to the tympanic membrane

External occipital protuberance: source of attachment for the trapezius muscles and the ligamentum nuchae

Foramen lacerum: passageway for the superficial petrosal nerve and the internal carotid artery

Foramen magnum: the large opening in the occipital bone where the medulla oblongata, the occipitoaxial ligaments, the vertebral arteries, the anterior and posterior spinal nerves, and the accessory nerves pass through

Foramen ovale: the opening in the sphenoid bone through which the mandibular and lesser petrosal nerves, accessory meningeal artery, and several small veins and lymph vessels pass

Foramen rotundum: passageway for the maxillary nerve

Foramen spinosum: passageway for the middle meningeal blood vessels and a branch of the mandibular nerve

Hard palate (roof of the mouth): formed by the horizontal portion of the palatine bones and the palatine processes of the maxillary bones

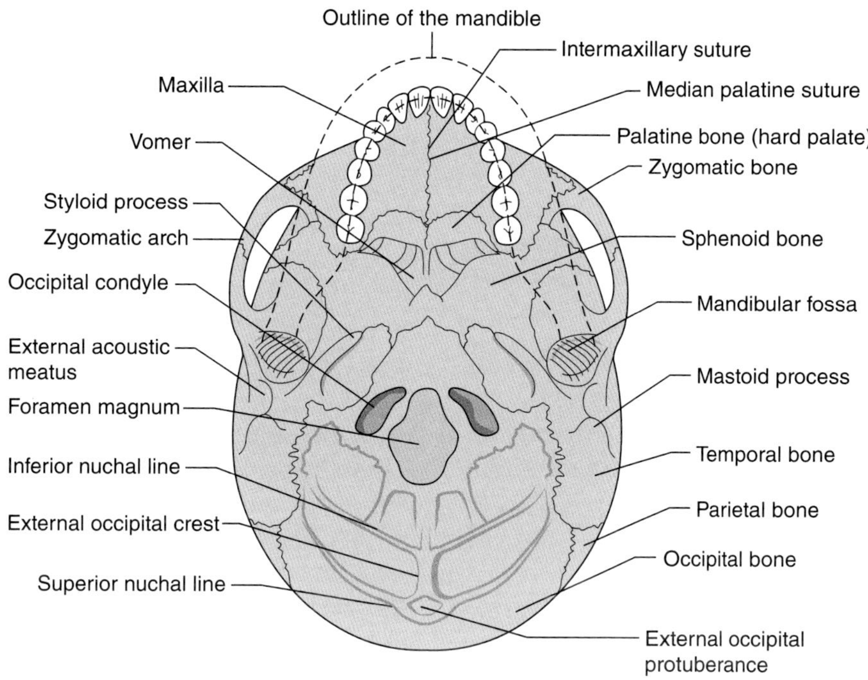

Figure 7.3 Inferior structures of the skull.

Hypoglossal canal: area on the occipital bone where a meningeal artery and the hypoglossal nerve leave the skull

Hypophyseal fossa: the area where the pituitary gland is located

Incisive canal: passageway for the nasopalatine nerves and the palatine artery

Inferior nuchal line: site of attachment for the semispinalis capitis muscle

Inferior orbital fissure: passageway for the infraorbital blood vessels, the zygomatic branch of the maxillary nerve, and branches of the sphenopalatine ganglion

Inferior tympanic canaliculus: area for passage of the tympanic branches of the glossopharyngeal nerves

Infraorbital foramen: opening in the maxillary bone for the infraorbital branch of the maxillary nerve and the infraorbital branch of the internal maxillary artery

Internal occipital crest: source of attachment for a portion of the inner layer of the dura mater (one of the coverings of the brain)

Jugular foramen: opening in the temporal bone for passage of the inferior petrosal sinus; the transverse sinus (internal jugular vein); meningeal branches of the pharyngeal and occipital arteries; and the 9th, 10th, and 11th cranial nerves

Jugular process: source of attachment for the lateral atlantooccipital ligament and the rectus capitis lateralis

Lacrimal fossa or groove: location of the lacrimal sac

Mandibular foramen: opening in the mandibular canal through which pass the inferior alveolar artery and nerve

Mandibular fossa: area where the head of the mandible articulates with the temporal bone

Mandibular head: forms the TMJ with the mandibular fossa on the temporal bone

Mastoid canal: passageway for the auricular branch of the vagus nerve

Mastoid foramen: passageway for the emissary vein and artery

Mastoid process: source of attachment for the splenius capitis and the sternocleidomastoid muscles

Mental foramen: passageway through the mandible for the mental nerve and blood vessels

Mylohyoid groove: area on the mandible for passage of the mylohyoid nerves and artery

Neck of the mandible: area for the attachment of the external pterygoid muscle

Occipital condyle: point of articulation between the skull and the superior articular process of the atlas

Optic foramen: passageway for the optic nerve

Orbit: eye socket housing the eye

Ramus of the mandible: bony attachment for the internal pterygoid and masseter muscles

Styloid process: bony attachment for the stylohyoid, styloglossus, and stylopharyngeus muscles and the stylomandibular ligament

Stylomastoid foramen: hole where the facial nerve and the stylomastoid artery pass through the temporal bone

Superior nuchal line: area of attachment for the trapezius, sternocleidomastoid, occipitalis, and splenius capitis muscles

Superior orbital fissure: passageway for the trochlear, oculomotor, and ophthalmic portions of the trigeminal and abducens nerves

Supraorbital notch: passageway in the frontal bone for the supraorbital branches of both the ophthalmic artery and ophthalmic nerve

Temporal fossa: area of attachment for the temporalis muscle

Zygomatic arch: articulation between the temporal process of the zygomatic bone and the zygomatic process of the temporal bone that provides the source of attachment for the masseter muscle and provides space for the temporalis muscle to attach to the coronoid process

The bones of the ear include the malleus (hammer), the incus (anvil), and the stapes (stirrup; figure 7.4). The malleus is attached to the tympanic membrane (eardrum) and has a head, neck, handle, and anterior and lateral processes. The handle and lateral process attach to the tympanic membrane, and the anterior process serves as an attachment site for ligamentous fibers. The malleus head connects to the body of the incus. The incus consists of a body and a long and short crura. The long crus attaches to the head of the stapes. The stapes consists of a head, neck, body, and an anterior and posterior crura. These three bones are suspended, by ligaments, in the tympanic cavity of the middle ear; they transmit sound from the tympanic membrane to the cochlea of the inner ear.

The ear consists of three parts: the external ear, the middle ear, and the internal ear. The external ear is divided into the pinna or auricle, the external meatus (auditory canal), and the tympanic membrane. The middle ear contains the three ear bones, the upper end of the Eustachian tube, and the round window (fenestra rotunda) and oval window (fenestra ovalis), which separate the middle ear from the inner ear. The three ear bones do not articulate with any bones of the skull. The inner ear (figure 7.5) contains the cochlea and the vestibule (consisting of the utriculus, the sacculus, and the semicircular canals, which are integral parts of the equilibrium mechanism). The following sequence is the basic explanation of how sound is heard: Sound waves are collected by the outer ear and conducted through the auditory canal to the tympanic membrane. The tympanic membrane transfers the waves to the middle ear structures (the malleus, incus, and stapes), which conduct the waves to the cochlea in the inner ear via the middle ear windows. Cochlear nerves, along with nerves from the vestibule, combine to form the vestibulocochlear nerve (or eighth cranial nerve) and transmit impulses to the medulla oblongata (see chapter 10).

The hyoid bone, having no articulations with other skull bones, is suspended under the tongue between the larynx and the mandible. Based on our bone classifications, defined in chapter 1, the hyoid bone is considered an irregular bone; however, some classify it as a sesamoid bone due to its free-floating nature. It serves as a point of attachment for a number of muscles of the neck and tongue (figure 7.6).

The mandible forms the lower portion of the jaw and articulates with the temporal bone of the skull (figure 7.7). Initially, it consists of two halves before birth, and these two halves are united after birth at an area known as the *mandibular symphysis*.

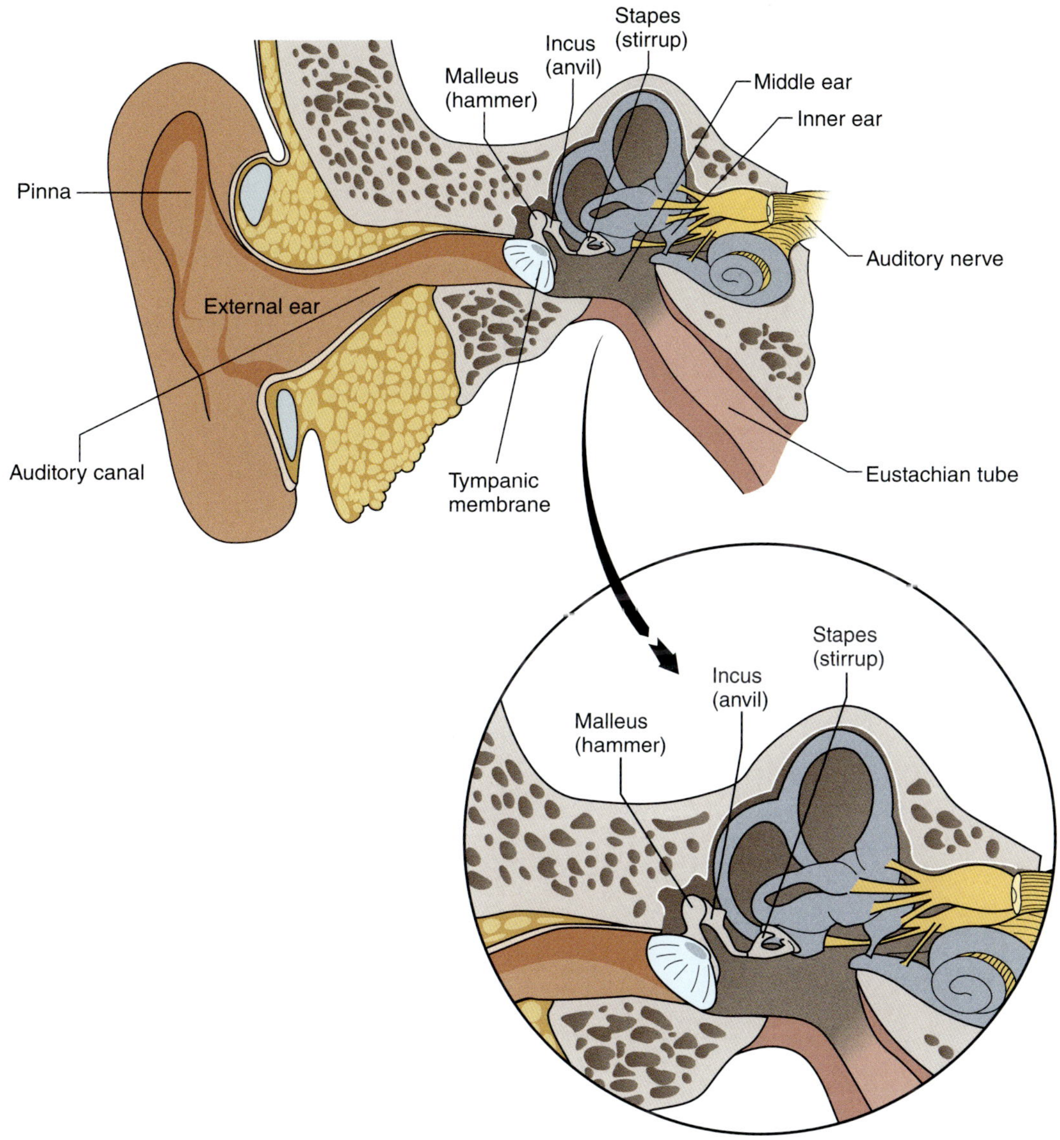

Figure 7.4 Structures and bones of the ear.

Joints of the Head

Joints of the skull are classified as intercranial (between cranial bones), interfacial (between facial bones), and craniofacial (between facial and cranial bones). There are also craniovertebral joints between the skull and the cervical vertebrae, which are presented in chapter 8. Bones of the skull articulate with each other through immovable joints known as *sutures*. These sutures are easily seen in the skulls of young babies (see the Focus On sidebar on fontanels), but once the bones mature, the sutures (seams) are fused. Most skull sutures are named after the bones they connect. The skull has 34 such sutures, as outlined in the following list. The suture is listed first, followed by the bones it connects. Items marked with an asterisk (*) indicate a suture that is labeled in figure 7.1, 7.2, or 7.3.

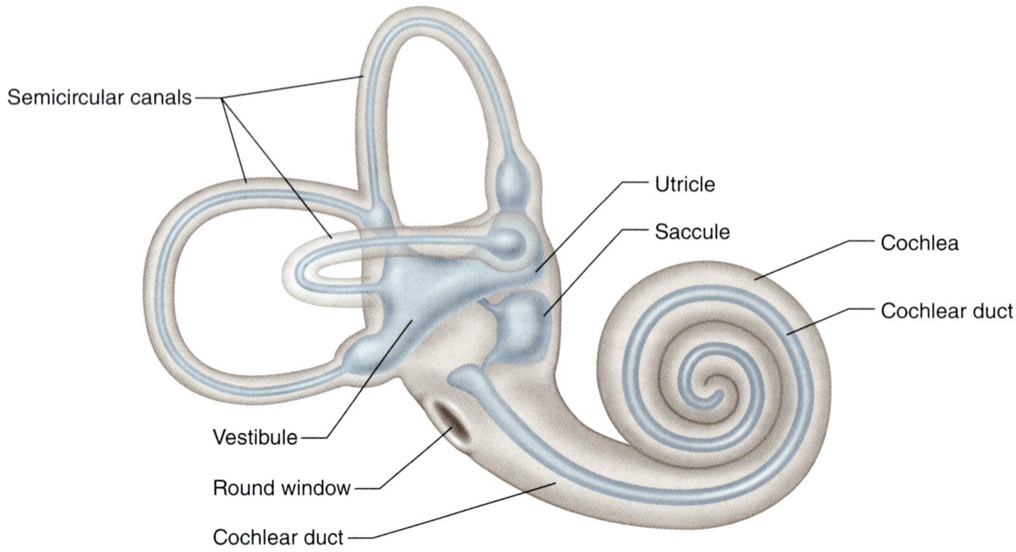

Figure 7.5 Structures of the inner ear.

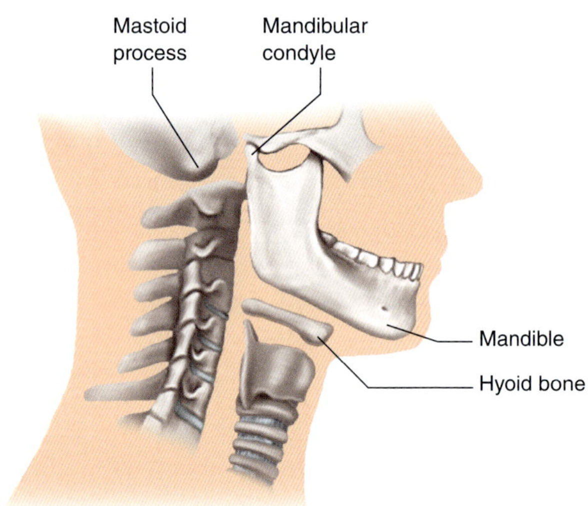

Figure 7.6 The hyoid bone.

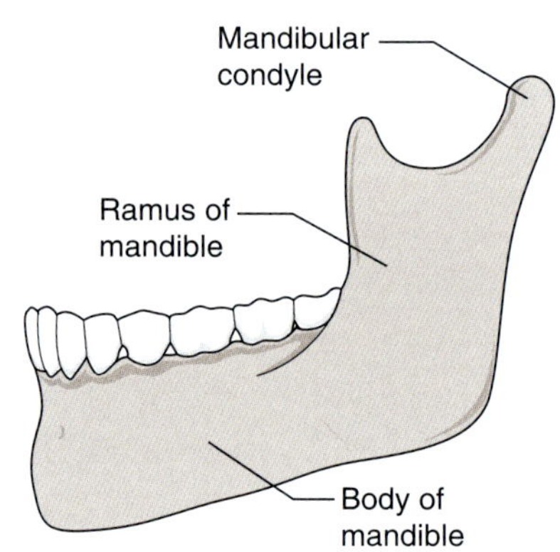

Figure 7.7 The mandible.

*Coronal: between the frontal and parietal bones

Ethmoidomaxillary: between the ethmoid and maxillary bones

Frontal: between the frontal bones until about age 6 to 10

Frontoethmoidal: between the frontal and ethmoid bones

Frontomaxillary: between the frontal and maxillary bones

*Intermaxillary: between the maxillary bones

Internasal: between the nasal bones

Lacrimoethmoidal: between the lacrimal and ethmoid bones

Lacrimofrontal: between the lacrimal and frontal bones

Lacrimomaxillary: between the lacrimal and maxillary bones

*Lambdoidal: between the occipital and parietal bones

Mastosquamosal: between the mastoid process and the squamous aspect of the temporal bone

Maxillozygomatic: between the maxillary and zygomatic bones

Median palatine: between the palatine process of the maxilla and the palatine bone

Nasofrontal: between the nasal and frontal bones

Nasomaxillary: between the nasal and maxillary bones

Occipitomastoid: between the occipital bone and the mastoid process
Palatoethmoidal: between the orbit of the palatine bone and the ethmoid bone
Palatomaxillary: between the orbit of the palatine bone and the maxillary bone
Parietomastoid: between the parietal bone and the mastoid process
Petrooccipital: between the petrous aspect of the temporal bone and the occipital bone
Petrosquamosal: between the petrous and squamous aspects of the temporal bone
*Sagittal: between the parietal bones
Sphenoethmoidal: between the wing of the sphenoid bone and the ethmoid bone
Sphenofrontal: between the wing of the sphenoid bone and the frontal bone
Sphenoorbital: between the body of the sphenoid bone and the orbital process of the palatine bone
Sphenoparietal: between the wing of the sphenoid bone and the parietal bone
Sphenopetrosal: between the sphenoid bone and the petrous aspect of the temporal bone
Sphenosquamosal: between the wing of the sphenoid bone and the squamous aspect of the temporal bone
Sphenozygomatic: between the wing of the sphenoid bone and the zygomatic bone
Squamosal: between the lower portion of the parietal bone and the squamous aspect of the temporal bone
Temporozygomatic: between the zygomatic process of the temporal bone and the temporal process of the zygomatic bone
Transverse palatine: between the palatine processes of the maxillary bones and the palatine bones
Zygomaticofrontal: between the zygomatic and frontal bones

All of the identified sutures (joints) are considered nonmoving joints, but the skull does contain one movable joint: the TMJ. The articulation between the head of the mandible and the mandibular fossa of the temporal bone forms a diarthrodial joint.

Ligaments of the Head

The mandible joint consists of an articular disc, a joint capsule, and a temporomandibular ligament (figure 7.8). Support for the joint is also provided by the stylomandibular ligament and sphenomandibular ligament (see figure 7.8). Movement in the joint is a rolling and gliding motion that results in elevation, depression, protrusion, and medial and lateral motion of the mandible.

Sinuses

Within the bones and bony markings are several air-filled cavities known as *sinuses* (figure 7.9). These spaces are lined with mucous membranes. All are either directly or indirectly connected

FOCUS ON

Fontanels

In an infant, there are four prominent areas where bones of the skull have not yet grown together. These soft spots between the bones are known as *fontanels*. The largest fontanel is the anterior (bregma) fontanel, which is the point between the coronal and sagittal sutures. An infant's pulse can be felt and often visibly observed in this area. The posterior (lambda) fontanel is the point where the lambdoidal and sagittal sutures meet. The lateral or sphenoidal (pterion) fontanel is the point where the greater wing of the sphenoid bone meets the sphenoidal angle of the parietal bone. The mastoid (asterion) fontanel is the point where the lambdoidal, occipitomastoid, and parietomastoid sutures meet. Under normal growth conditions, these fontanels "close," or the sutures are completely formed, by age 2.

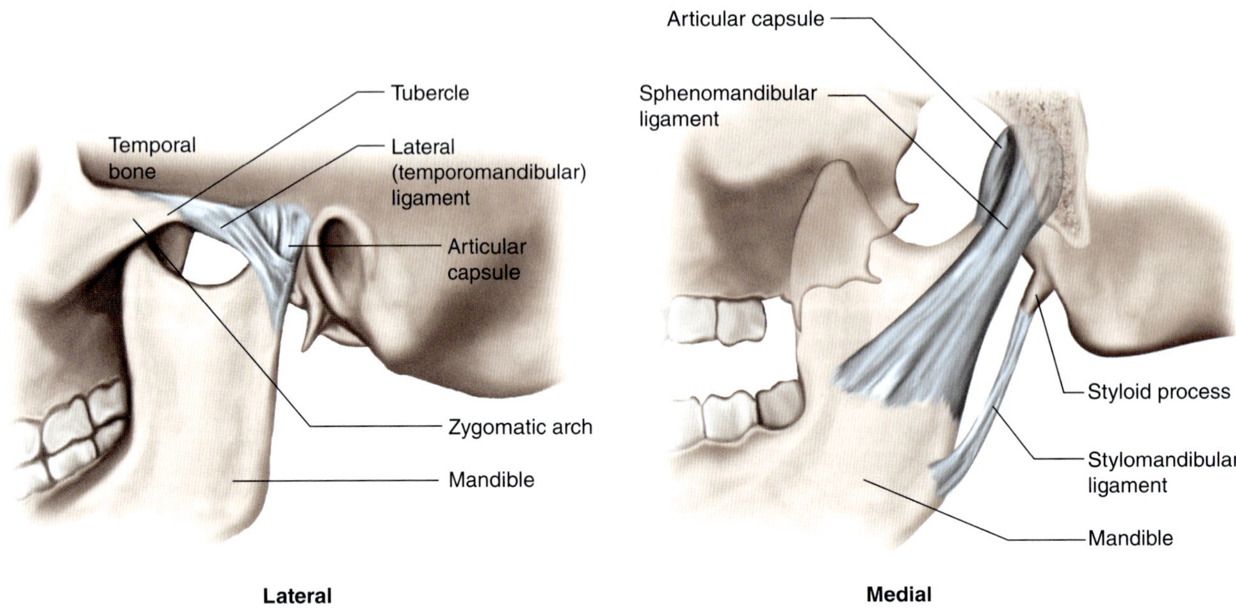

Figure 7.8 Temporomandibular joint ligaments.

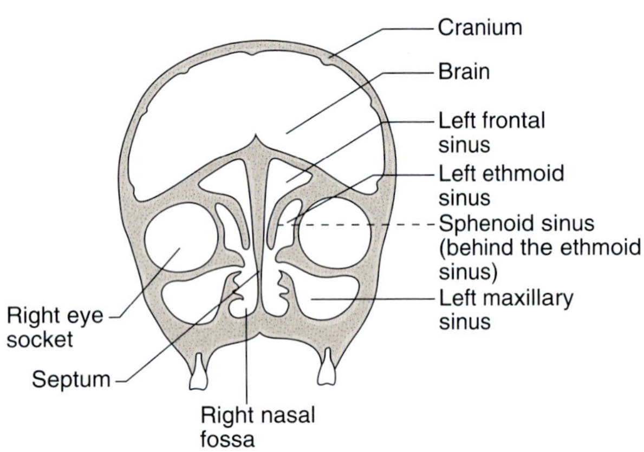

Figure 7.9 Sinuses of the skull.

to the nasal cavity and are thought to act as resonating cavities for the voice. Inflammation of the mucosa of any of the sinuses is known as *sinusitis*. There are four major sinuses:

The ethmoid sinuses are located in the ethmoid bone between the eyes and the middle turbinate (the bony walls of the nasal cavity form four pairs of projections known as *turbinates*: supreme, superior, middle, and inferior), and they communicate with the nasal cavity. A paper-thin bone (lamina papyracea) separates these sinuses from the eye socket contents.

The frontal sinuses are two triangular-shaped cavities (left and right) located at the base of the frontal bone (forehead); they communicate both directly and indirectly with the nasal cavity.

Located in the cheeks, the large maxillary sinuses are cavities within the maxillary bone, communicating with the middle meatus of the nasal cavity. The sphenoid sinus cavities are located in the anterior body of the sphenoid bone behind the ethmoid sinuses, opening into the nasal cavity above the superior nasal concha. The carotid arteries and the optic nerve lie in this area.

Fundamental Movements and Muscles of the Head

You might think that with the TMJ being the only movable joint of the head, there would be a minimal number of muscles. However, there are numerous muscles specific to movements within the skull, as well as many muscles attached to the skull that facilitate movement of the cervical spine (neck). The muscles of the head are divided into six groups: cranial, facial, eyeball, tongue, soft palate, and pharynx (figures 7.10 and 7.11).

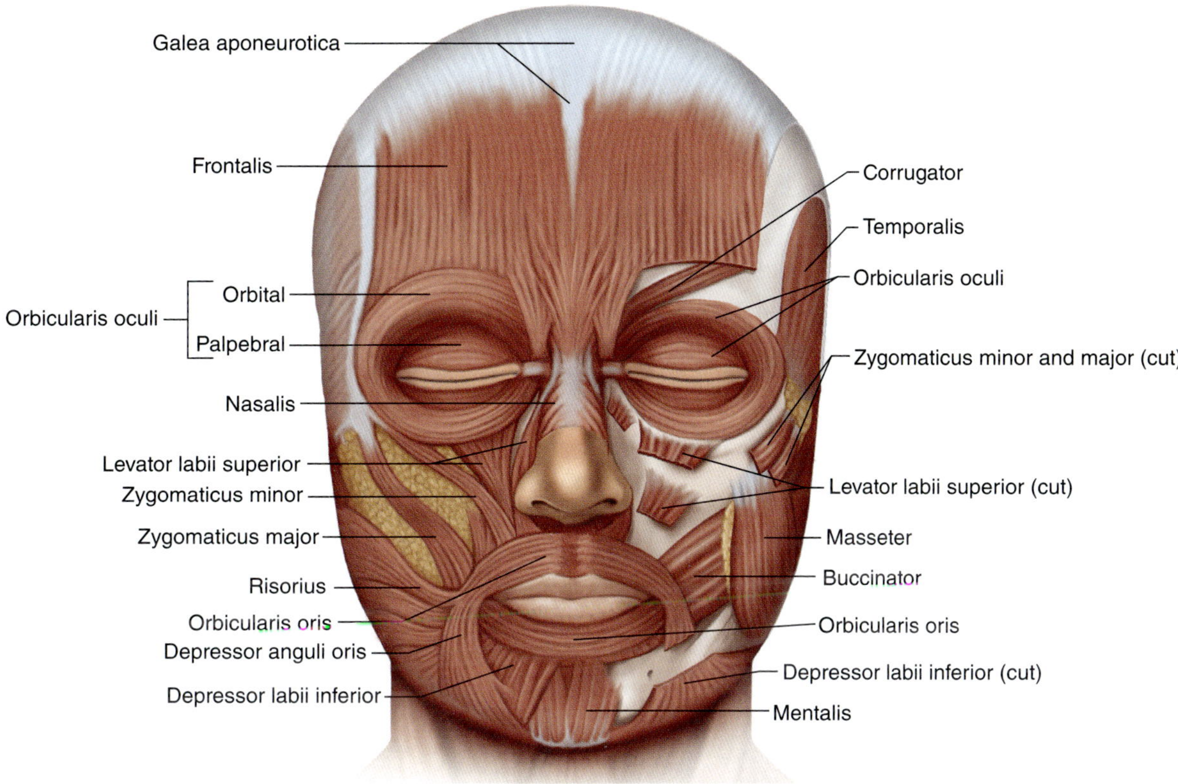

Figure 7.10 Superficial and deep muscles of the head.

- *Cranial muscles:* The cranial muscles are divided into three groups: (1) epicranius, consisting of the frontalis and occipitalis muscles; (2) external ear, consisting of the anterior, posterior, and superior auricularis muscles; and (3) middle ear, consisting of the stapedius and tensor tympani muscles.
- *Facial muscles:* The facial muscles (muscles of expression) are divided into four groups: (1) eyelid muscles, consisting of the corrugator supercilii, levator palpebrae superioris, and orbicularis oculi muscles; (2) nose muscles, consisting of the caput angulare, depressor alaeque nasalis, dilator naris, nasalis, and procerus muscles; (3) mouth muscles, consisting of the buccinator, caninus, mentalis, orbicularis oris, quadratus labii (inferior and superior), caput angulare, caput infraorbitale, caput zygomaticum, risorius, triangularis, and zygomaticus muscles; and (4) mastication muscles, consisting of the pterygoid (internal and external), masseter, and temporalis muscles.

Hands On

Either facing a partner or using a mirror, perform the following activities, considering which of the facial muscles are used: (1) Raise your eyebrows, (2) wrinkle your nose, and (3) form a kiss with your lips. Place your fingers over the area of the TMJ, open and close your mouth, and feel the muscles of mastication contracting. What facial muscles contribute to this action?

- *Eyeball muscles*: A cross-sectional view of the eye shows the following structures (figure 7.12). The cornea is the transparent layer just under the eyelid, and the iris is an elastic membrane located between the cornea and the lens. The iris is the portion of the eye that is colored (e.g., blue, green, or brown); it has an opening in its center known as the *pupil*. The space between the iris and the cornea is known as the *anterior chamber*, and the space between the iris and the lens is known as the *posterior chamber*. Both chambers are filled with a fluid known as *aqueous humor*.

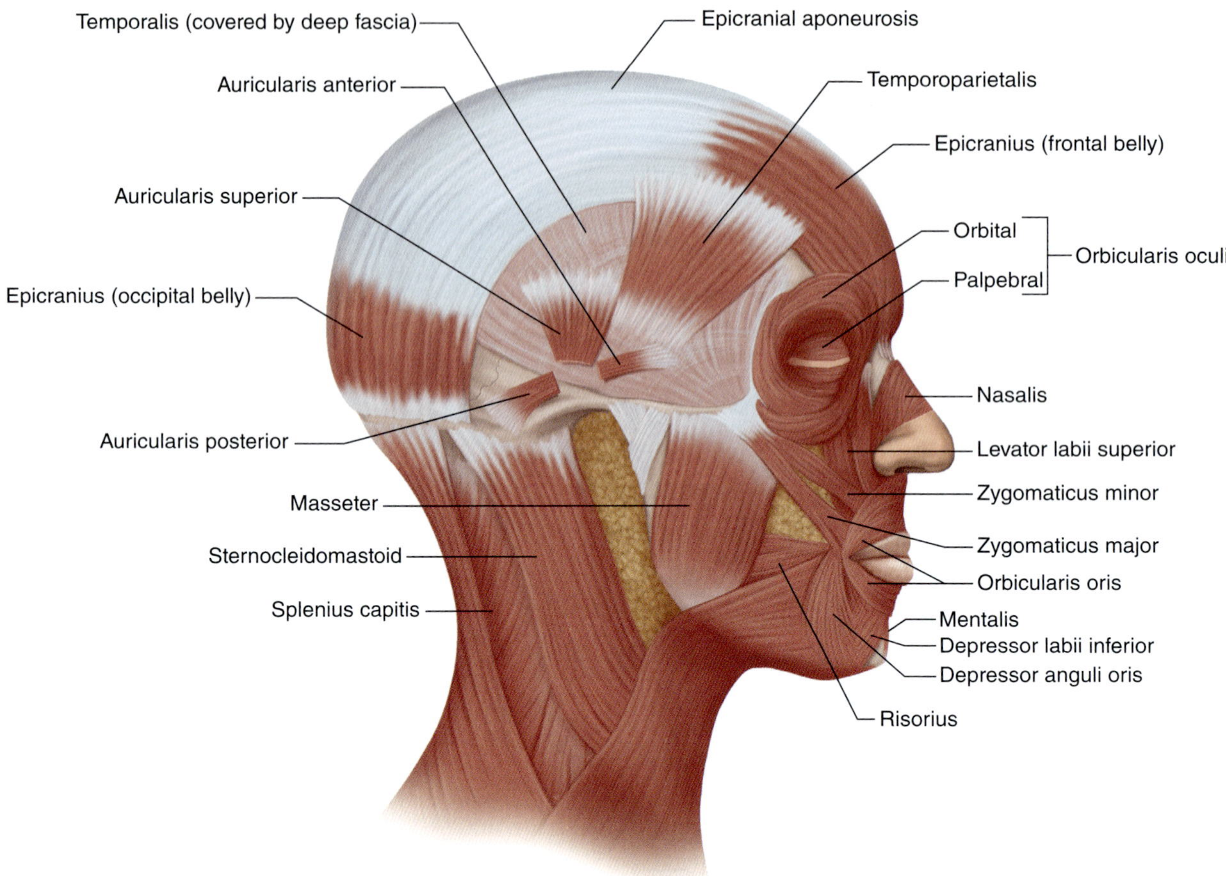

Figure 7.11 Lateral and posterior muscles of the head.

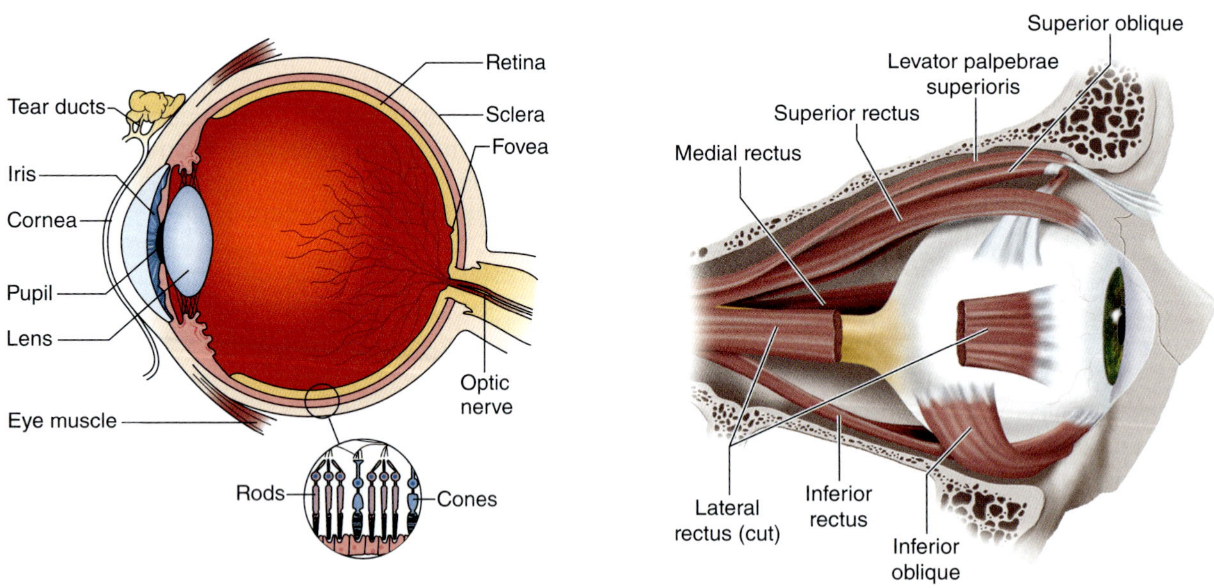

Figure 7.12 Muscles and structures of the eyeball.

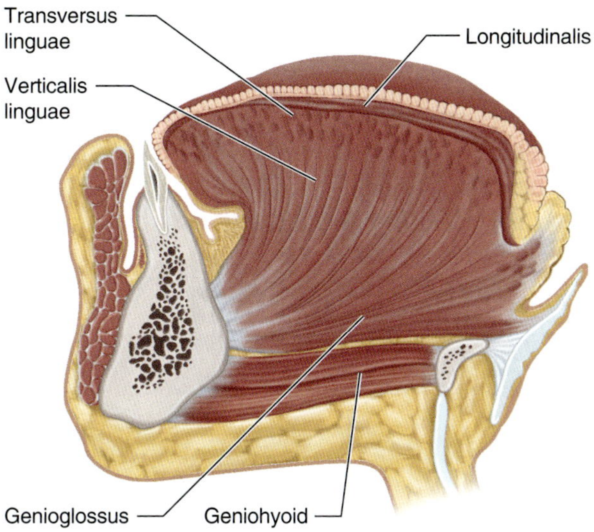

Figure 7.13 Muscles of the tongue.

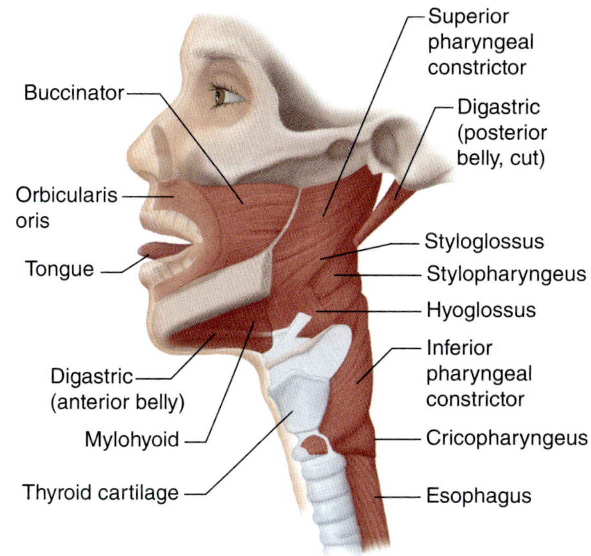

Figure 7.14 Superficial muscles of the pharynx and larynx.

A suspensory ligament suspends the lens in the eyeball. Posterior to the lens lies a three-layered wall consisting of the outer layer (sclera), an inner layer (retina), and a layer between these two (choroid). The area between the lens and the retina is filled with a fluid known as the *vitreous humor*. The muscles of the eyeball are divided into two groups: (1) extrinsic, consisting of the obliquus (superior and inferior), orbitalis, and four recti muscles (superior, inferior, lateralis, and medialis) and (2) intrinsic, consisting of the ciliary and two muscles of the iris, the dilator pupillae and the sphincter pupillae.

Hands On

Either facing a partner or using a mirror, perform the following activities, considering which of the eyeball muscles are used: (1) Look upward, (2) look downward, and (3) look side to side. Cover one eye with your hand, and then quickly uncover it; observe the pupil. Did it increase or decrease in size?

- *Tongue muscles:* The muscles of the tongue (figure 7.13) are divided into two groups: (1) extrinsic, consisting of the chondroglossus, genioglossus, hyoglossus, and styloglossus muscles and (2) intrinsic, consisting of the longitudinalis (superior and inferior) muscles, transversus linguae, and verticalis linguae.

Hands On

Either facing a partner or using a mirror, perform the following activities, considering which of the tongue muscles are used: (1) Stick out your tongue, and (2) wiggle your tongue left and right.

- *Soft palate muscles:* The muscles of the soft palate consist of the glossopalatinus, uvulae, levator veli palatini, tensor veli palatini, pharyngopalatinus, and salpingopharyngeus muscles.
- *Pharynx and larynx muscles:* The muscles of the pharynx consist of the stylopharyngeus and the constrictor pharyngis muscles (superior, inferior, and medius). While typically considered muscles of the neck, the muscles of the larynx are included in this grouping of muscles of the head since their actions relate only to the larynx and not to the movement of the cervical spine. The muscles of the larynx are divided into two groups: extrinsic and intrinsic. The extrinsic muscles are considered muscles of the hyoid and are also divided into two groups: (1) the infrahyoid group, consisting of the levator glandulae thyroideae, sternohyoid, sternothyroid, thyrohyoid, and omohyoid (anterior and posterior bellies) and (2) the suprahyoid group, consisting of the geniohyoid, mylohyoid, stylohyoid, and digastric (anterior and posterior bellies; figure 7.14). The intrinsic muscles of the larynx

are the aryepiglotticus, arytenoid oblique and transverse, cricoarytenoid posterior and lateralis, cricothyroid, thyroarytenoid, thyroepiglotticus, ventricularis, and vocal muscle.

Figure 7.14 also illustrates the thyroid cartilage of the larynx. A bump develops on this cartilage that is anatomically known as the *laryngeal prominence* but is commonly referred to as the *Adam's apple*. Particularly in males, as the thyroid cartilage grows together with the larynx, this prominence grows in size. This growth typically coincides with puberty and creates the deepening of the male voice. This process is much more diminished in females but can occur.

POP QUIZ

Based on this photo, the hockey player's helmet would help reduce the risk of which of the following injuries?

- **a.** skull fracture
- **b.** mandible fracture
- **c.** maxilla fracture
- **d.** hyoid fracture

Eliot J. Schechter/NHLI via Getty Images

Visit the web resource to complete the activities for this chapter.

CHAPTER 8

The Spinal Column and Pelvis

The spinal column (figure 8.1) is a stack of 33 bones, called *vertebrae*, held together by ligaments and muscles, with cartilaginous discs (primarily water and protein) between the bones. The 33 vertebrae are divided into 5 distinct sections. All the vertebrae have many common characteristics, but each group has unique features designed for specific purposes. The most superior group is known as the *cervical (neck) spine* and contains 7 vertebrae. The next group is known as the *thoracic (chest) spine* and contains 12 vertebrae. The next group is known as the *lumbar (low back) spine* and contains 5 vertebrae. The next group is known as the *sacral spine* and contains 5 vertebrae fused together into one structure known as the *sacrum*. The last, or most distal, group is known as the coccygeal spine and contains 4 vertebrae fused together into one structure known as the *coccyx*.

The lateral view of the spinal column reveals four curvatures: anterior (convex) curves in the cervical and lumbar spine and posterior (concave) curves in the thoracic and sacrococcygeal spine. These curvatures may increase or decrease as the body's center of gravity shifts (e.g., with pregnancy, weight gain, weight loss, or trauma). These changes in curvatures are the result of one of the spinal column's functions: to maintain, in the upright body position, the brain over the body's center of gravity.

Over- or underdevelopment of the musculature on either side of the spinal column, structural deformities, or other causes can result in excessive curvatures of the spinal column. Three of the more common conditions resulting from excessive curvature are kyphosis (figure 8.2), excessive posterior curvature of the thoracic spine (hunchback or round shoulders); lordosis (figure 8.3), excessive anterior curvature of the lumbar spine (swayback; often found in individuals with excessive abdominal weight and pregnant women); and scoliosis (figure 8.4), excessive lateral curvature of the spinal column, usually in the thoracic spine but sometimes to a lesser extent in the cervical and lumbar spine (often found in individuals involved in unilateral activities that lead to muscular development on one side of the trunk, such as baseball pitchers,

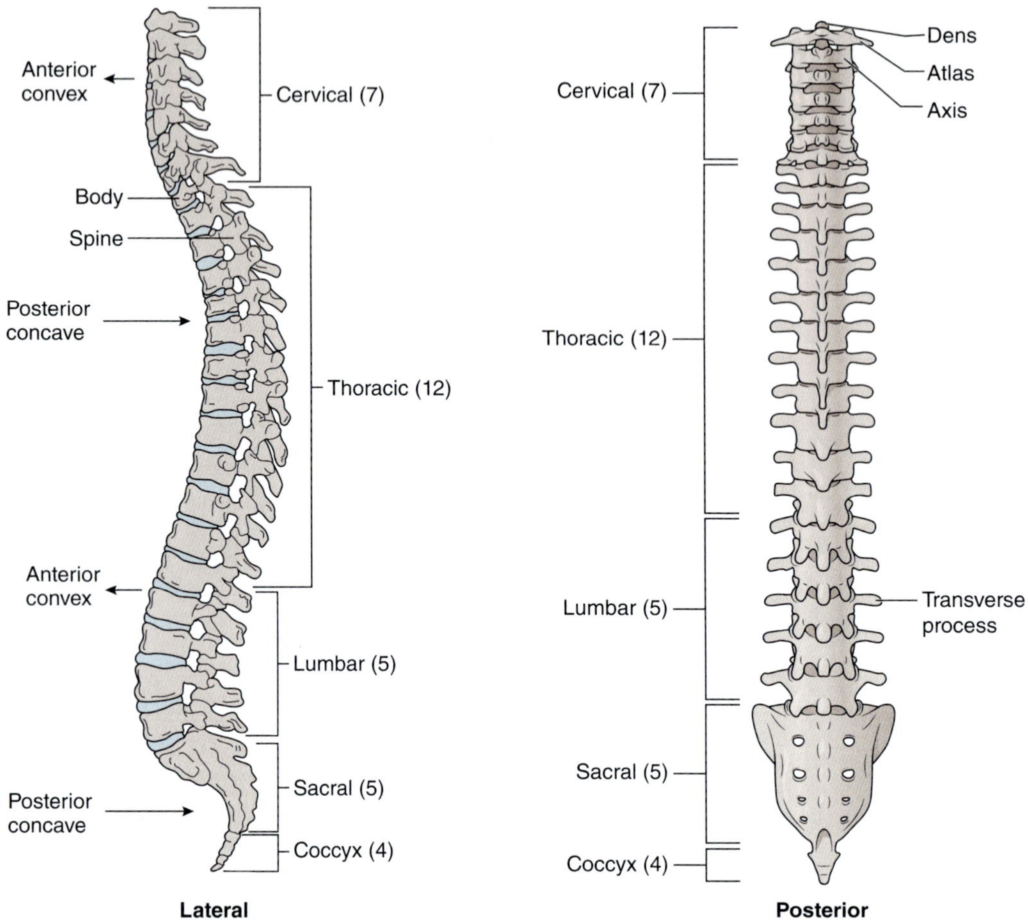

Figure 8.1 The spinal column.

Figure 8.2 Thoracic kyphosis.

Figure 8.3 Lumbar lordosis.

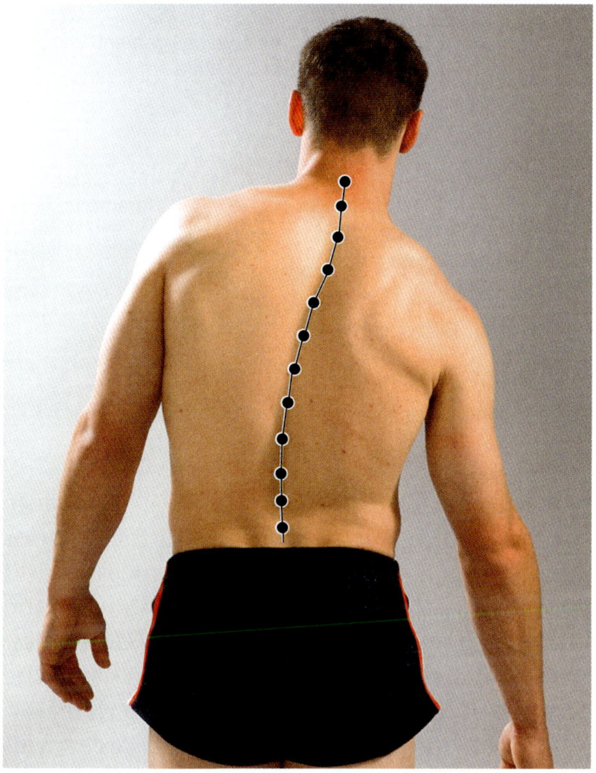

Figure 8.4 Scoliosis.

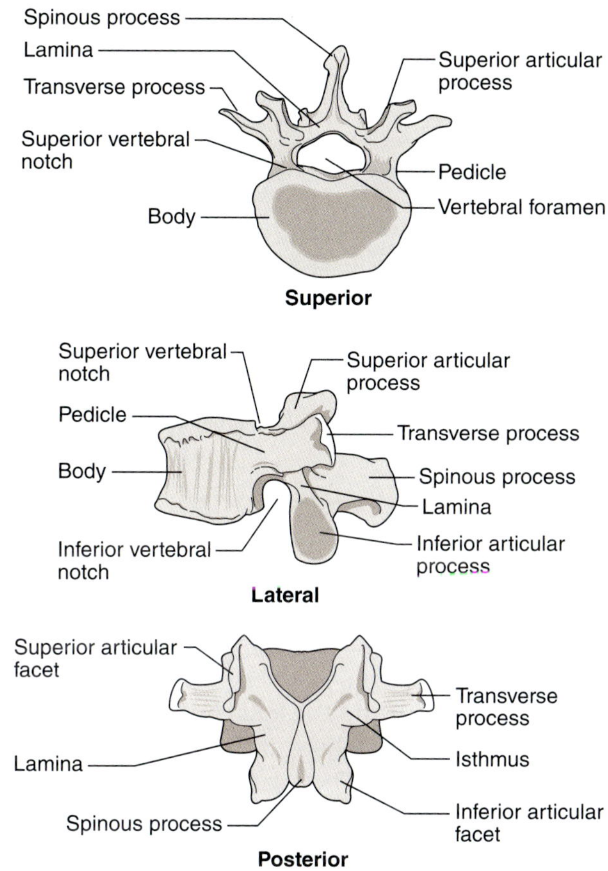

Figure 8.5 A typical vertebra.

football quarterbacks, golfers, and mail carriers). If you have the opportunity, observe individuals performing such activities. Look for the trunk leaning to one side laterally, differences in shoulder height, and whether there is an abnormal curvature of the spine in the shape of the letter *C* or the letter *S*. Remember that one of the spinal column's functions is to keep the brain over the body's center of gravity. If the *C*, or lateral, curvature becomes too excessive, the spinal column will attempt to bring the brain back over the center of gravity by curving to form an *S*.

Bones of the Spinal Column

Although all five sections of the spinal column have common structures, the vertebrae of each are spaced uniquely and may be shaped slightly differently, depending on their functions in a particular anatomical area. All vertebrae have a body, two transverse processes laterally (serving as sources for ligamentous and muscular attachments), a spinous process (except for the atlas (C1)—serving as another source for ligamentous and muscular attachments), and a vertebral foramen (where the spinal cord and nerve roots pass; figure 8.5). With the exception of the atlas (C1), each vertebra has the following features. Superior and inferior articular facets are where a vertebra articulates with the vertebrae above and below it. A lamina forms the posterior aspect of the vertebral foramen. (Remember the area formed by the lamina because it will be reintroduced later regarding the term *laminectomy*.) Pedicles form the lateral sides of the vertebral foramen. The intervertebral foramen between the vertebrae allows the nerve branches from the spinal cord to pass through. The isthmus (also called the *pars interarticularis*, or neck) is the bony area between the superior and inferior articular facets.

The seven cervical vertebrae are numbered from the most superior to the most inferior as C1, C2, and so on. The first cervical vertebra, the atlas (C1), and the second cervical vertebra, the axis (C2), are shaped differently from the

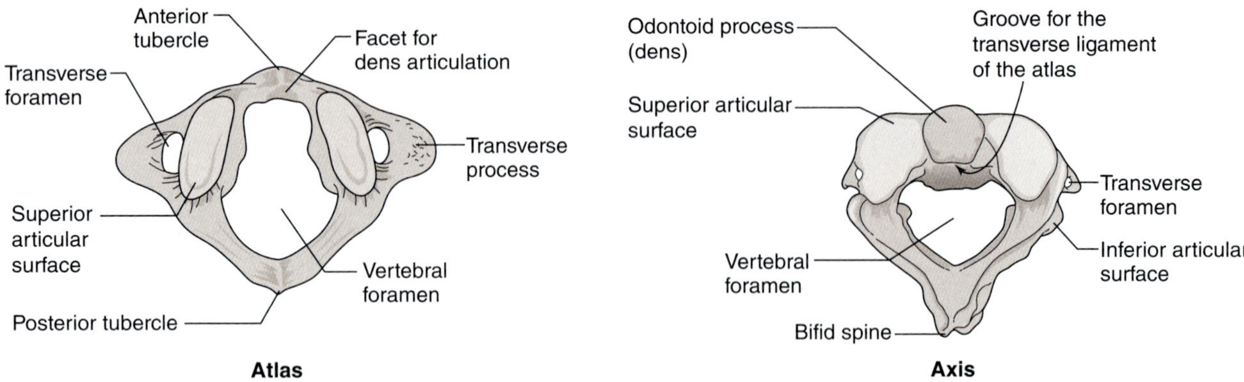

Figure 8.6 Superior view of the first cervical vertebra (atlas) and the second cervical vertebra (axis with the dens).

other five cervical vertebrae (C3-C7) to permit the head to rotate.

The atlas has no significant body but has two large articular facets that provide the surface where the skull and the spinal column articulate (figure 8.6). This is a very firm articulation between the skull and the first cervical vertebrae, creating a situation where, if the skull moves, the atlas moves with it. The atlas slides over the axis and rests on top of the two large superior articular surfaces of the axis, between which lies a rather large bony process from the axis body known as the *dens*, or *odontoid process* (see figure 8.6). It is the joint between the atlas (C1), attached to the skull, and the axis (C2) where rotation of the head occurs. Note also that the cervical vertebrae have a bifid (split) spinous process and a foramen in each transverse process to provide for the passage of blood vessels through the cervical spine (figure 8.7). These two features are unique to the cervical vertebrae. One other difference to note is the rather long and prominent spinous process of the seventh cervical vertebra (C7). This prominence is easily palpated.

Hands On

Run your finger down the posterior aspect of your cervical spine until you find a large protrusion. This is the C7 spinous process, which serves as an anatomical landmark for determining the spinous processes of the other cervical vertebrae and where the thoracic spine begins.

The 12 thoracic vertebrae have bony features similar to all other vertebrae, with a few unique features. Note the longer and more vertical spinous processes (see figure 8.7). Also note the articulating surfaces (fovea) on the anterior lateral aspects of the transverse processes and on the superior and inferior portions of the posterior lateral aspects of the vertebral bodies. These notches provide the articulation site of the 12 pairs of ribs with the 12 thoracic vertebrae.

The five lumbar vertebrae are the largest vertebrae (see figure 8.7). They have no foramina through their transverse processes, nor are there any articular facets (fovea) on their bodies. Although separate at birth, the five sacral vertebrae (S1-S5) fuse together to form a large triangular-shaped bone known as the *sacrum* (figures 8.8 and 8.9) during the growth process. The two large articular surfaces formed on the lateral aspects of the sacrum are where the spinal column articulates with the bones of the pelvis, forming the pelvic girdle.

The coccyx (the final four vertebrae), like the sacrum, is originally four vertebrae that, during the growth process, fuse to form one structure (figure 8.10). It serves as a source of attachment for ligamentous and muscular structures. Some people refer to the coccyx as humans' "vestigial tail," in reference to the evolution of humans from species that had tails.

Ligaments of the Spinal Column

There is only one movable joint in the skull, the temporomandibular joint, which enables the opening and closing of the mouth; any movement of the head is the result of movement at

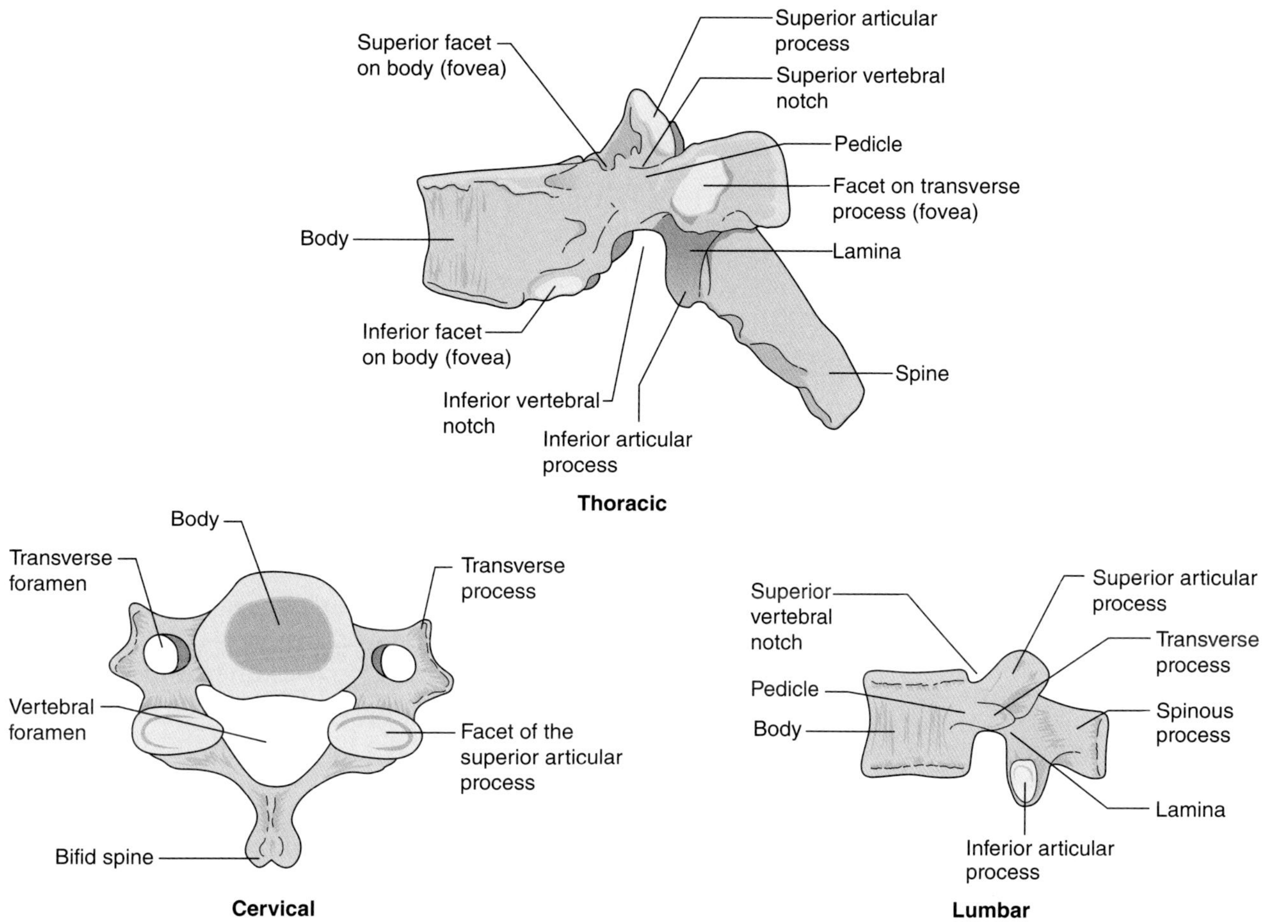

Figure 8.7 The thoracic, cervical, and lumbar vertebrae.

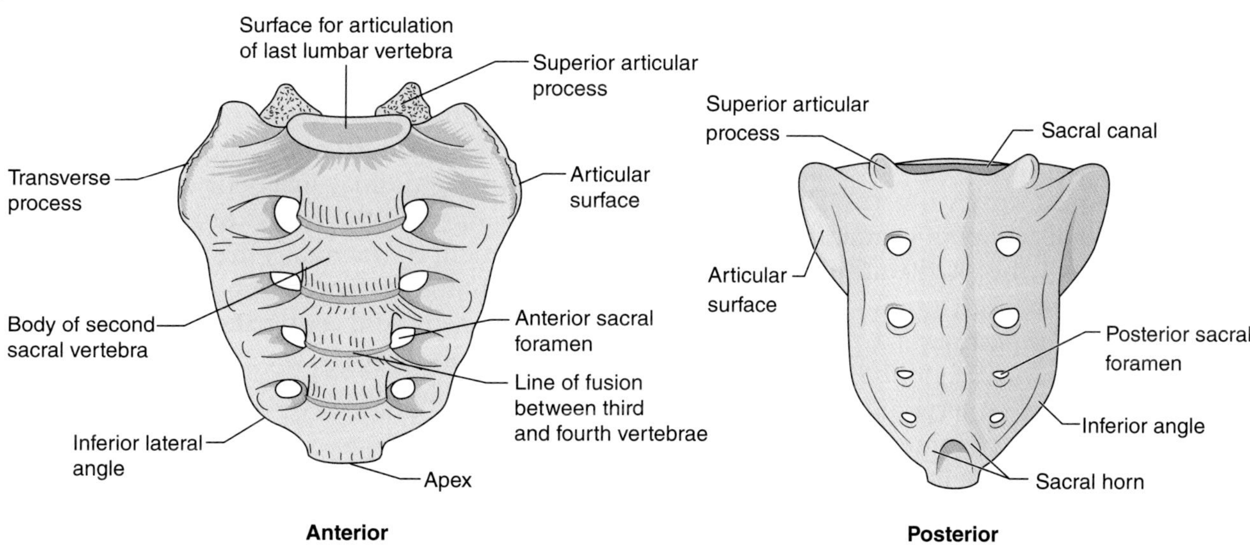

Figure 8.8 The sacrum.

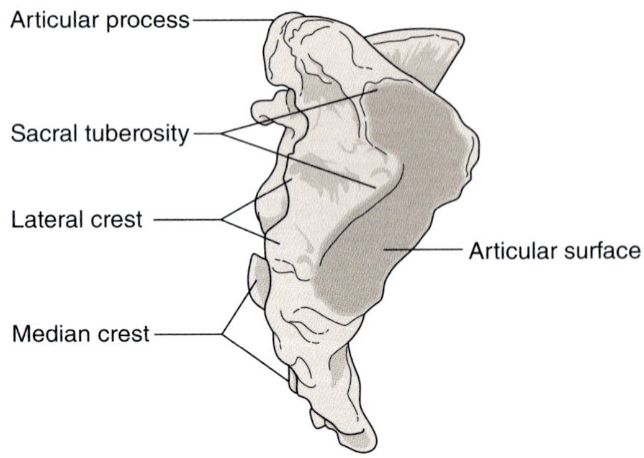

Figure 8.9 Lateral view of the sacrum.

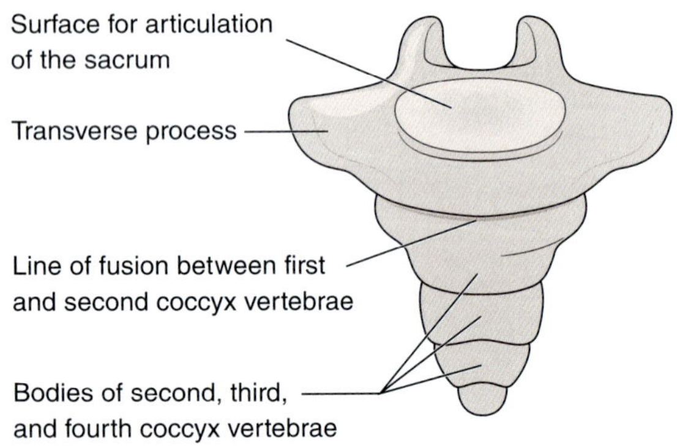

Figure 8.10 Anterior view of the coccyx.

the joints between the occipital bone of the skull and the first and second cervical vertebrae. There are approximately 25 ligaments between the skull and the cervical spine (figures 8.11 and 8.12). With the exception of one ligament discussed later, we will not devote time to the individual ligaments but simply note the groups they belong to. Atlantooccipital ligaments attach the occipital bone of the skull to the first cervical vertebra (atlas, or C1). Occipitoaxial ligaments (also known as the *membrana tectoria*) attach the occipital bone of the skull to the dens (odontoid process) of the second cervical vertebra (axis, or C2), and atlantoaxial ligaments attach the atlas and the axis. A group of ligaments not involved in the movement of the spinal column includes the costovertebral ligaments (six per rib), which articulate the 12 pairs of ribs of the thorax with the 12 thoracic vertebrae.

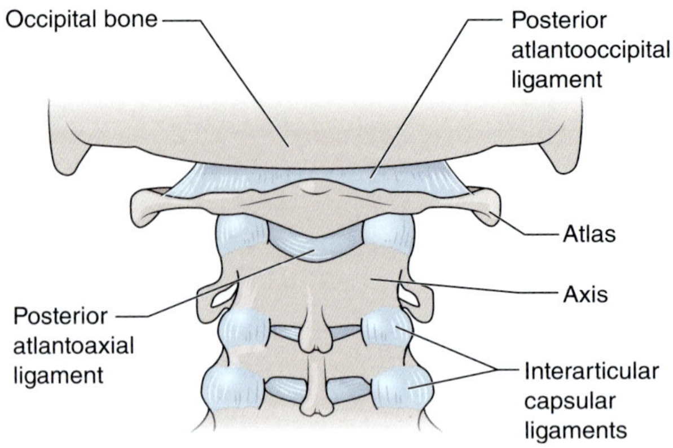

Figure 8.11 Atlantooccipital and atlantoaxial ligaments of the cervical spine.

One of the atlantoaxial ligaments, the transverse ligament (figure 8.13), runs from one transverse process of the atlas (C1), across the vertebral foramen, to the other transverse process and holds the dens (odontoid process) of the axis (C2) in place. Failure of this ligament to restrain movement of the dens of C2 beneath C1 could cause the dens to do irreparable damage to structures in the vertebral foramen at the C1-C2 level of the spinal column. If damaged or torn, the dens of C2 may tear the transverse ligament of C1 and compress the spinal cord against C1, disrupting nerves to the heart and lungs.

From the most superior to the most inferior aspects of the spinal column are a number of ligaments that play a major role in the movement of the joints between the vertebrae (figure 8.14). Two ligaments known as the interbody ligaments run the entire length of the spinal column. The anterior longitudinal ligament runs along the anterior aspect of the bodies of all 33 vertebrae. This ligament is structurally the weakest of all the spinal column ligaments. The posterior longitudinal ligament runs along the posterior aspect of the bodies of all 33 vertebrae. The posterior longitudinal ligament forms the anterior wall of the spinal canal.

The ligamentum flavum runs between the laminae of successive vertebrae (see figure 8.14). The interspinous ligament runs between the spinous processes of successive vertebrae. Running between the dorsal tips of each vertebra's

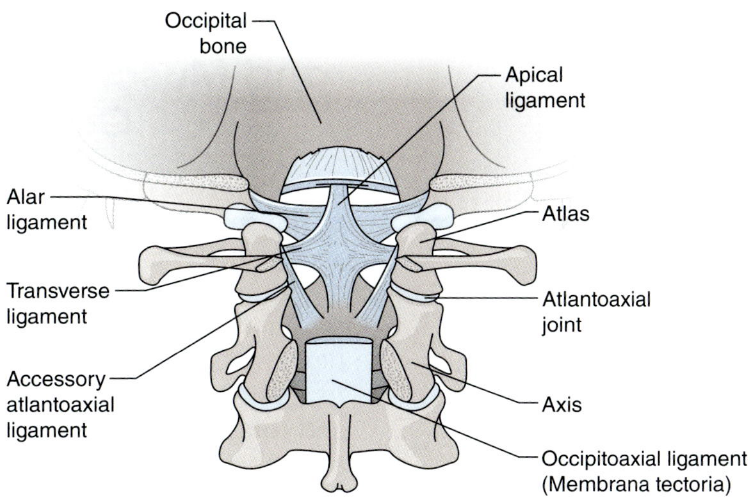

Figure 8.12 Occipitoaxial and atlantoaxial ligaments of the cervical spine.

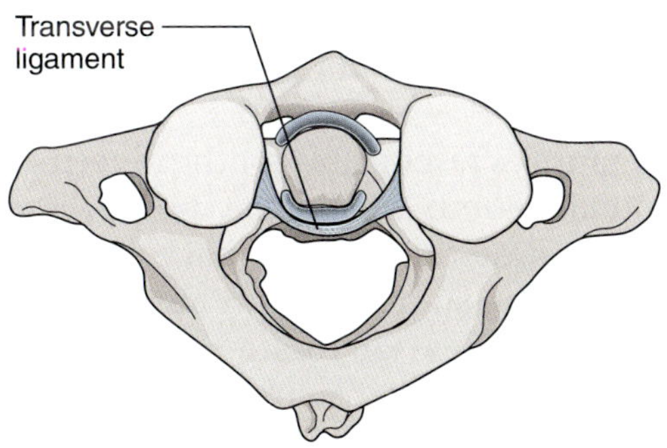

Figure 8.13 The transverse ligament of the atlas (C1).

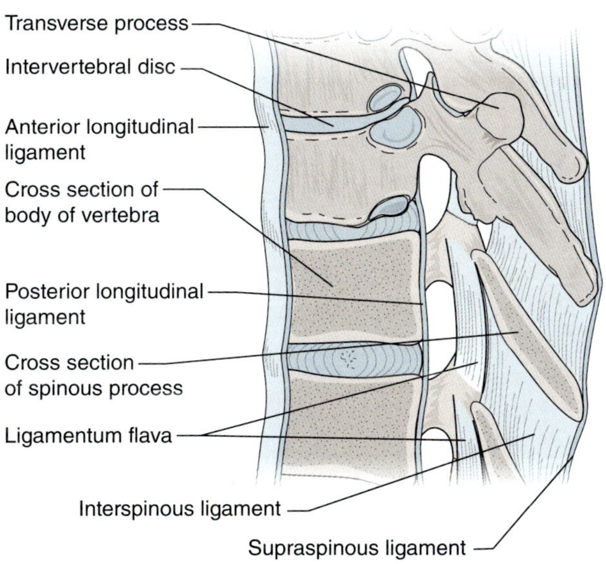

Figure 8.14 Ligaments of the spinal column.

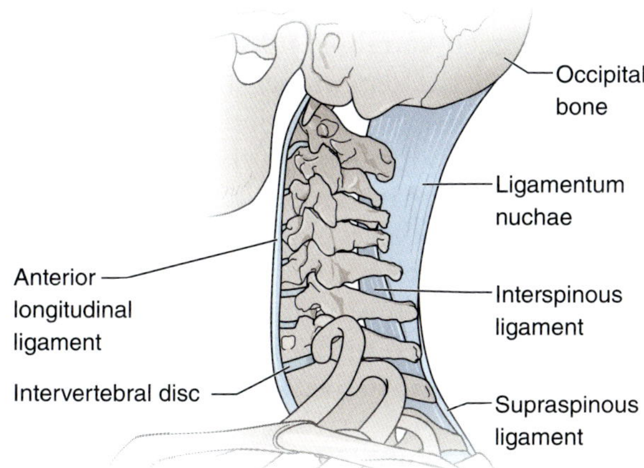

Figure 8.15 The ligamentum nuchae of the cervical spine.

spinous process, from the coccyx to the external occipital protuberance of the occipital bone, is the supraspinous ligament. Between the external occipital protuberance and the spinous process of the seventh cervical vertebra, the supraspinous ligament is known as the *ligamentum nuchae* (figure 8.15).

One additional ligament of the spinal column is the iliolumbar ligament, which runs between the transverse processes of the fifth lumbar vertebra to the ilium of the pelvis (figure 8.16). The joints and ligaments between the spinal column and the pelvis are presented later in this chapter.

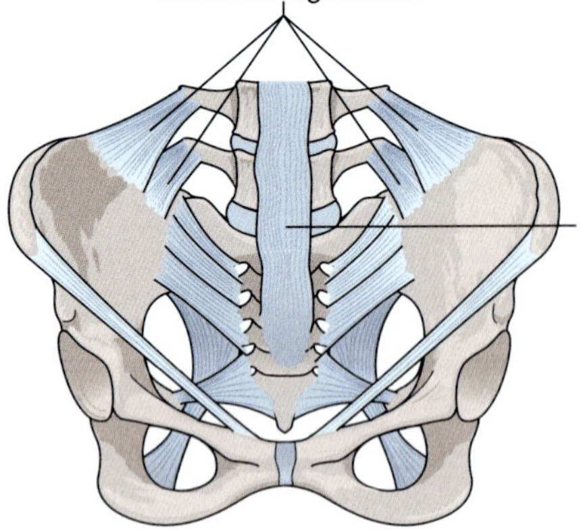

Figure 8.16 Anterior view of the lumbosacral joint and iliolumbar ligament.

One other structure of the spinal column is the intervertebral disc (figure 8.17). These cartilaginous (primarily water and protein) discs lie on the bodies of each vertebra and serve as spacers (to help separate the vertebrae and allow nerve roots to pass from the spinal canal to other structures of the body) and shock absorbers for the spinal column. The disc has two distinct portions: The inner portion is known as the *nucleus pulposus*, and the outer portion is known as the *annulus fibrosus*. The annulus fibrosus consists of fibrous tissue, whereas the nucleus pulposus consists of soft, pulpy, elastic tissue.

Fundamental Movements and Muscles of the Spinal Column

Movements of the joints between the vertebrae of the spinal column occur in all three planes about all three axes. The fundamental movements of the spinal column (the cumulative actions of all the joints of the 33 vertebrae) are flexion, extension, lateral flexion (as opposed to abduction and adduction in the extremities), and rotation (figures 8.18 and 8.19). The greatest amount of movement of the spinal column takes place in the cervical and lumbar regions. Movement is more restricted in the thoracic region because of the attached ribs. There is no movement in the sacral and coccygeal regions because the vertebrae in these regions are fused together into the five-bone sacrum and the four-bone coccyx.

The muscles responsible for the movements of the spinal column are, for the most part, either anterior or posterior to the spinal column. Those anterior to the spinal column are muscles that flex, laterally flex, or rotate the spine. Those posterior to the spinal column are muscles that extend, laterally flex, or rotate the spine. Some

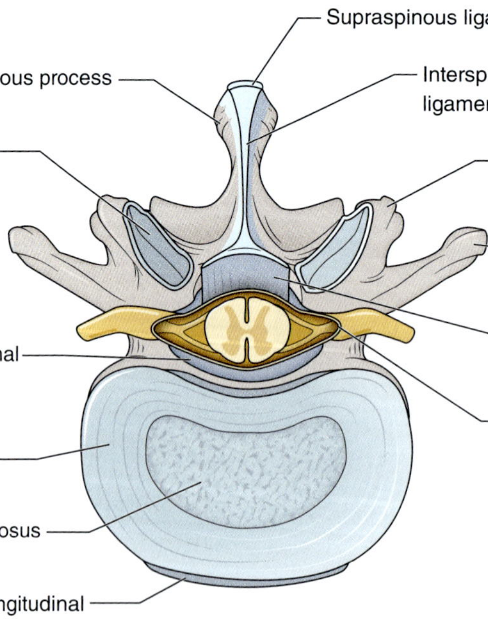

Figure 8.17 The intervertebral disc.

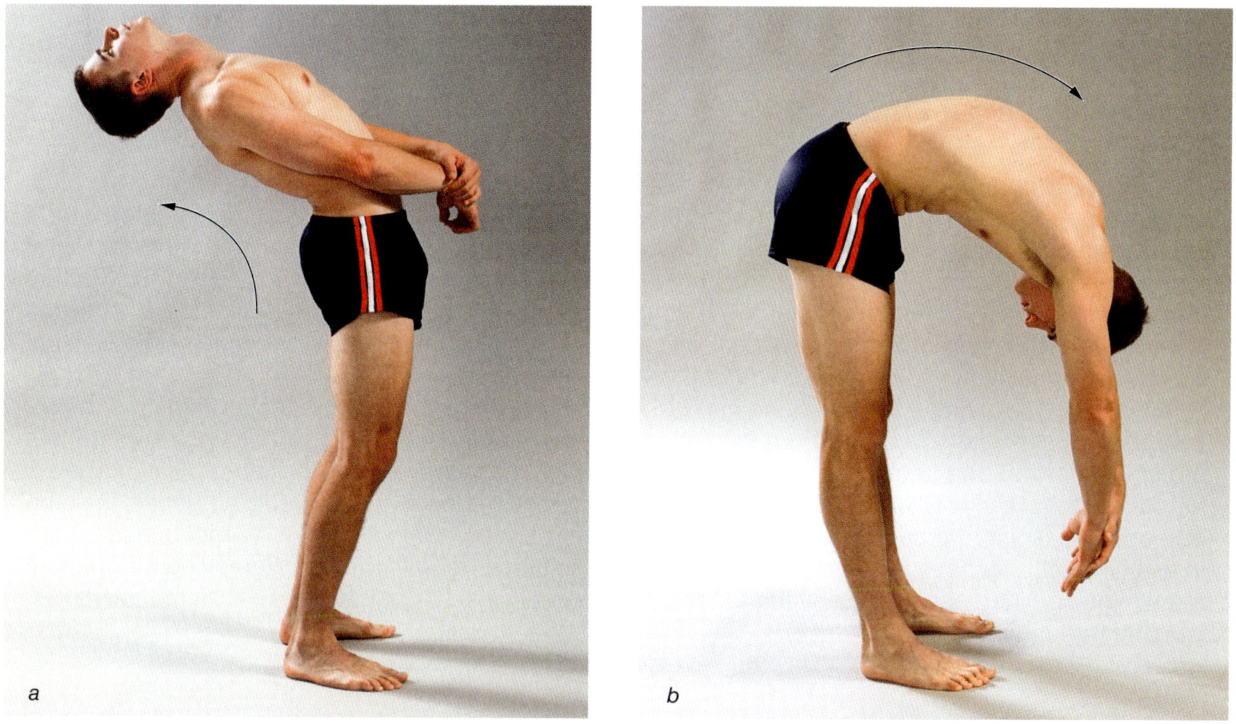

Figure 8.18 *(a)* Extension and *(b)* flexion of the spine.

Figure 8.19 *(a)* Rotation and *(b)* lateral flexion of the spine.

FOCUS ON

Intervertebral Disc

The terms *slipped disc* and *herniated disc* are frequently heard in reference to someone having back pain. A slipped disc means the intervertebral disc has protruded into the space occupied by either the spinal cord or its nerve roots, compressing the nerve and causing pain and disability. A more technical term would be *protruded disc*. The term *herniated disc* refers to the actual tearing of the disc structure that allows a portion of the disc material to escape and cause pressure on the spinal cord or its nerve roots. Both the protruded and herniated discs can occur as a result of excessive motion of the spinal column (hyperflexion, hyperextension, lateral flexion, or rotation).

A herniated disc in the lumbar spine can be detected from a spinal tap, or lumbar puncture, a procedure in which fluid is removed from the space beneath the arachnoid membrane (which surrounds the spinal cord) of the lumbar region of the spinal cord. Elevated levels of a telltale protein indicate a herniated disc. Disc problems need to be treated by neurosurgeons and orthopedic surgeons. In some of these situations, the surgeon might find the need to perform a laminectomy. As mentioned earlier in this chapter, the lamina forms the posterior aspect of the vertebral foramen. Removing a portion of the lamina can increase the size of the foramen and decrease some of the pressure on the nerve root passing through the foramen.

spinal column muscles are specific to a particular area of the spinal column (cervical or lumbar), whereas others have branches in multiple areas of the spinal column.

In the cervical region, the major anterior muscles and muscle groups are the sternocleidomastoid, the prevertebrals (rectus capitis anterior, rectus capitis lateralis, longus capitis, and longus colli), and the scaleni (scalenus anterior, scalenus medius, and scalenus posterior; figures 8.20 and 8.21). The sternocleidomastoid muscle, as its name indicates, attaches to the sternum, the clavicle, and the mastoid process of the skull just posterior to the ear.

Hands On

Referring to figure 8.22, palpate the sternal and clavicular portions of the sternocleidomastoid muscle. Using simple reasoning, and considering there is both a right and a left sternocleidomastoid muscle, answer the following question: If this muscle lies anterior and slightly lateral to the cervical spine, what actions is it capable of performing? (Answer: flexion, lateral flexion, and rotation of the cervical spine)

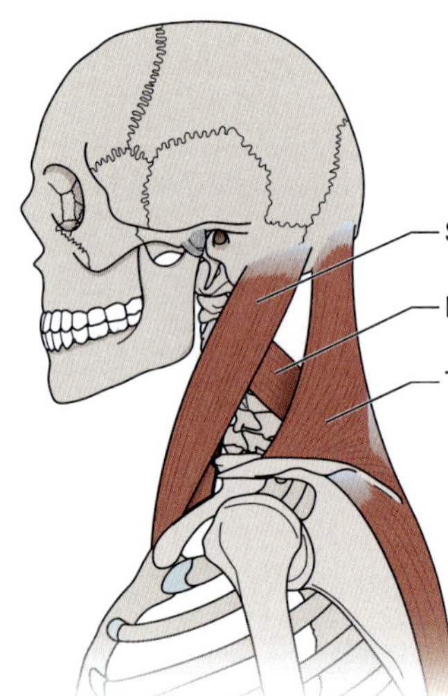

Figure 8.20 The sternocleidomastoid.

The recti muscles (rectus capitis anterior and rectus capitis lateralis), the longus capitis, and the longus colli flex and rotate the cervical spine. The three scalene muscles (scalenus anterior, scalenus medius, and scalenus posterior) run from the cervical vertebrae to the first three ribs.

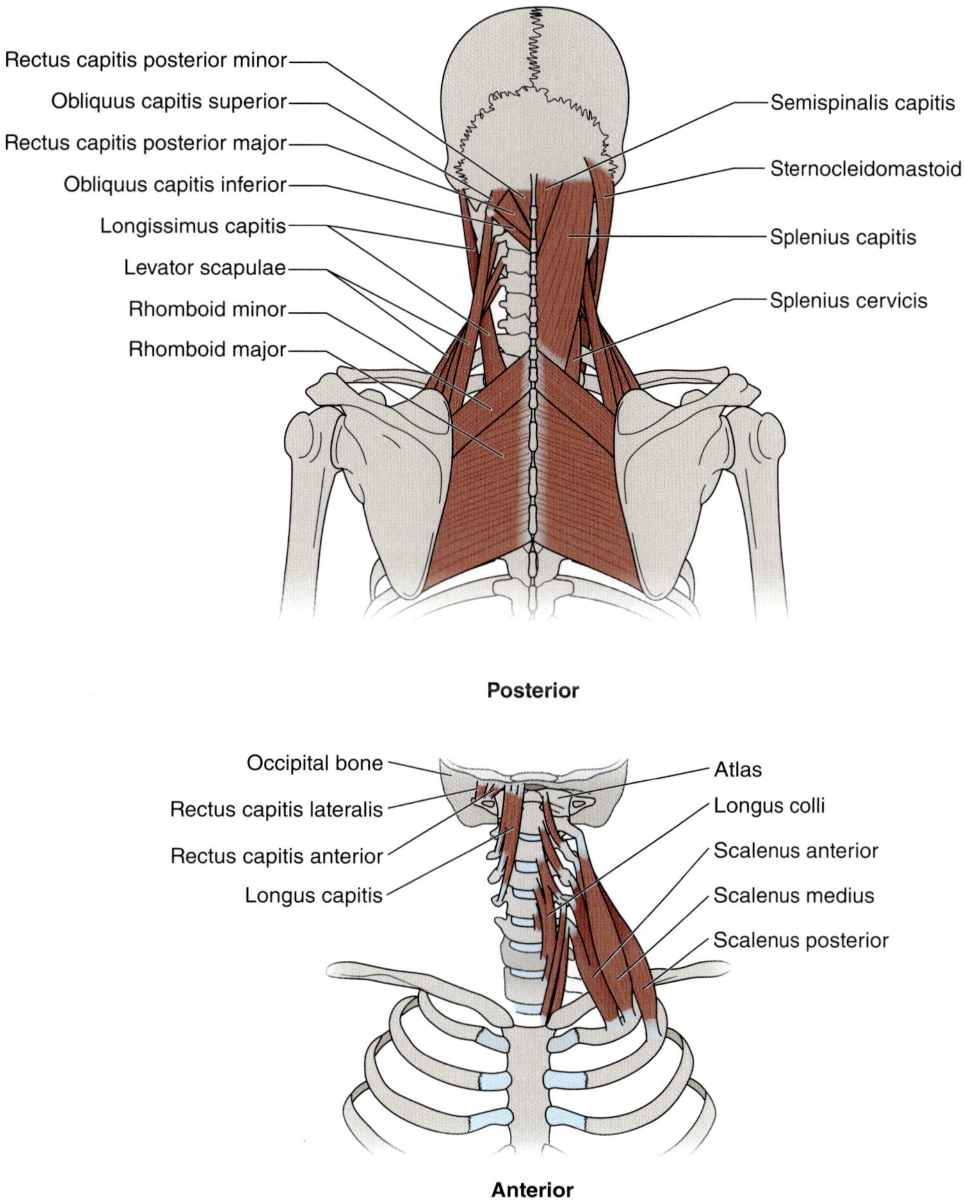

Figure 8.21 Posterior and anterior muscles of the cervical spine.

The scaleni not only laterally flex the cervical spine but, during forced respiration, also elevate the first three ribs to allow the lungs to expand.

Hands On

Isolating all three of the scalene muscles is difficult, but you should be able to locate the group by having your partner laterally flex their neck against your resistance (figure 8.23).

Restricted to the posterior aspect of the cervical spine are the rectus (rectus capitis posterior major and minor), obliquus (obliquus capitis superior and inferior), and splenius (splenius capitis and cervicis) muscle groups (see figure 8.21). These muscles are involved in the extension, lateral flexion, and rotation of the cervical spine.

Muscles of the shoulder girdle (see chapter 3) are also involved in cervical spine movement.

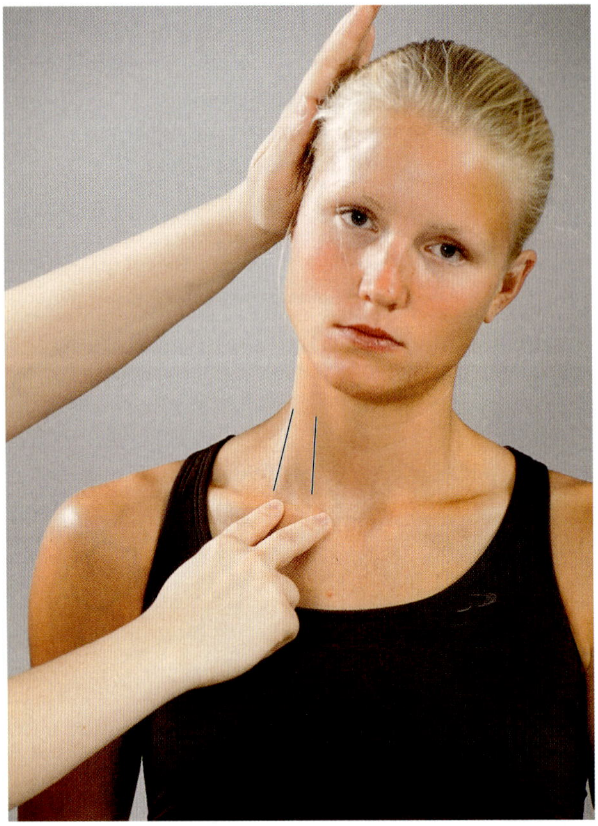

Figure 8.22 Locating the sternal and clavicular parts of the sternocleidomastoid.

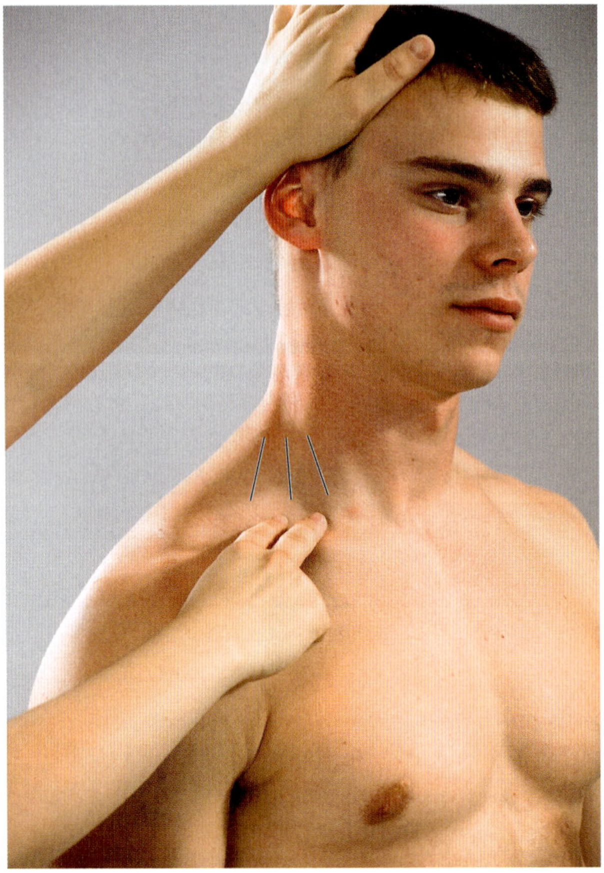

Figure 8.23 Locating the scalene muscles.

The levator scapulae, the trapezius, and the most superior portion of the rhomboids have attachments to the cervical spine and assist with its extension, lateral flexion, and rotation (figure 8.24). Other muscles are also involved in cervical spine movement, but they are not specific to the cervical spine and are presented later in this chapter.

The muscles anterior to the thoracic and lumbar regions of the spinal column are often classified as the abdominal muscles: the rectus abdominis, the internal and external obliques, and the transversus abdominis (figure 8.25).

Hands On

Having your partner perform and hold an abdominal curl should reveal both the rectus abdominis (the "abs") and external oblique muscles (figures 8.26 and 8.27). Additionally, put on a coat, put your hands inside the coat pockets, and observe the direction your four fingers are pointing. The fingers are pointing in the same direction as the external oblique muscles align.

The rectus abdominis originates on the pubic symphysis (see later in this chapter; see figure 8.25) and inserts on the fifth, sixth, and seventh ribs and the xiphoid process of the sternum (see chapter 9). It flexes the lumbar and thoracic spine. The external oblique runs between the inferior edges of the last eight ribs and the outer edge of the middle half of the crest of the ilium. It flexes and contralaterally (to the opposite side) rotates the lumbar and thoracic spine. The internal oblique runs between the outer edge of the middle two-thirds of the crest of the ilium and the seventh, eighth, and ninth ribs. It flexes and ipsilaterally (to the same side) rotates the lumbar and thoracic spine. The transversus abdominis runs between the lower six ribs and the middle half of the internal edge of the crest of the ilium, the linea alba (formed by overlapping fascia of the abdominal muscles), and the pubic bone. Its primary function involving the spinal column is to assist with the ipsilateral rotation of the thoracic and lumbar spine.

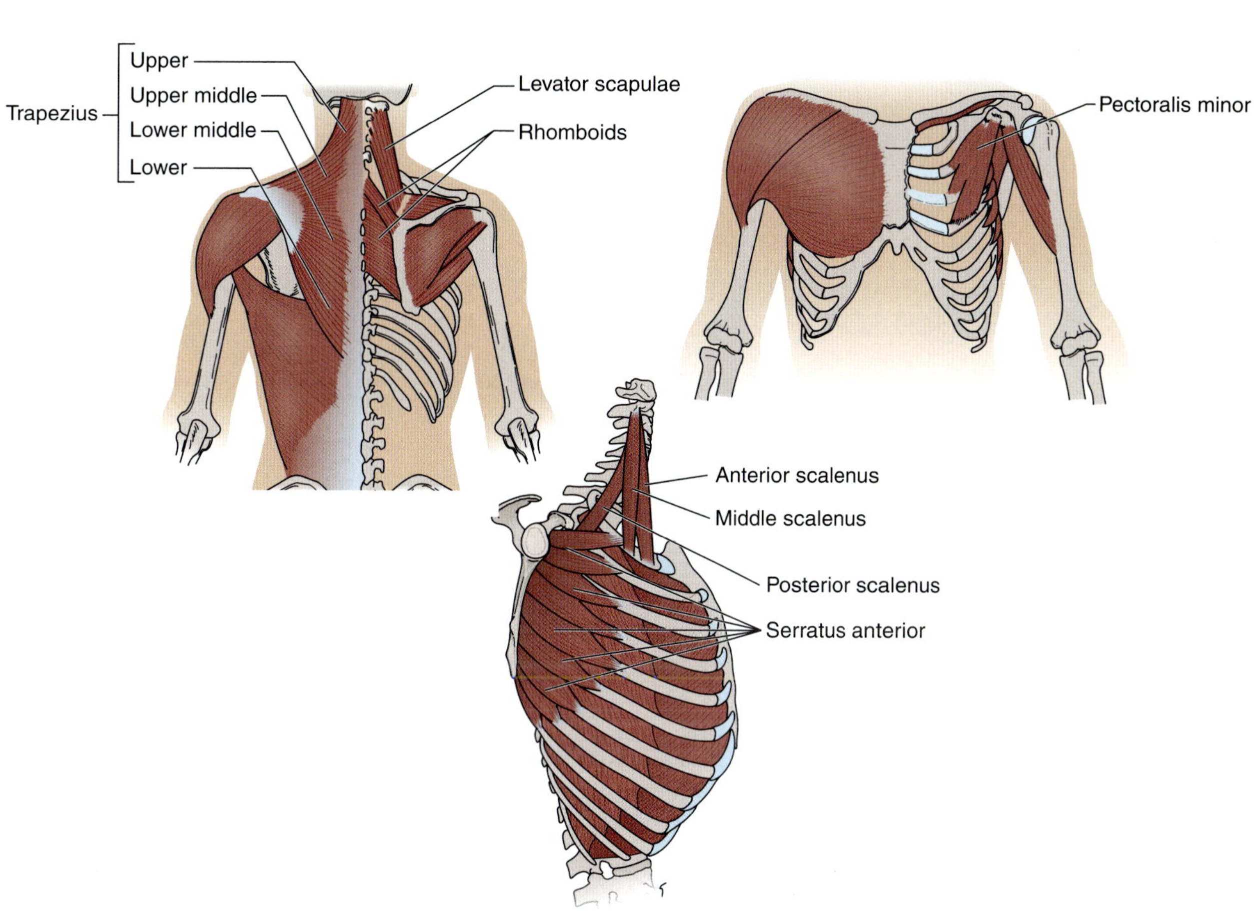

Figure 8.24 Superficial muscles acting on the scapula.

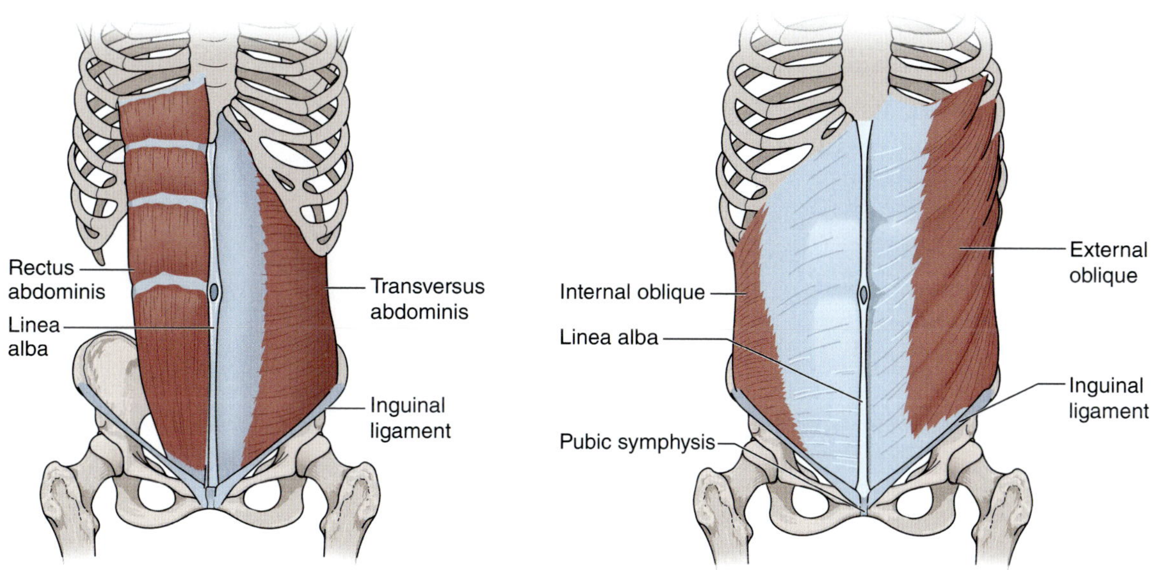

Figure 8.25 The linea alba, rectus abdominis, internal oblique, external oblique, and transversus abdominis.

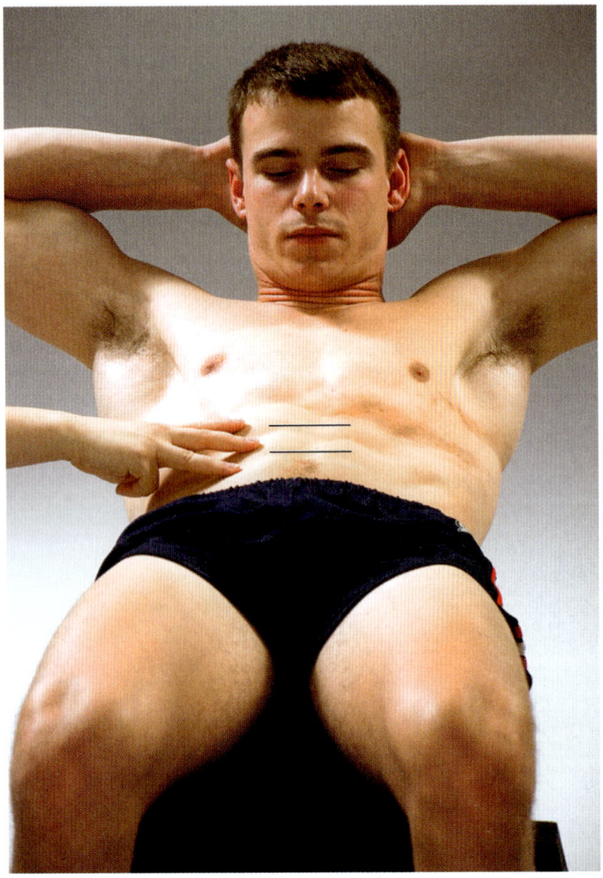

Figure 8.26 Locating the rectus abdominis.

Figure 8.27 Locating the external oblique.

POP QUIZ

Observe this individual performing a Russian twist exercise. As they rotate to the right side, which abdominal muscles are being used to rotate the lumbar and thoracic spine?

The posterior muscles of the spinal column are actually groups of muscles that cover two or more areas of the spinal column (figure 8.28). The spinalis group contains the spinalis dorsi (also called *spinalis thoracis*), the spinalis cervicis, and the spinalis capitis. The spinalis group of muscles originates from the second lumbar vertebra through the seventh cervical vertebra and inserts on the thoracic and cervical vertebrae and the occipital bone of the skull. Because their position is so closely aligned with the spinal column, their function is primarily extension. Another group

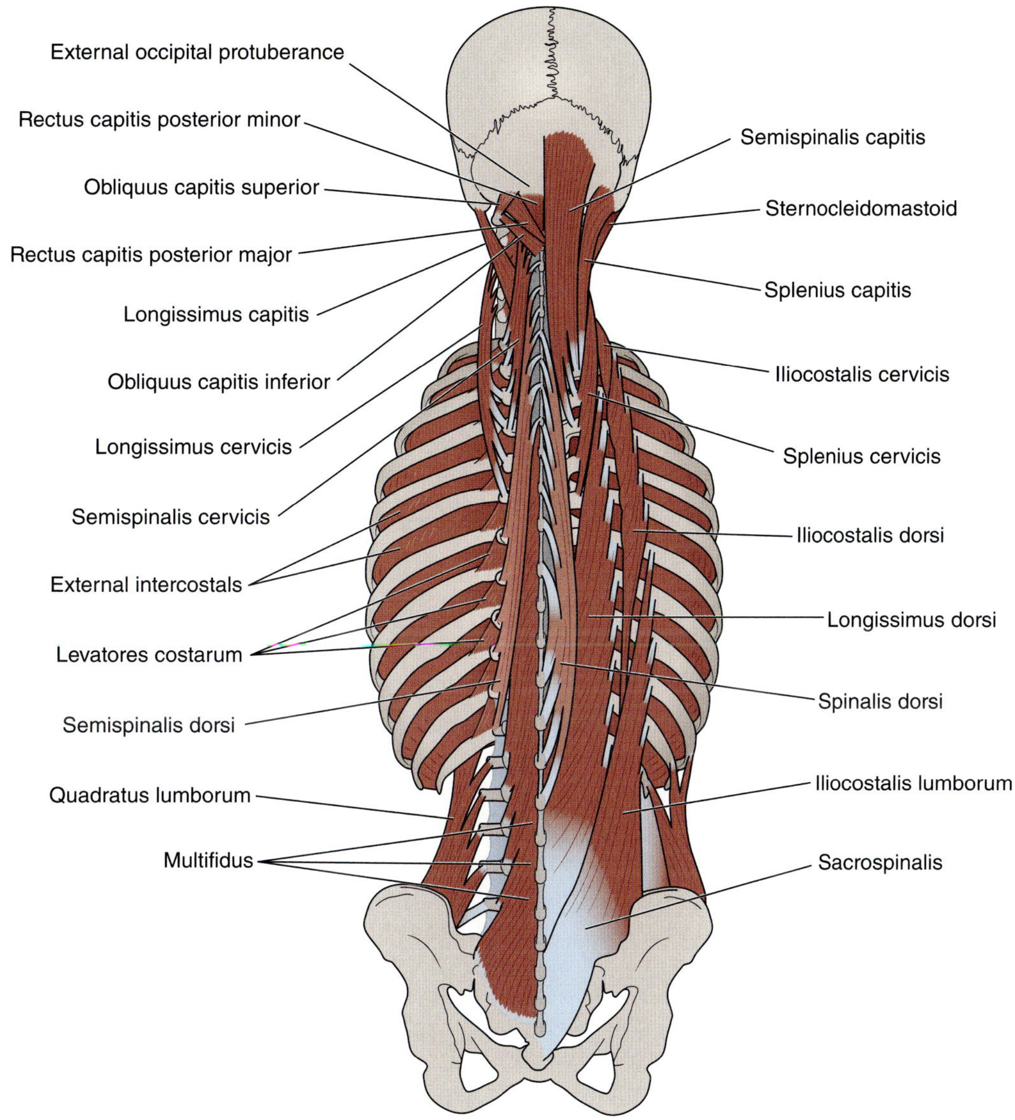

Figure 8.28 Posterior muscles of the spinal column.

of posterior spinal column muscles is the semispinalis, consisting of the semispinalis dorsi (or semispinalis thoracis), the semispinalis cervicis, and the semispinalis capitis. This group originates from the transverse processes of the seventh thoracic vertebra through the last four cervical vertebrae (C4-C7) and inserts on the spinous processes of the upper thoracic, all cervical vertebrae, and the occipital bone of the skull. The semispinalis group extends the cervical and thoracic spine and, particularly in the cervical section, is also involved in lateral flexion and rotation. A third group of posterior spinal column muscles is the iliocostalis group, which includes the iliocostalis lumborum, the iliocostalis dorsi (or iliocostalis thoracis), and the iliocostalis cervicis. This group runs from the posterior aspect of the crest of the ilium and the 3rd through 12th ribs (lumborum and dorsi portions) and the transverse processes of the 4th, 5th, 6th, and 7th cervical vertebrae. All three sections of this group extend the spine, with the upper two sections also involved in lateral flexion and rotation. The fourth group of posterior spinal column muscles is the longissimus group, which includes the longissimus capitis, longissimus cervicis, and longissimus dorsi (or longissimus thoracis). This muscle group runs from the posterior aspect of the crest of the ilium and the transverse processes of the lumbar, thoracic, and lower cervical ver-

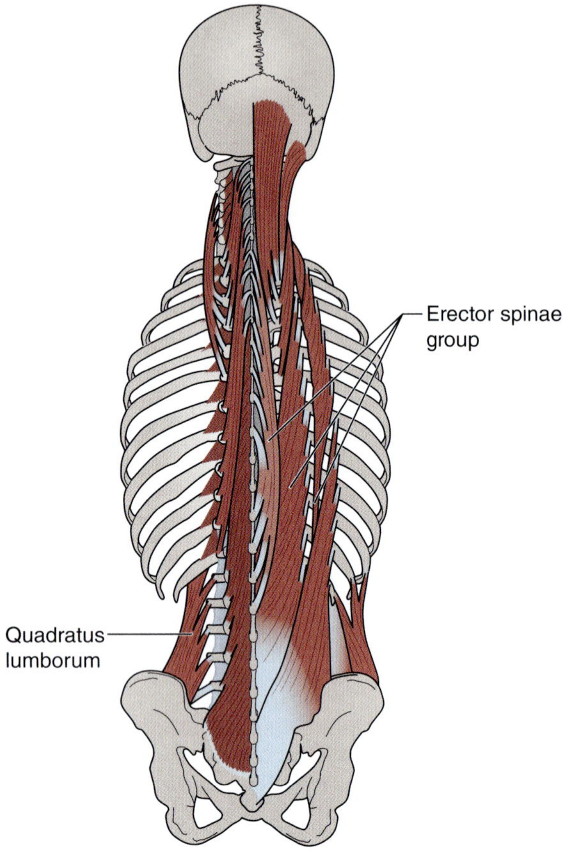

Figure 8.29 Posterior view of the erector spinae group.

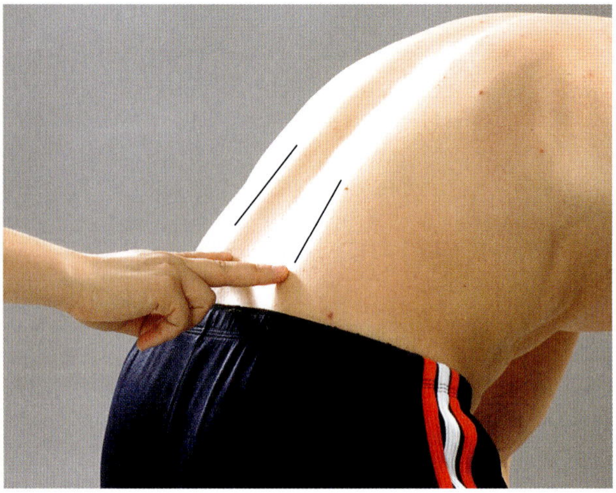

Figure 8.30 Locating the erector spinae group.

tebrae to all thoracic vertebrae and to the cervical vertebrae up to the second. This group extends, laterally flexes, and rotates the lumbar, thoracic, and cervical sections of the spinal column.

Three additional muscles of the posterior aspect of the spinal column are the sacrospinalis (erector spinae), the quadratus lumborum, and the multifidus. The sacrospinalis originates on the spines of the lumbar and sacral vertebrae and splits into three branches that are part of other muscle groups previously described: the iliocostalis lumborum, the longissimus dorsi, and the spinalis dorsi.

Hands On

Ask your partner to flex their trunk at the waist and hold this position. The erector spinae muscle group can be observed running parallel to the spinal column (figures 8.29 and 8.30). Ask your partner to return to standing and feel the erector spinae muscle group contract as they extend their trunk.

These branches (and the sacrospinalis) extend and laterally flex the lumbar and thoracic spine. The quadratus lumborum runs from the posterior aspect of the crest of the ilium and transverse processes of the lower lumbar vertebrae to the transverse processes of the upper four lumbar vertebrae and the lower aspect of the 12th rib (see figure 8.29). Although this muscle does contribute to the extension of the lumbar spine, its primary function is lateral flexion of the lumbar spine. The multifidus is a posterior muscle of the spinal column that covers all movable sections of the column. The multifidus originates from the sacrum; the posterior superior aspect of the iliac spine; and the lumbar, thoracic, and lower four cervical vertebrae and inserts on the spines of the lumbar, thoracic, and all but the first cervical vertebrae (see figure 8.28). This muscle extends and rotates the spinal column.

As did the cervical region of the spinal column, the thoracic and lumbar spine also have muscles with primary functions in the shoulder joint and shoulder girdle (figure 8.31). The trapezius (with attachments to T1-T12), the rhomboids (with attachments to T1-T5), and the latissimus dorsi (with attachments to T1-T12 and L1-L5) also contribute to lateral flexion, rotation, and extension of the spinal column.

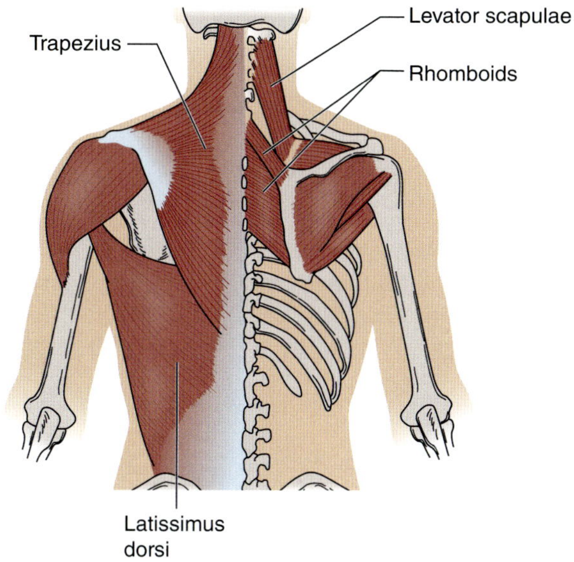

Figure 8.31 Posterior view of the superficial muscles of the shoulder that also act on the spinal column.

Bones of the Pelvis

Depending on your point of view, the pelvis could consist of 3, 7, or 11 bones (figure 8.32). The posterior aspect of the pelvis is the sacrum, the five sacral vertebrae fused together (figure 8.33). On both sides of the sacrum are two large bones often referred to as the *innominate bones*. Actually, each innominate bone is three separate bones that have fused together: the ilium, the ischium, and the pubic bone. Note the lines of fusion illustrated on the lateral view of the innominate bone in figure 8.32. The ilium bones articulate with the sacrum posteriorly, and the pubic bones articulate with each other at the pubic symphysis to form the pelvis. The male and female pelvises differ to allow for childbearing by the female (figure 8.34). The female pelvis is proportionally

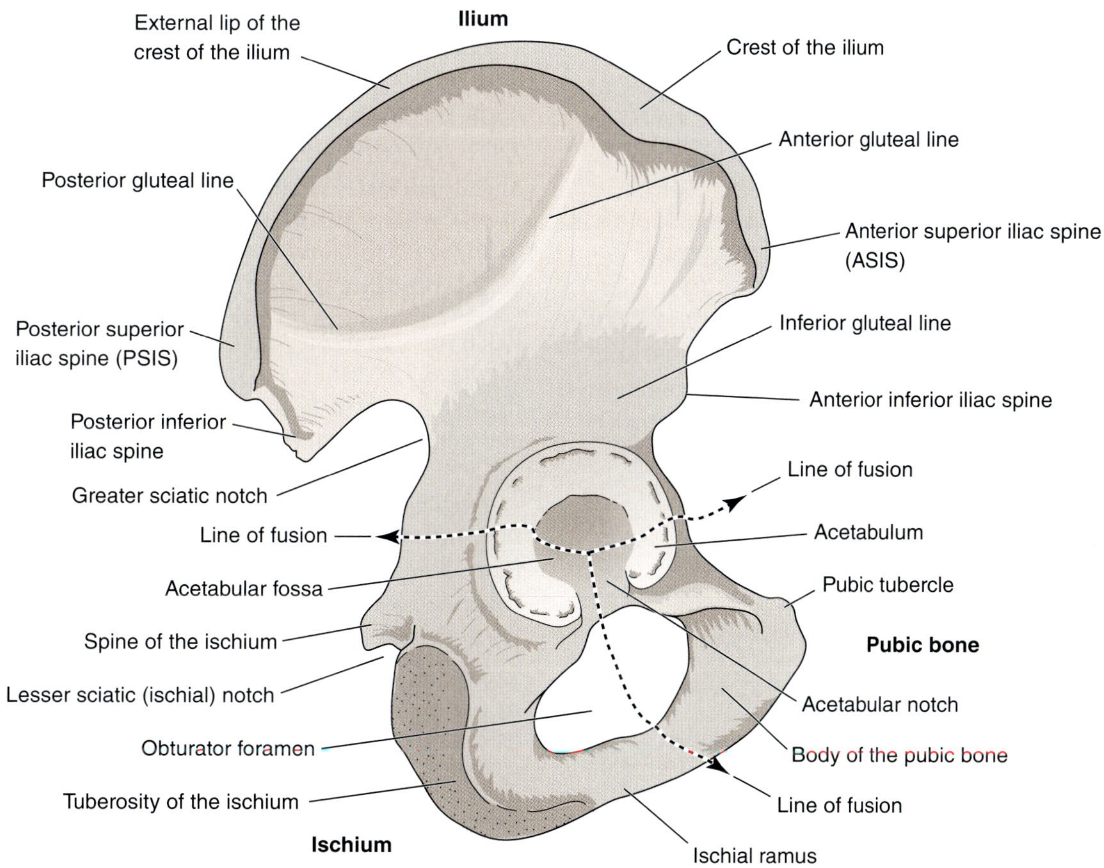

Figure 8.32 Lateral view of the innominate bone (ilium, ischium, and pubic bone).

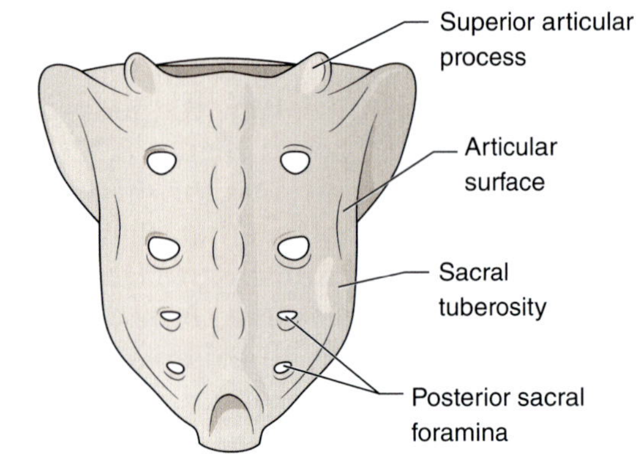

Figure 8.33 Posterior view of the sacrum.

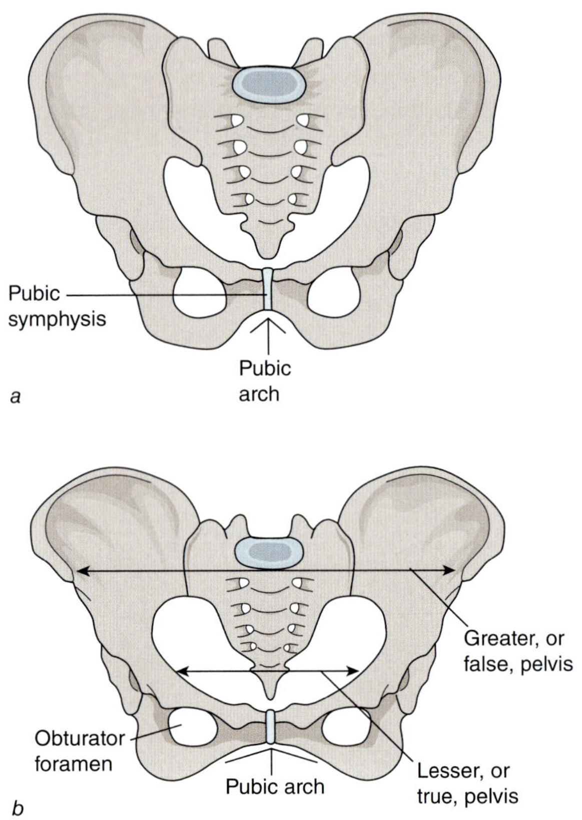

Figure 8.34 Anterior views of the *(a)* male and *(b)* female pelvis.

wider and flatter and is tilted forward to a greater degree to allow for this function.

The ilium, a wide, winglike structure, is the most superior part of the innominate bone and is the bone that articulates the pelvis with the spinal column through the sacrum.

At the most anterior and most posterior aspects of the ilium are bony prominences known as the *anterior superior iliac spine* (ASIS) and posterior superior iliac spine (PSIS). Running between the ASIS and PSIS is a large ridge of bone serving as a major source of muscular attachments known as the *crest of the ilium* (see figure 8.32).

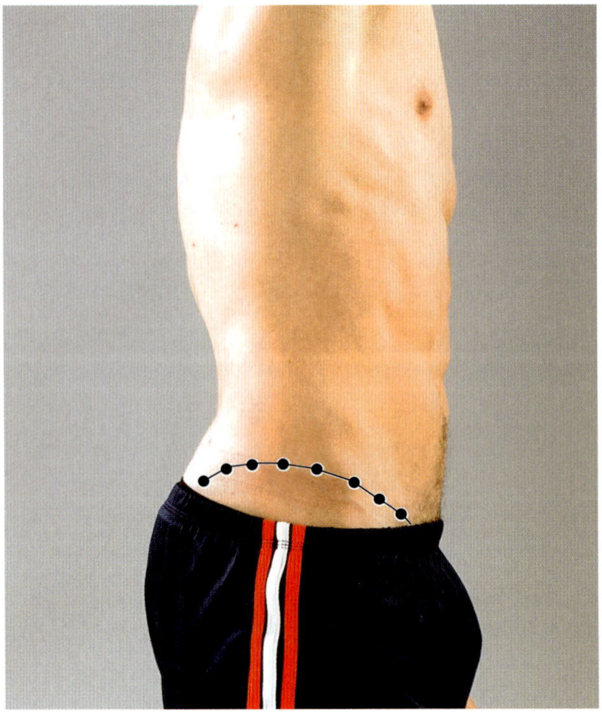

Figure 8.35 Locating the crest of the ilium.

Hands On

Place your hands on your hips and feel your iliac crests (figure 8.35). While you are feeling your iliac crests, follow the crest to its anterior and posterior ends. These end points are your ASIS and PSIS (figure 8.36).

Beneath both the ASIS and PSIS are smaller bony prominences known as the *anterior inferior iliac spine* and the *posterior inferior iliac spine*. Just inferior to the posterior inferior iliac spine is a large notch known as the *greater sciatic notch*.

The ischium is the most posterior of the three bones of the innominate bone and is distal to the ilium. Most posterior on the ischium is a bony prominence known as the *spine of the ischium*. Beneath the spine of the ischium is a notch in the bone known as the *lesser sciatic (ischial) notch*. At the very distal aspect of the ischium is a large bony prominence known as the *ischial tuberosity*. The ischial tuberosity serves as the source of attachment for the lower-extremity muscle group commonly known as the *hamstrings*.

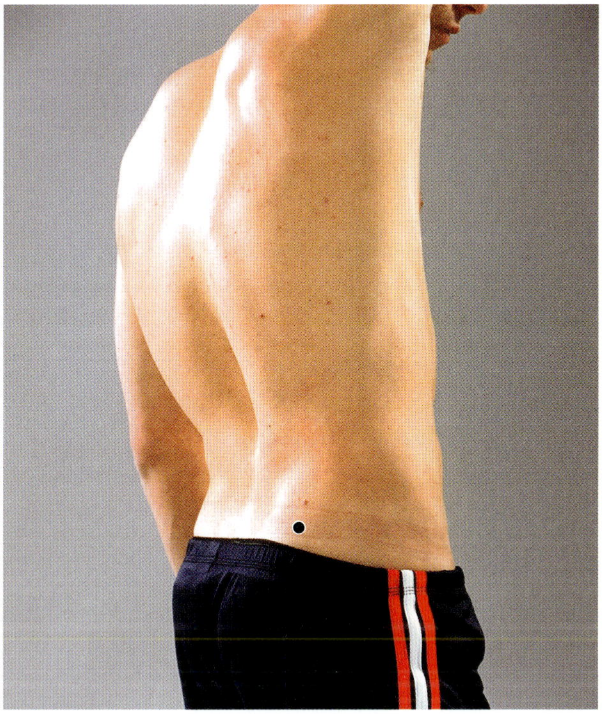

Figure 8.36 Locating the posterior superior iliac spine.

Hands On

The ischial tuberosities are what you sit on. While standing, you can apply pressure to your gluteal areas and palpate your ischial tuberosities.

The pubic bone is the most anterior of the three bones making up the innominate bone. The pubic bones articulate with the ischium (inferior pubic ramus), the ilium (superior pubic ramus), and each other (body of the pubic bone). The articulation between the two pubic bones is known as the *pubic symphysis* (see figure 8.34). Just lateral to the pubic symphysis, each pubic bone has a bony prominence on its superior surface known as the *pubic tubercle*.

Hands On

Apply pressure at the most distal aspect of your abdomen and then move your hand distally to palpate your pubic symphysis.

On the lateral aspect of the innominate bone, the three bones (ilium, ischium, and pubic bone) form a deep socket known as the *acetabulum*. This depression is the socket for the triaxial ball-and-socket hip joint. Dislocation of the hip joint often can include fractures in the area of the acetabulum. Beneath the acetabulum is a large foramen formed by the ischium and the pubic bone. This foramen is known as the *obturator foramen*.

Ligaments of the Pelvis

The iliolumbar ligament runs between the fifth lumbar vertebra and the crest of the ilium (figure 8.37). Two ligaments articulate the sacrum with the ilium: the anterior sacroiliac and the posterior sacroiliac. The anterior sacroiliac ligament runs between the anterior surface of the sacrum and the anterior surface of the ilium (see figure 8.37). The posterior sacroiliac ligament (see figure 8.37) has three sections: the short sacroiliac, the long sacroiliac, and the interosseous. The short sacroiliac ligament runs between the posterior ilium and the lower portions of the sacrum. The long sacroiliac ligament runs

FOCUS ON

Crest of the Ilium

The crest of the ilium is the source of many muscle attachments, both origins and insertions. When a person contuses this area, it is very painful to attempt to move the lower spinal column and the hips, and even to breathe heavily because of the muscular attachments. This type of trauma to the soft tissue attaching to the crest of the ilium is often referred to as a "hip pointer." Sometimes, injuries such as a hip pointer can be the result of improperly fitted equipment or failure to use equipment. All coaches should have instruction in the proper use and fitting of equipment. Knowledge of specific anatomical structures and how equipment protects those structures can reduce the incidence and severity of some injuries.

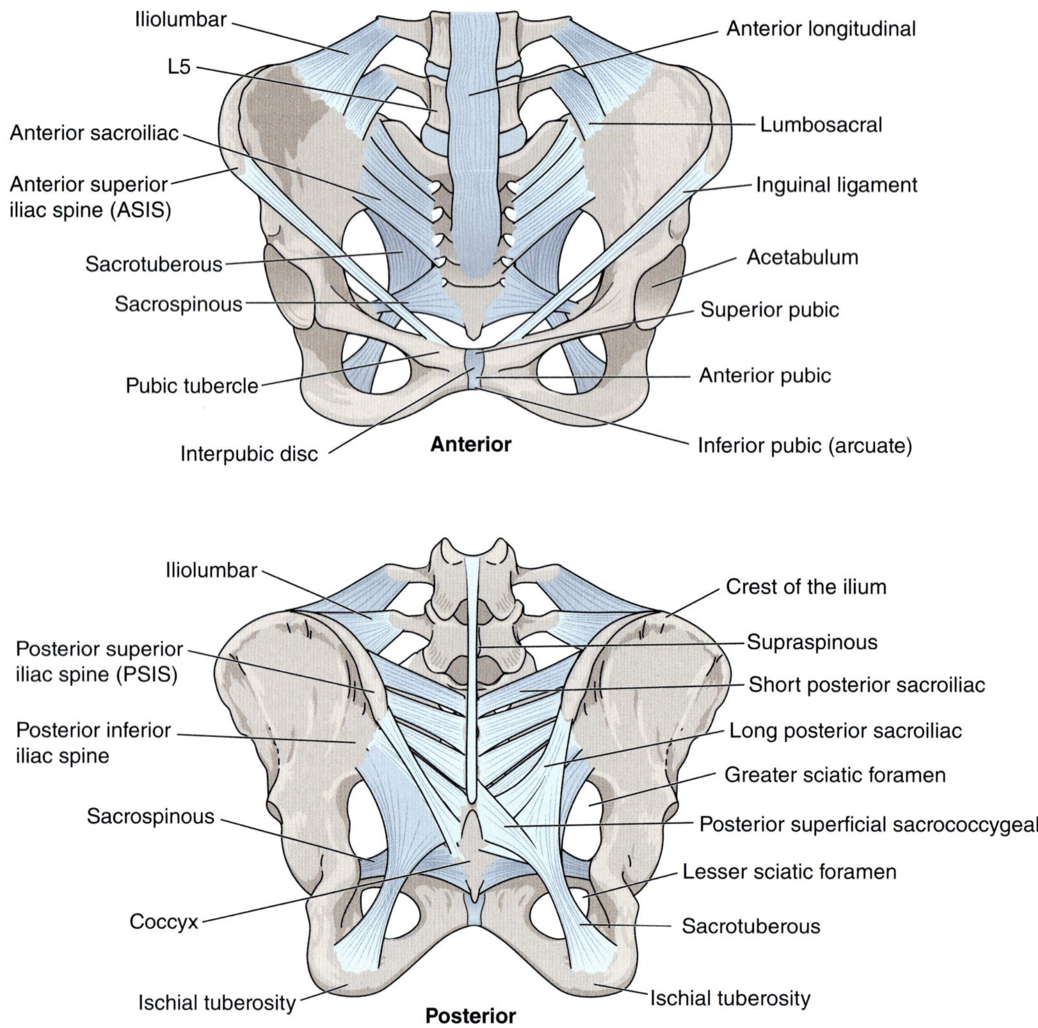

Figure 8.37 Landmarks and ligaments of the pelvis.

FOCUS ON

Inguinal Hernia

Excessive stress on the muscles that attach to the inguinal ligament can result in an *inguinal hernia.* The term **hernia** indicates an abnormal protrusion of an organ. Straining the musculature that attaches to the inguinal ligament can cause a portion of the abdominal contents to protrude through the strained muscle tissue, resulting in an inguinal hernia.

Improper lifting techniques can cause hernias and are also a leading contributor to low-back muscular strain. Proper lifting techniques can be taught in weight-training programs and in academic courses in kinesiology and biomechanics.

between the posterior superior spine of the ilium and the third and fourth vertebrae of the sacrum. The interosseous ligament is made of short fibers that connect the posterior aspects of the sacroiliac joint.

The ligaments of the pubic symphysis are the anterior pubic, the inferior (arcuate) pubic, the posterior pubic, and the superior pubic. The function of each of these ligaments is to articulate the anterior, posterior, superior, and inferior aspects of the two pubic bones to form the pubic symphysis. Additionally, there is an interpubic fibrocartilage disc between the pubic bones (see figure 8.37).

Running between the anterior superior spine of the ilium and the pubic tubercle is a long ligament known as the *inguinal ligament*, which serves as a major source of muscular attachments (see figure 8.37). Two ligaments that stabilize the pelvis are the sacrospinous and the sacrotuberous ligaments (see figure 8.37). The sacrospinous ligament runs from the sacrum and coccyx to the spine of the ischium. The sacrotuberous ligament runs between the posterior inferior spine of the ilium, the sacrum and coccyx, and the ischial tuberosity.

Fundamental Movements and Muscles of the Pelvis

Movement is not normal in the joints of the pelvic bones. The sacroiliac joint and the joints between the ilium, the ischium, and the pubic bone are essentially fused. The pubic symphysis has minimal movement of a gliding nature.

A combination of movements of the spinal column and the hip joint results in pelvic motion: backward tilt, forward tilt, lateral tilt, and rotation. From the anatomical position (figure 8.38*a*), the backward tilt of the pelvis involves the pubic symphysis moving upward and the sacrum moving downward. This is accomplished by flexion of the lumbar spinal column and extension of the hip joints (figure 8.38*b*). Forward tilt of the pelvis involves the pubic symphysis moving downward and the sacrum moving upward. Forward pelvic tilt is accomplished by extension of the lumbar spinal column and flexion of the hip joints (figure 8.38*c*). Lateral tilt of the pelvis results when one ilium is elevated above the other ilium. This is accomplished by lateral flexion of the lumbar spinal column, abduction of one hip joint, and adduction of the other hip joint (figure 8.39). Rotation of the pelvis involves

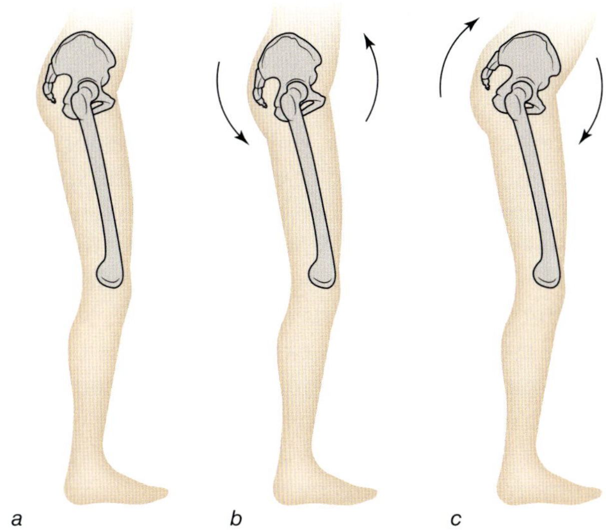

Figure 8.38 Movements of the pelvic girdle: *(a)* anatomical position, *(b)* backward tilt, and *(c)* forward tilt.

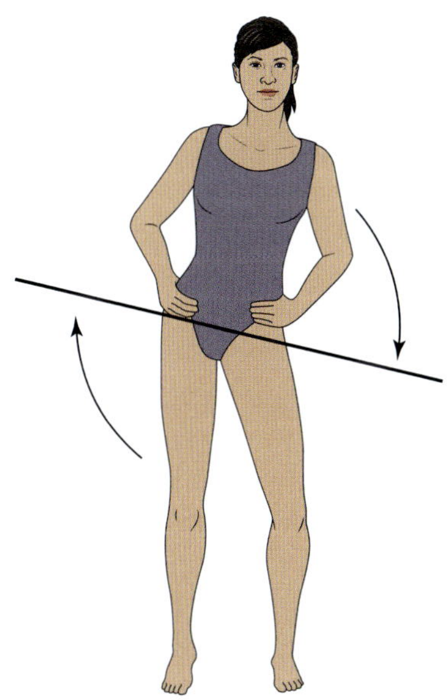

Figure 8.39 Lateral tilt of the pelvic girdle.

rotation of the lumbar spine, with internal rotation of one hip joint and external rotation of the other hip joint (figure 8.40).

The muscles of the spinal column are presented in this chapter, and the muscles of the hip joint are presented in chapter 11. The spinal column muscles produce the flexion, extension, lateral flexion, and rotation necessary for the pelvic movements, along with the muscles that produce flexion, extension, abduction, adduction, and internal and external rotation of the hip joint. From this information, you should be able to understand which muscles produce which movements of the pelvis through the spinal column and hip joint movements.

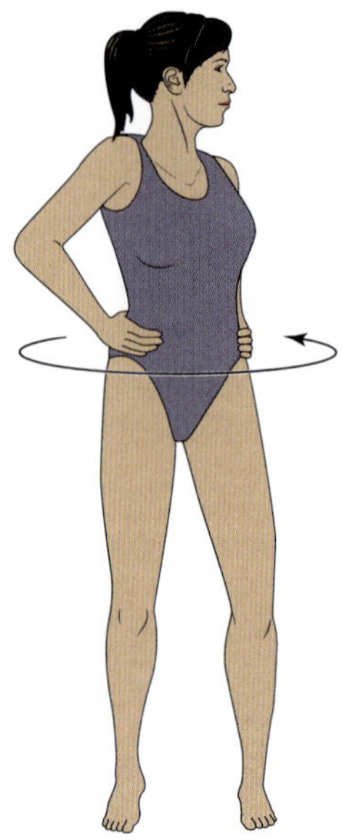

Figure 8.40 Rotation of the pelvic girdle.

Visit the web resource to complete the activities for this chapter.

CHAPTER 9

The Thorax

The thorax (figure 9.1) is a structure formed by bones that create a large compartment known as the *thoracic cavity* (chest cavity), which houses the lungs and heart. Linings of the lungs and the inner walls of the cavity create a vacuum, and movement of the thorax creates changes in pressure within the thoracic cavity, which causes air to enter or be expelled from the lungs. Normal inspiration and expiration of air by the lungs are commonly known as *quiet respiration*, whereas movement of air in and out of the lungs during physical exertion is known as *forced respiration*. Movement at the joints of the thorax allows for the expansion and contraction of the thoracic cavity, and muscles attached to the bones of the thorax create that movement. For muscles to create movement, oxygen must be supplied to the tissues, and carbon dioxide needs to be removed. The heart and lungs are responsible for providing oxygen and removing carbon dioxide. Although respiration and circulation are more suitable discussions for human physiology, a brief presentation of the anatomy of these two major structures within the thorax (heart and lungs) is appropriate in this chapter.

Bones of the Thorax

The anterior bone of the thorax is known as the *sternum* (figure 9.2). The sternum protects the structures beneath it; serves as a source of muscular attachment for muscles of the thorax, the neck, and the abdomen; and provides attachment for costal cartilage. The sternum consists of three parts: the manubrium, the body, and the xiphoid process. The manubrium has a suprasternal (jugular) notch at its superior edge (figure 9.3). The line where the inferior edge of the manubrium attaches to the body of the sternum is known as the *angle of the sternum*. The manubrium also has two clavicular notches and two first costal notches where the sternum articulates with the clavicles (sternoclavicular joints) and the first

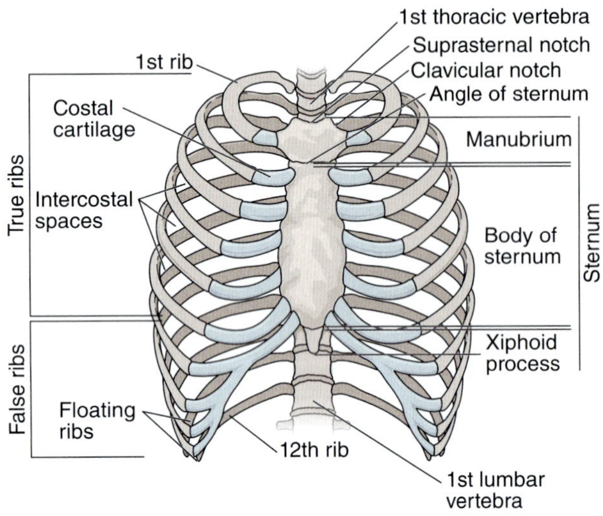

Figure 9.1 Anterior view of the bones of the thorax and their landmarks.

ribs of the thorax. The body of the sternum has costal notches along the lateral sides where the ribs articulate. The most inferior portion of the sternum, the xiphoid process, is cartilaginous in early life and becomes bone in adulthood.

The thorax has 12 pairs of ribs: The first (superior) seven pairs are known as true ribs, and the last (inferior) five pairs are known as false ribs. The first seven ribs attach to the sternum (the first to the manubrium and the second through seventh to the sternal body) through costal cartilage, which extends from 1 to 3 in. (2.5-7.6 cm), depending on the length of the rib, between the sternal end of the rib and the sternum. The first three false ribs (ribs 8-10) are attached indirectly to the sternum by costal cartilage that attaches to the costal cartilage of the rib above it. The last two pairs of false ribs are also referred to as *floating ribs* because they have no anterior attachment.

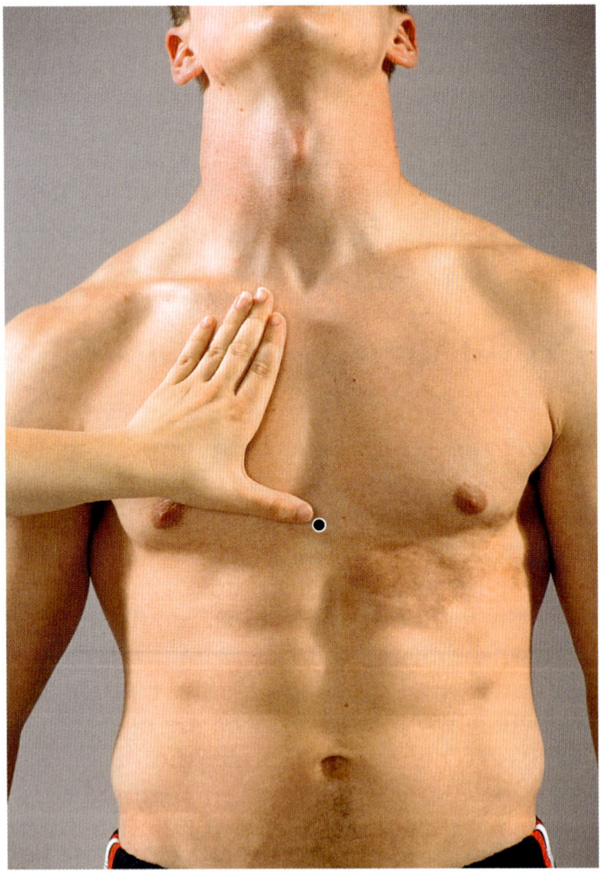

Figure 9.2 Locating the sternum.

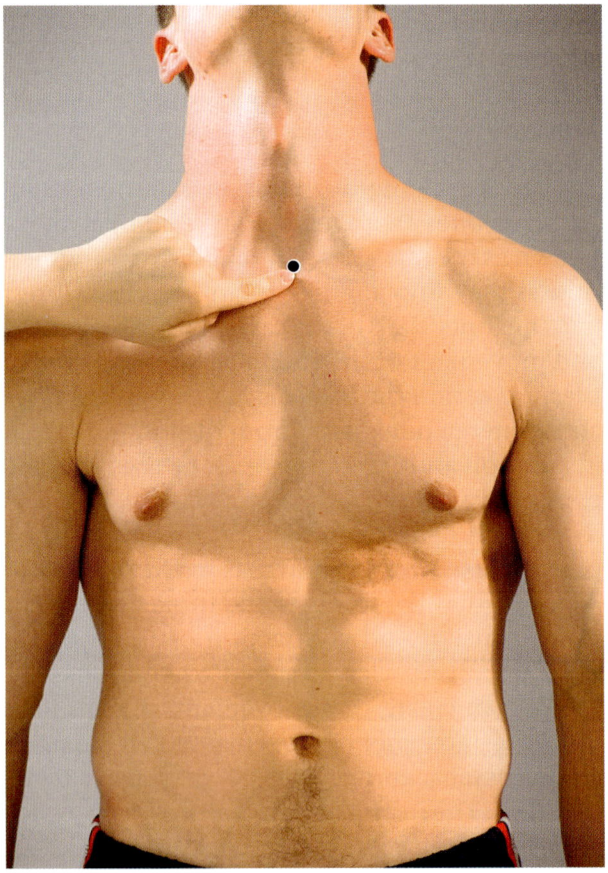

Figure 9.3 Locating the suprasternal notch.

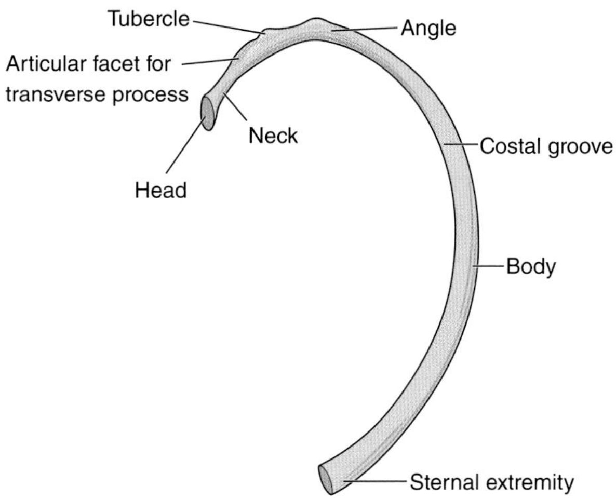

Figure 9.4 Anatomical landmarks of a typical rib.

The head and the neck of a rib (figure 9.4), with an articular facet and tubercle, appear at the end of the rib that articulates with a thoracic vertebra. Lateral to the head of the rib is the angle. This creates the broad back of the human body and gives humans the ability to lie down in a supine position without rolling to one side or the other.

In addition to the sternum and the 12 ribs, the 12 thoracic vertebrae of the spinal column are also considered bones of the thorax. The anatomy of the thoracic vertebrae is covered in chapter 8.

FOCUS ON

Cardiopulmonary Resuscitation (CPR) and the Xiphoid Process

CPR requires proper hand placement on the chest. Rescuers are taught to use the xiphoid process as a landmark to establish hand placement. Pressure applied to the xiphoid process not only produces ineffective compressions but also increases the risk of fracturing the xiphoid process. First-aid courses; CPR courses; emergency medical technician and paramedic courses; and programs in medicine, nursing, physical therapy, athletic training, and other allied health sciences require the acquisition of CPR skills. Knowledge of the anatomy of the thorax helps you implement these CPR skills properly and not expose the victim to further trauma.

Joints and Ligaments of the Thorax

The anterior articulations of the thorax (figure 9.5) include the chondrosternal joints, the costochondral joints, and the interchondral joints. The chondrosternal joints articulate the costal cartilages of the upper seven (true) ribs with the costal notches of the manubrium (first rib) and the body of the sternum (ribs 2-7). The ligaments include the costochondral (capsular), the costosternal, and the interchondral (interarticular). A gliding, rotating motion occurs at these joints, which allows external rotation and elevation of the rib. The costochondral joints are the articulations between the sternal end of the rib and the cartilage attaching the rib to the sternum (ribs 1-7) or to the cartilage of the rib above (ribs 8-10). There is no actual ligament associated with the costochondral articulation because the periosteum of the rib (a covering of connective tissue) in this case is actually continuous with the perichondrium of the cartilage (a thin, fibrous tissue that covers cartilage and provides it with nutrition). In athletics, the costochondral joint often comes under stress when the thorax is compressed beyond the normal range of motion of this joint. Slight rotation of the costochondral joint allows upward and outward movement of the rib.

All 12 ribs articulate with the 12 thoracic vertebrae through 2 articulations for each rib. The costovertebral articulations are known as the *corpocapitate* and *costotransverse capsular articulations*. The corpocapitate articulation is between the head of the rib and the body of the adjacent thoracic vertebra, and the ligaments involved are the costotransverse capsular and costovertebral radiate (figure 9.6). A gliding, rotational movement allows elevation and depression of the ribs. The costotransverse articulations are the joints between the tubercles of the first 10 ribs with the adjacent vertebra's transverse process. The 11th and 12th ribs do not have an articulation with the adjacent transverse processes. The only ligament of these costotransverse articulations is the capsular ligament.

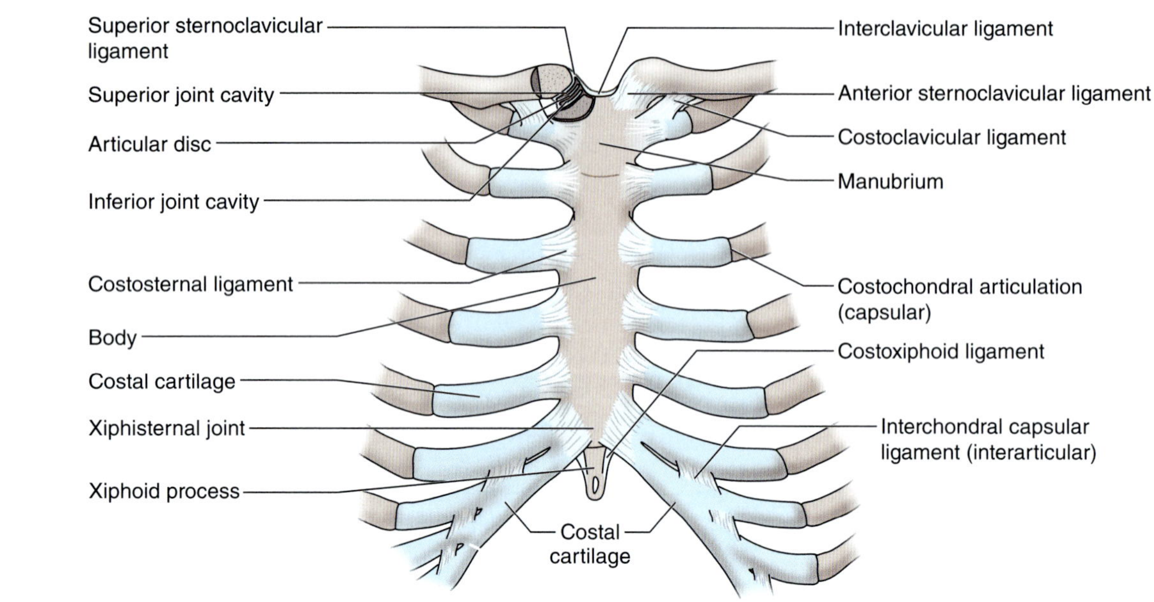

Figure 9.5 The costochondral joints and anterior ligaments of the thorax.

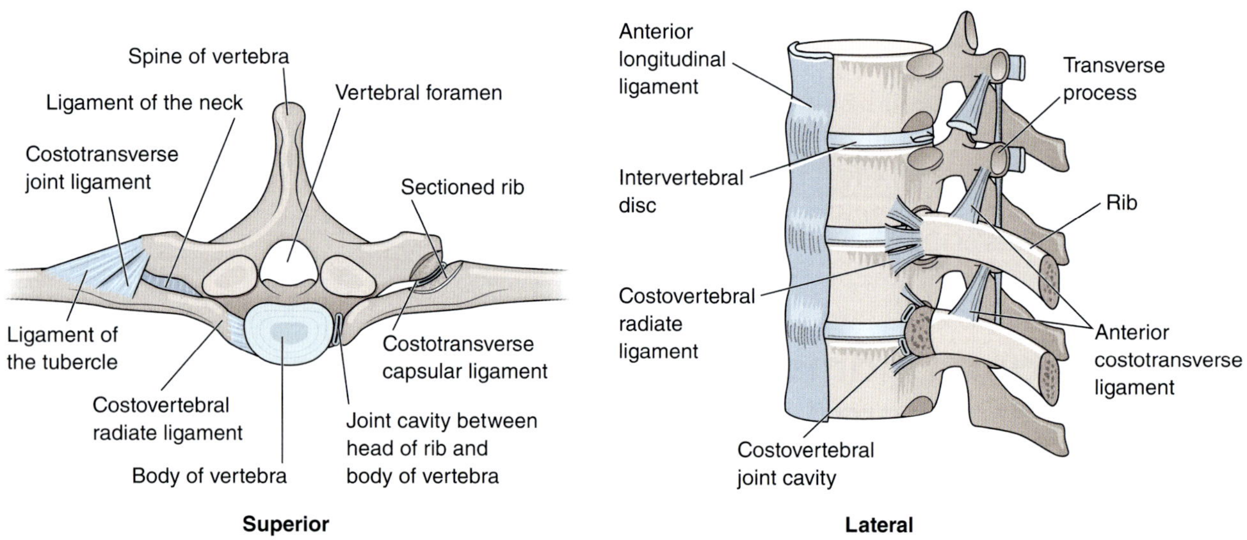

Figure 9.6 Posterior articulations and ligaments of the thorax.

FOCUS ON

Costochondral Sprain

The costochondral joint, like other joints, has a normal range of motion. When that range of motion is exceeded, a costochondral sprain is possible. The thorax of a football player being tackled may be compressed between the tackler and the playing surface, forcing the costochondral joints beyond their normal limits. Likewise, the thorax of a wrestler taken down from a standing position may be compressed between the opponent and the wrestling mat, causing injury to the costochondral joints. Some old-time wrestling coaches refer to this injury as a "rib-out." If the costochondral articulation is sprained, the rib and the cartilage may separate, allowing the bone to move more than normal and resulting in the appearance of a rib attempting to come through the skin at the site of the sprain. The mechanism that causes a costochondral sprain can also result in other forms of trauma (e.g., rib fractures or damage to internal organs of the thoracic cavity).

Recognizing a costochondral sprain requires study of human anatomy, athletic training, sports medicine, and coaching. Care of such a sprain is essential for proper healing. Although it would be ideal to completely immobilize the joint for several weeks (i.e., prevent breathing), this is obviously not possible. Other means of restricting movement need to be used, such as rib belts or taping or wrapping, which will reduce the movement of the costochondral articulation during respiration.

Fundamental Movements and Muscles of the Thorax

Respiration depends on movements of the thorax. As mentioned earlier, normal respiration is known as quiet respiration. The muscles of quiet respiration are considered specific to respiration (figure 9.7). Muscles discussed in other chapters become involved in forced respiration if they have any attachment to the bones of the thorax.

The muscles of the thorax include the diaphragm. This large, dome-like muscle runs from the xiphoid process, cartilage of the last seven

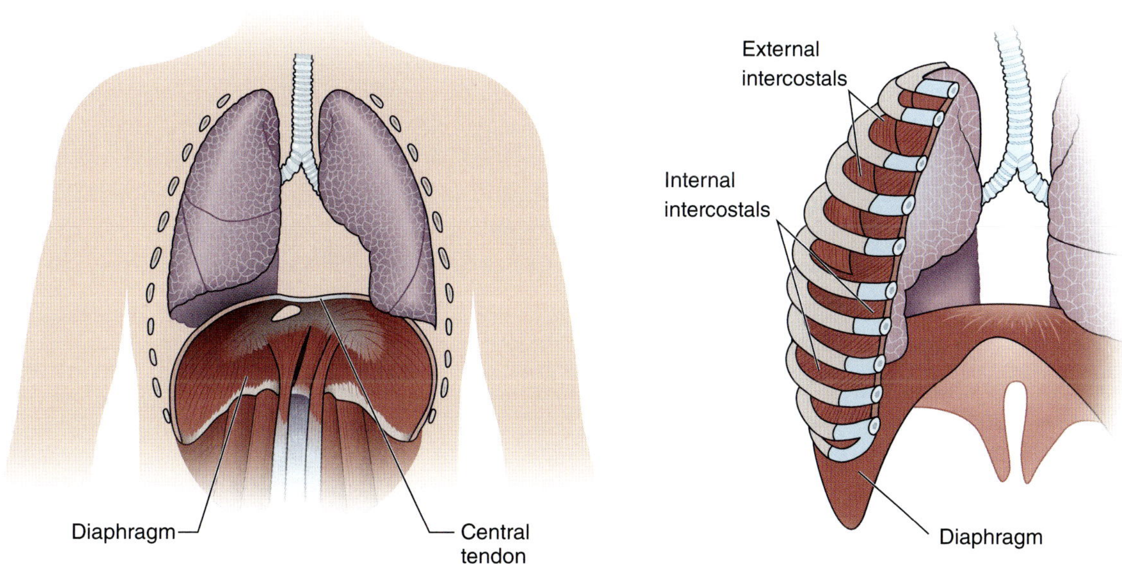

Figure 9.7 Main respiratory muscles.

ribs, and the lateral aspect of the first four lumbar vertebrae to a structure known as the *central tendon* (see figure 9.7). When the diaphragm contracts, its central tendon pulls downward, increasing space in the thoracic cavity and thus changing the pressure within it (figure 9.8). As air is drawn into the lungs and they expand, both the internal and external intercostal muscles (figure 9.9) draw the ribs together, allowing the lungs to expand outward and downward.

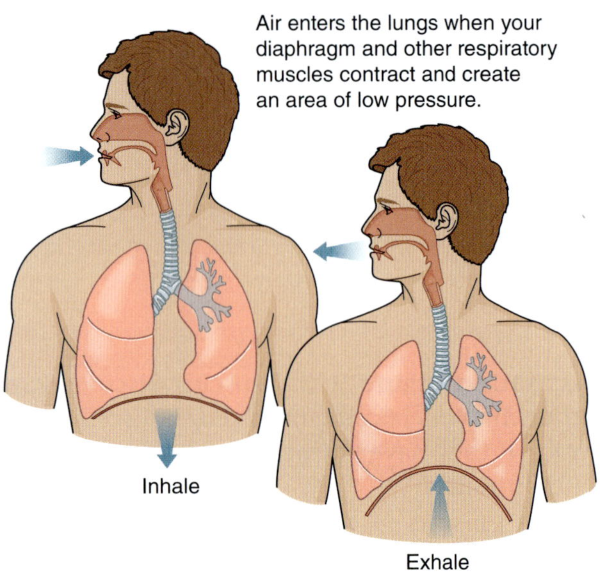

Figure 9.8 The diaphragm contracting and relaxing during inspiration and exhalation.

Hands On

Raise either arm above your head; place your fingers in the space between the anterior lateral aspect of the thorax to place pressure on the intercostal muscles (figure 9.10). The external intercostals (11 pairs), on the anterior aspect of the ribs, run between the lower border of a rib and the upper border of the rib below. Note the angle made by the direction of the external intercostal fibers.

Hands On

Place your open hands in the pockets of a jacket. Note the direction your fingers are pointing. Is there any similarity between the angle your fingers point and the angle the external intercostal fibers run?

The internal intercostals (11 pairs), on the posterior aspect of each rib, run between the upper border of a rib to the upper border of the rib below. The movement created by the contraction of both the external and internal intercostal muscles is a drawing together of the ribs. Observe that figures 9.9 and 9.12 illustrate the intercostal muscles from both a lateral and a posterior view. Look carefully at the angles of pull, and notice that they are the same in both views, which may be confusing. Simply remember that the angles of pull by the intercostal muscles remain the same, no matter which view of the thorax is presented.

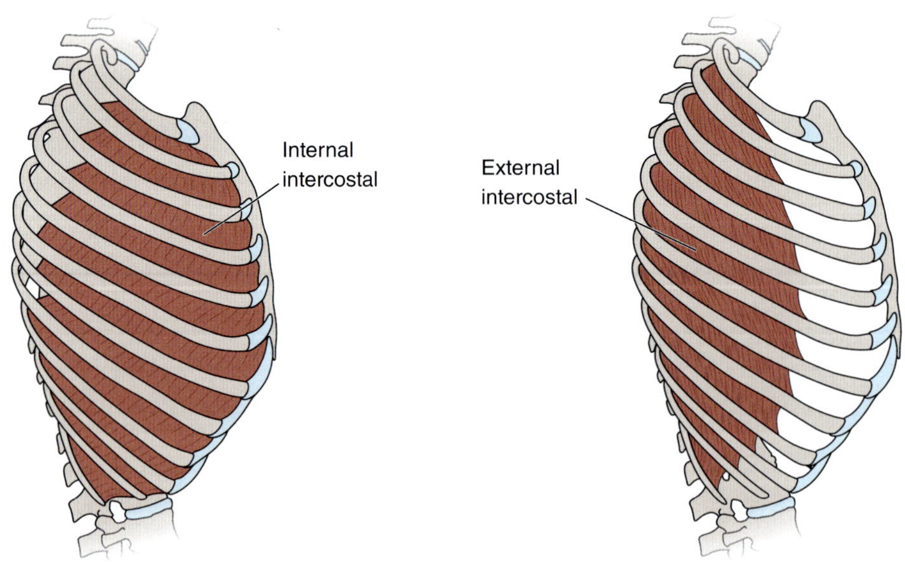

Figure 9.9 Lateral view of internal and external intercostal muscles.

Figure 9.10 Locating the intercostals.

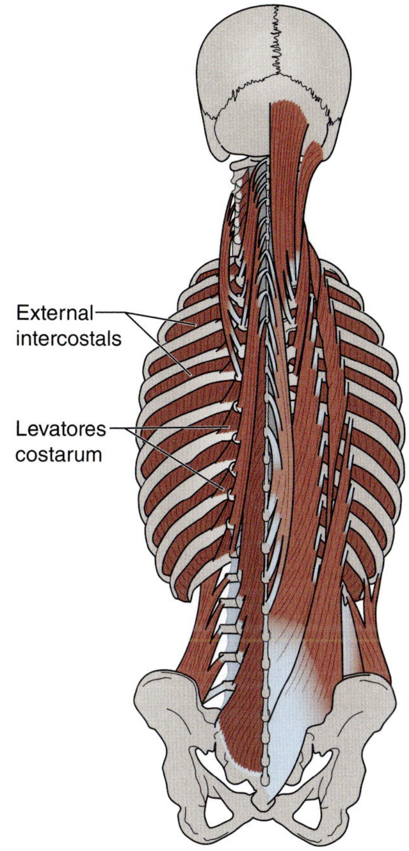

Figure 9.11 The levatores costarum.

If the scalene muscles of the cervical spine contract to fix (stabilize) the first rib, the external intercostal muscles elevate the ribs to allow more volume in the thoracic cavity for the lungs to fill. If the last rib is fixed by the quadratus lumborum muscle, the lower ribs are drawn closer together to decrease the volume of the thoracic cavity.

Other thoracic muscles include the levatores costarum (12 pairs), which run from the transverse processes of the vertebrae C7 through T11 to the angle of the rib below (figure 9.11). Contraction assists with elevation of the rib. The subcostal muscles (10 pairs) run between the inner surface of a rib near its angle to the inner surface of the second or third rib below (figure 9.12). Contraction assists with elevation of the rib. The serratus posterior muscle has inferior and superior portions (figure 9.13). The inferior portion runs from the spinous processes of the last two thoracic and first three lumbar vertebrae to the inferior border of the last four ribs, just lateral to their angles, and depresses these ribs. The superior portion of the serratus posterior runs from the spinous processes of the

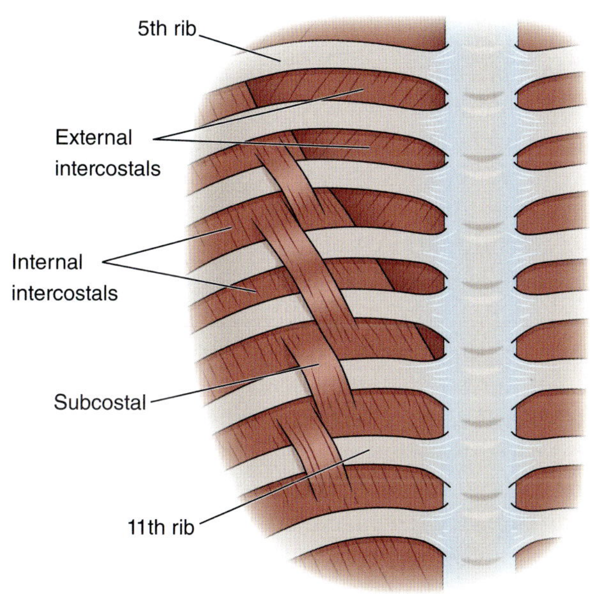

Figure 9.12 The subcostal muscles.

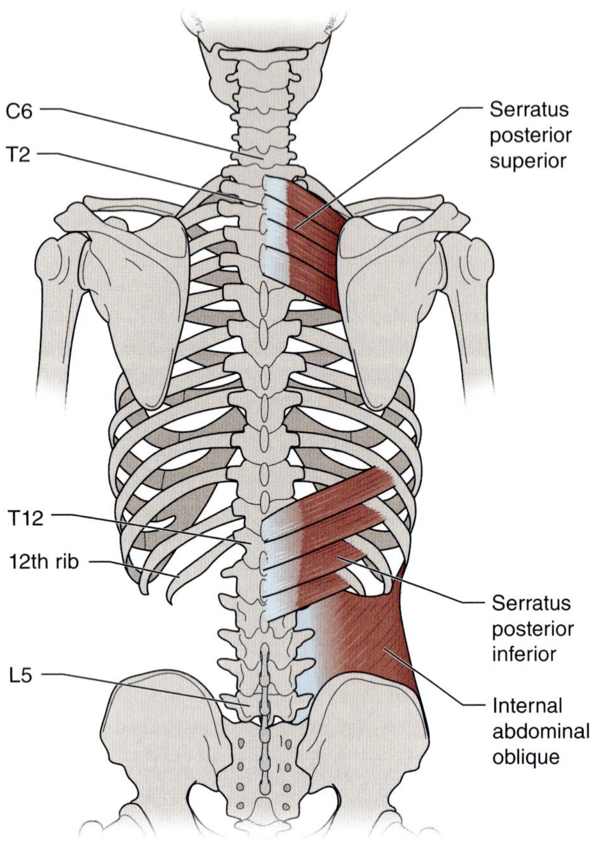

Figure 9.13 Superior and inferior portions of the serratus posterior.

seventh cervical and first three thoracic vertebrae to just lateral to the angle of the second, third, fourth, and fifth ribs and elevates these ribs. The last muscle considered a muscle of the thorax is the transversus thoracis (figure 9.14). It runs from the lower third of the posterior aspect of the sternum, the xiphoid process, and the costal cartilage of the fourth through seventh ribs to the inner aspect of the costal cartilage of the second to sixth ribs and elevates the ribs.

Numerous other muscles have one of their attachments to one of the bones of the thorax and usually are not involved in quiet respiration but can be involved in forced respiration, when lung capacity needs to increase to meet demands and therefore needs additional space within the thoracic cavity to expand. Table 9.1 describes the muscles involved in quiet respiration and those involved in forced respiration.

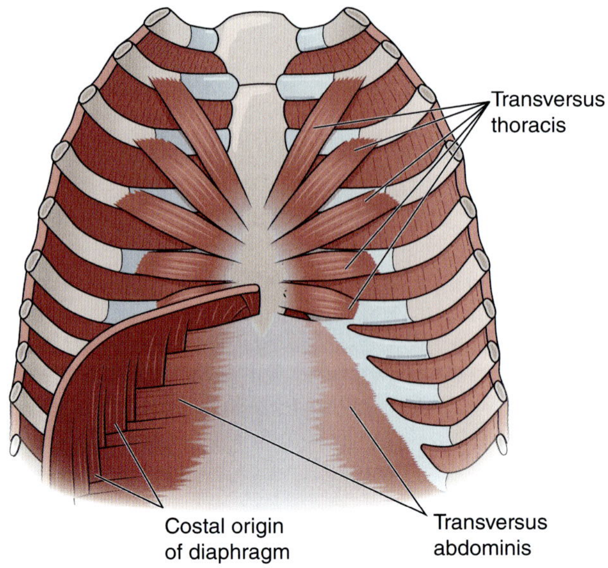

Figure 9.14 The transversus thoracis.

The information in table 9.1 implies that quiet inspiration requires the diaphragm to contract and, as the lungs fill with air, the external and internal intercostal muscles contract to elevate the ribs to create more space for the expanded lungs. When the diaphragm and intercostal muscles relax and return to their starting positions, quiet expiration occurs. In forced inspiration, to make space for greater filling of the lungs, additional muscles that have an attachment to the thorax and are capable of expanding the thorax are called into action. When as much air as possible needs to be expelled from the lungs, any muscle capable of squeezing the thoracic cavity as small as possible is called into action during forced expiration.

Hands On

First, have your partner lie flat on a table or the floor, and simply observe the movement of the thorax during quiet respiration (inspiration and expiration). What muscles created both movements of quiet respiration? Then, while in a standing position, have your partner inhale as much air as possible and then exhale as much air as possible. Visibly or preferably with your hands on your partner's thorax, feel the muscles of the thorax work to create maximal forced expiration.

Table 9.1 Muscles of Respiration

	QUIET RESPIRATION		FORCED RESPIRATION	
Muscle	**Inspiration**	**Expiration**	**Inspiration**	**Expiration**
Diaphragm	•	Passive	•	
External intercostals	•	Passive	•	•
Internal intercostals	•	Passive	•	•
Latissimus dorsi			•	
Levatores costarum			•	
Pectoralis major and minor			•	
Rhomboid major and minor			•	
Scaleni anterior, middle, and posterior			•	
Serratus anterior			•	
Serratus posterior (superior aspect)			•	
Sternocleidomastoid			•	
Subcostals			•	
Trapezius			•	
Obliques, external and internal			•	
Rectus abdominis				•
Sacrospinalis group				•
Serratus posterior (inferior aspect)				•
Transversus abdominis				•
Transversus thoracis				•

Structures Within the Thorax

Any discussion of the thorax needs to include the anatomy of the two major structures housed within it: the heart and the lungs.

Heart Structures

The heart is the key organ in the circulatory system. It receives deoxygenated blood from the body and pumps it to the lungs to pick up oxygen. The oxygenated blood returns from the lungs to the heart, and the heart then pumps this blood to the other organs and tissues. Various anatomical structures accomplish this blood flow to, through, and from the heart and lungs.

The heart (figures 9.15 and 9.16) lies just above the diaphragm and between the lungs in the area known as the *mediastinum*. The heart is a hollow organ with thick muscular walls and is wider at the upper end (base) and narrower and more pointed at the lower end (apex). The heart is wrapped in a membrane known as the *pericardium*. A little larger than the size of a fist, the heart contains four specific compartments, or chambers (figure 9.17).

The two upper chambers are the right atrium and left atrium. These two chambers are separated by a wall of tissue known as the *interatrial septum*. The atrial cavity of these chambers is enlarged by an extension of the atrial wall known as the *auricle* (earlike projection). The two lower chambers, larger in size compared to the atria, are known as the *left ventricle* and the *right ventricle*.

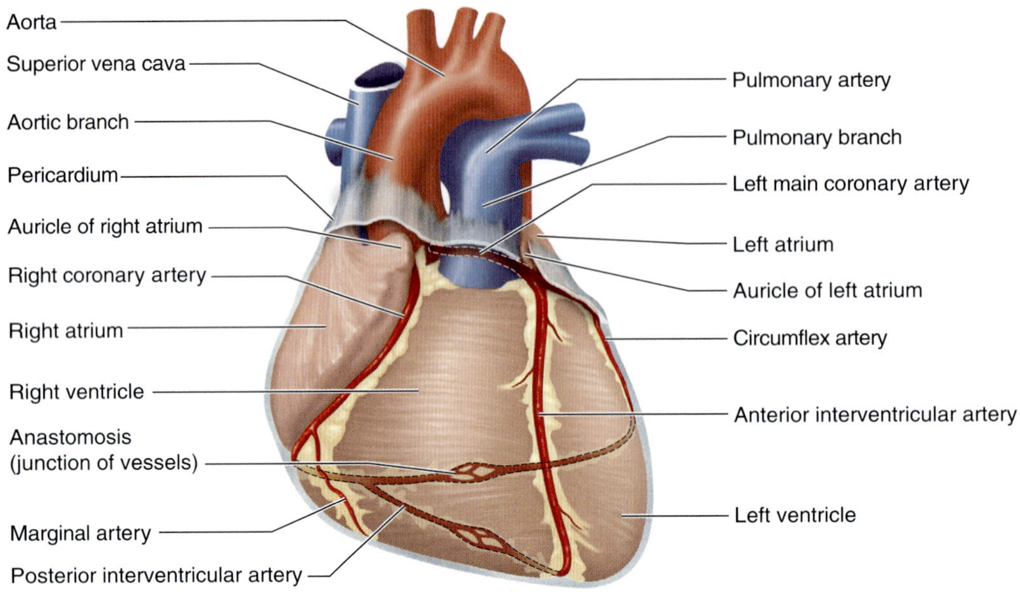

Figure 9.15 Anterior heart.

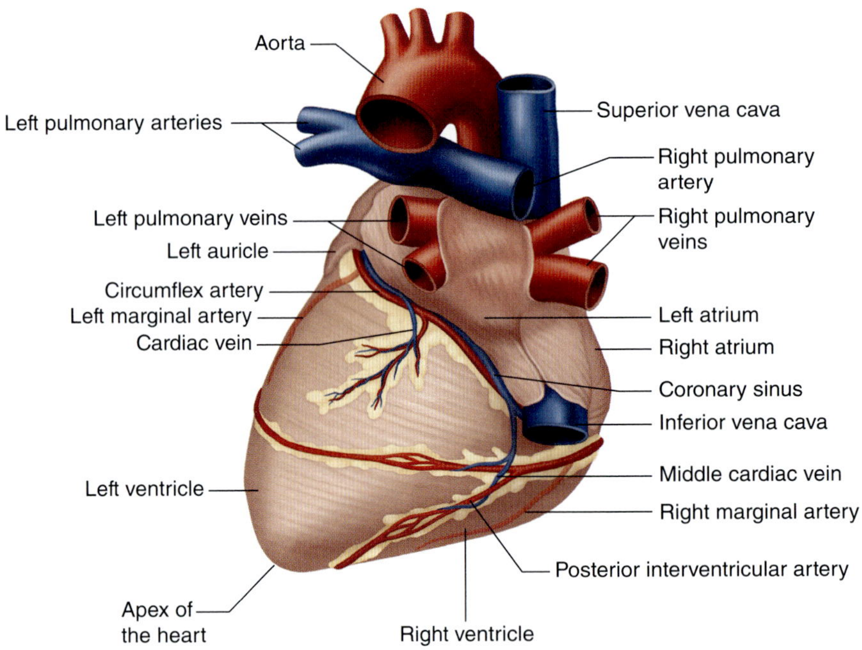

Figure 9.16 Posterior heart.

These chambers are separated by a wall known as the *interventricular septum*. Additionally, there is a wall (septum) separating the atria from the ventricles: the atrioventricular septum. Within the atrioventricular septum is an opening between the right atrium and right ventricle, and another opening between the left atrium and left ventricle. The opening between the right atrium and right ventricle (atrioventricular orifice) has a valve with three cusps (flaps) covering the orifice, known as the *tricuspid valve*. The opening between the left atrium and left ventricle has a

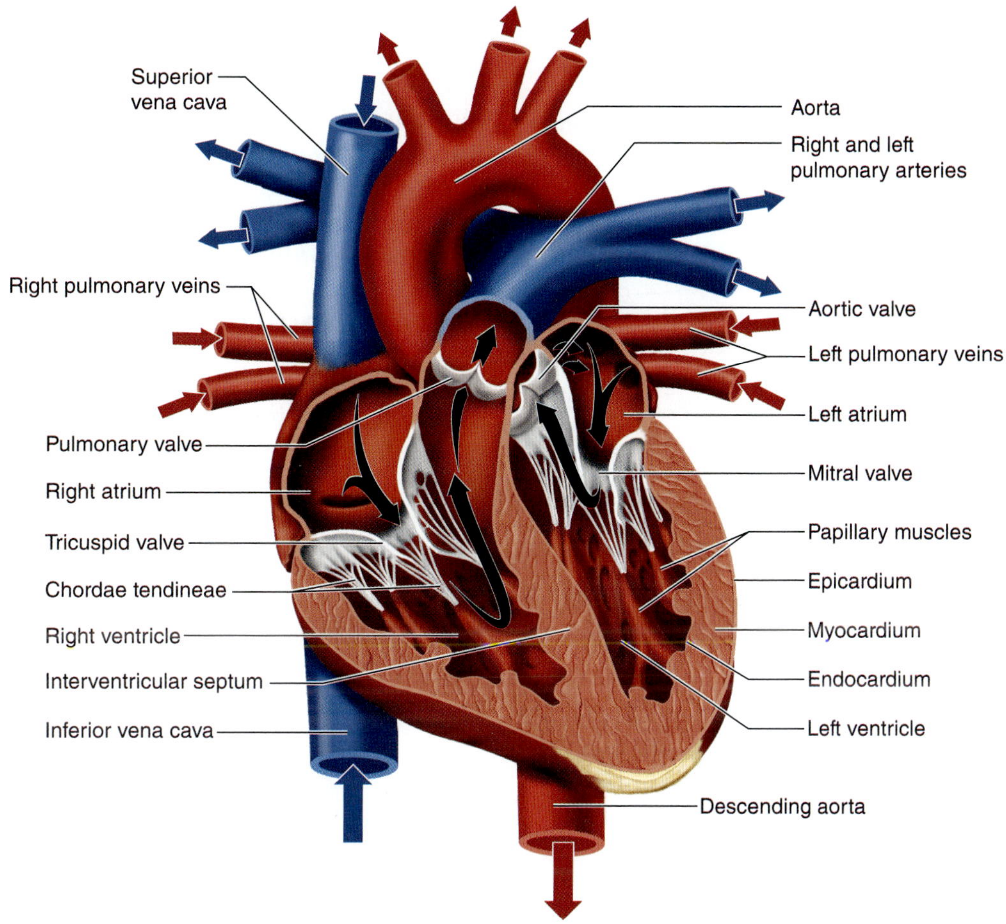

Figure 9.17 Heart anatomy.

valve with two cusps (flaps), known as the *bicuspid valve* (also known as the *mitral valve*). Both valves are designed to prevent the backward flow of blood from the ventricles into the atria. The chordae tendineae are tendon-like cords that attach to the cusps (flaps) of each valve. These cords are also attached to the ventricular walls by small muscles known as the *papillary muscles*. The papillary muscles are responsible for maintaining appropriate tension on the valves to prevent blood from flowing back into the atria from the ventricles.

Under normal conditions, the opening and closing of the heart valves produce a "lubb-dup" sound. When the valves do not function normally, sounds such as whirring or hissing may occur. Two common defects are known as *regurgitation* and *stenosis*. Failure of a heart valve to open fully produces a sound of rushing blood during and just after the "lubb" sound. This is known as *stenosis*. The rushing of blood backward, regurgitation, is the result of a valve failing to completely close and can be heard during and after the "dup" sound. The puncta maxima (greatest or best points) refers (collectively) to the six points on the thoracic wall where the sounds of the opening and closing of the various heart valves are best heard (figure 9.18). Although the points may not seem to be directly over the heart, they are the best places to put the stethoscope to hear a particular valve.

The ventricular walls are three layers thick. A thin outer layer of cells is known as the *epicardium*. An equally thin inner layer of cells on the ventricular walls is known as the *endocardium*. Between these two walls is a thicker muscular

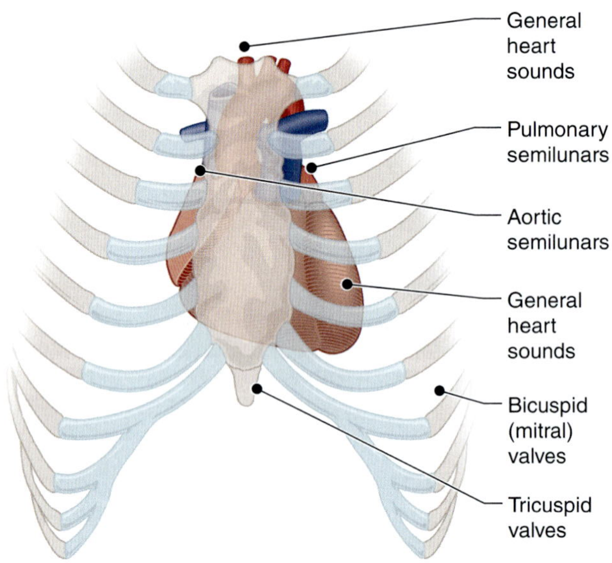

Figure 9.18 Location of different heart sounds via the puncta maxima.

wall of tissue known as the *myocardium*. Because of the amount of force needed to push the blood throughout the body, the left ventricular myocardium is thicker than the right ventricular myocardium. The left ventricle forces blood from the heart throughout the entire circulatory system, whereas the right ventricle forces blood from this chamber to only the nearby lungs.

Airway and Lung Structures

As air enters the mouth and nose, it passes through structures known as the *respiratory tree* (figure 9.19). After being inhaled through the oral cavity (mouth) and the nasal cavity (nose), air passes through the pharynx, larynx, and trachea (see figure 9.19). The trachea divides into two bronchi; each bronchus divides into smaller passageways known as *bronchioles*. The bronchioles

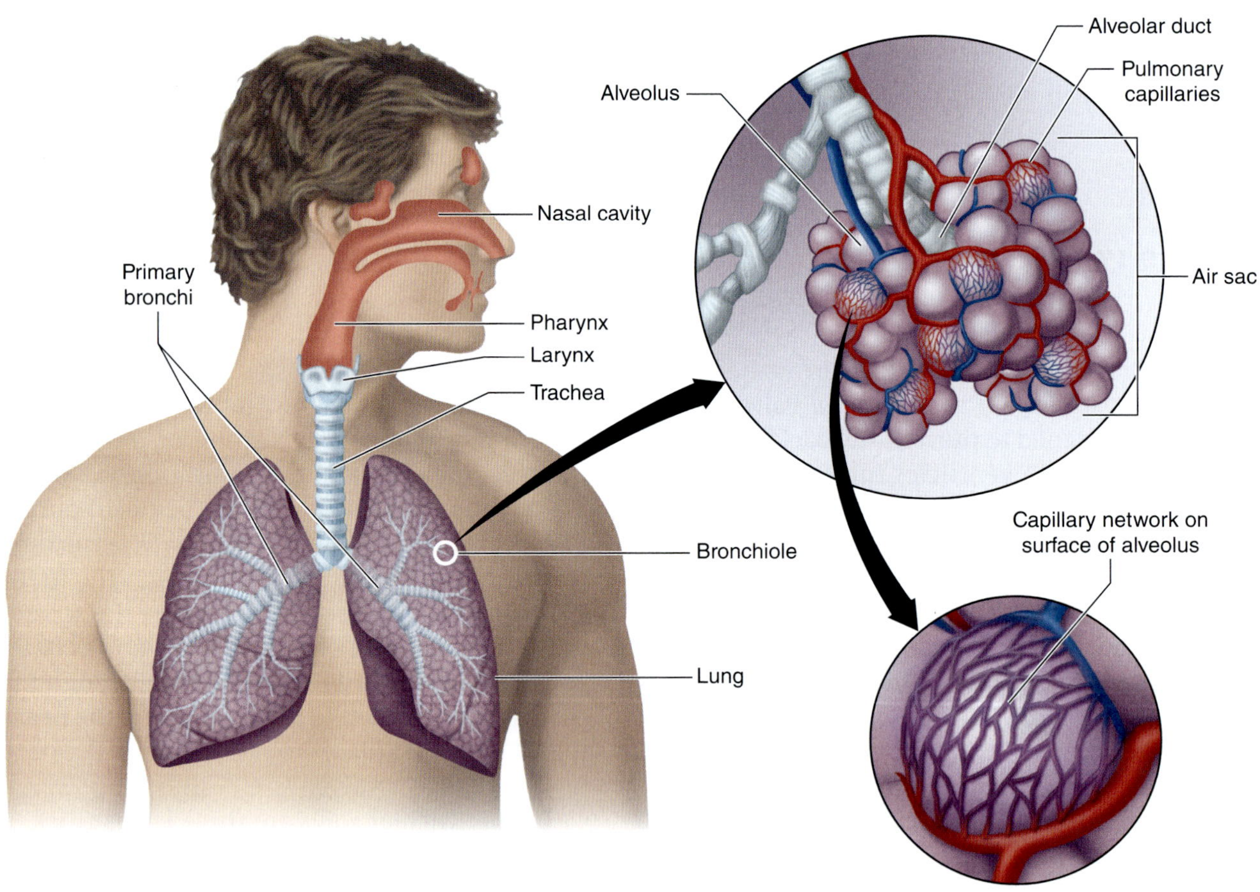

Figure 9.19 Respiratory tree.

Adapted by permission from W.L. Kenney, J.H. Wilmore, and D.L. Costill, *Physiology of Sport and Exercise*, 5th ed. (Human Kinetics, 2012), 165.

POP QUIZ

Observe the location of the pads worn by the lacrosse goalie. Which of the following structures are being protected by the chest pads?

a. sternum

b. ribs

c. heart

d. spine

divide into smaller tube-like structures, respiratory bronchioles, which end as small alveolar ducts. These ducts lead into expansions known as *atria*. From the atria, air then flows into structures known as *air sacs*.

Within the thorax is the thoracic cavity, which contains the pleural cavities (left and right). The pleural cavities are separated by the space known as the *mediastinal septum* (figure 9.20). Once the trachea divides into the bronchioles, the bronchioles enter the pleural cavities and ultimately terminate at the air sacs in either the right or left lung. Both lungs are covered by a membrane known as the *pleural membrane*. This membrane has two layers: the parietal pleura, which lines the walls of the pleural cavity, and the visceral membrane, which lines each lung. A fluid is present between the membrane layers for lubricating purposes to facilitate the movement between the lungs and the pleural walls during the process of respiration.

Although alike in most ways, the right and left lungs do slightly differ anatomically. The lungs have the following anatomical similarities: an apex at the narrow superior portion; a base at the wide inferior portion; costal surfaces, which lie next to the ribs; costal grooves, which are caused by the ribs; and mediastinal surfaces, which are where the lungs touch the mediastinal septum. Located on the mediastinal surfaces is the hilum,

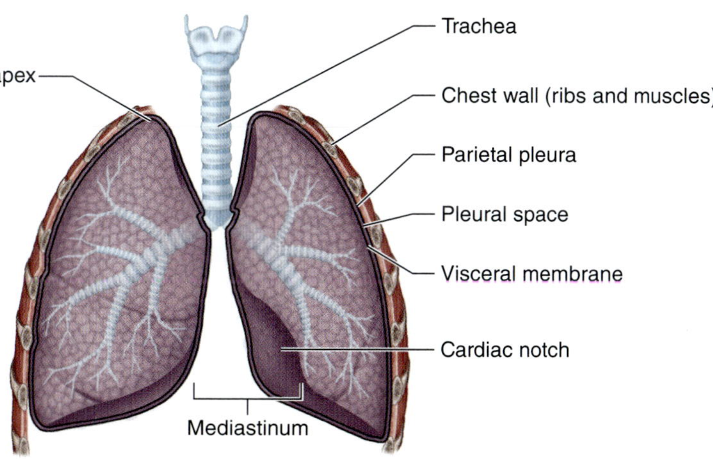

Figure 9.20 Airway and lung anatomy.

which is a depression midway between the apex and base where the bronchial and pulmonary arteries and veins enter and leave the lungs. The major anatomical difference between the right and left lungs is the number of lobes in each lung. The right lung has three lobes (lower, middle, and upper), and the left lung has only two lobes (upper and lower). The left lung also has an area known as the *cardiac notch*, which creates space for the heart. The right lung is shorter than the left lung because most of the liver lies under the right side of the diaphragm, which pushes the right lung higher in the thoracic cavity than the left lung.

Visit the web resource to complete the activities for this chapter.

CHAPTER 10

Nerves and Blood Vessels of the Head, Spinal Column, Thorax, Heart, and Lungs

This chapter starts with an overview of the anatomical structures of the brain, including the cerebrum, brain stem, and cerebellum. The chapter then presents the nerves and blood vessels of the head and brain before shifting focus to the thorax and the two major organs contained there: the heart and lungs. For movement to occur, muscles need oxygen and other nutrients and a means of disposing of the by-products that result. The heart (circulatory system) and the lungs (respiratory system) are the organs mainly responsible for providing the necessary materials essential for muscle activity. The nerves, arteries, and veins of the thorax and its major organs are presented in this chapter.

The Brain

Before discussing the three main parts of the brain, the anatomical structures covering the brain need to be examined (figure 10.1).

Immediately beneath the scalp is the periosteum of the skull, and the epidural space is found just beneath this. The next layer is a brain covering called the *dura mater*. Beneath the dura mater is another space known as the *subdural space*. Under the subdural space is another brain covering known as the *arachnoid*. The space below the arachnoid, the subarachnoid space, contains cerebrospinal fluid. The subarachnoid space is between the arachnoid and pia mater coverings. Beneath the pia mater is the brain. These coverings are also referred to as *meninges*.

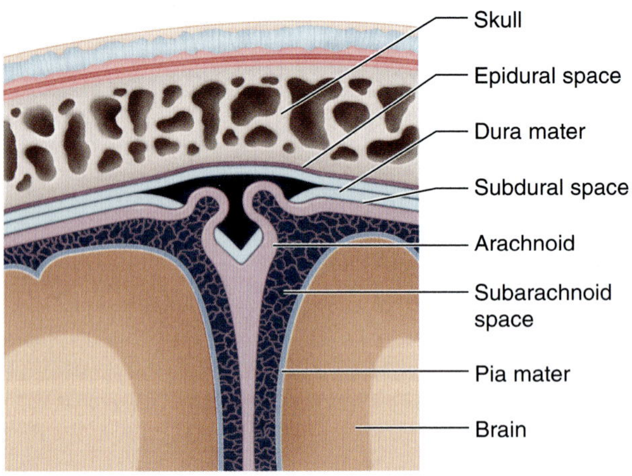

Figure 10.1 Cross section of the skull and brain.

FOCUS ON

Head Trauma

Most of the major blood vessels supplying the brain are found in the spaces between the brain's coverings. The epidural space primarily contains the arteries, and the subdural space primarily contains the veins. Frequently, the terms *epidural hematoma*, *subdural hematoma,* and *subarachnoid hemorrhage* are used to describe head trauma (figure 10.2). (A **hematoma** is defined as a well-formed blood tumor, while a **hemorrhage** is defined as a more diffuse discharge of arterial, venous, or capillary blood.) An epidural hematoma usually develops rapidly (it is arterial in nature) and can have the most serious consequences. Since the skull is above the epidural space, a hematoma in this area cannot expand anywhere but downward, exerting pressure on the brain. Within both the subdural and subarachnoid spaces, expansion of a hemorrhage (which is venous in nature) is slower and has more space (above and below) in which to expand before exerting pressure on the brain. These factors are major considerations when someone suffers head trauma, and part of the immediate treatment includes observation. Someone experiencing head trauma may appear symptom-free after an initial period of confusion, but a subdural hematoma or subarachnoid hemorrhage may take time to develop.

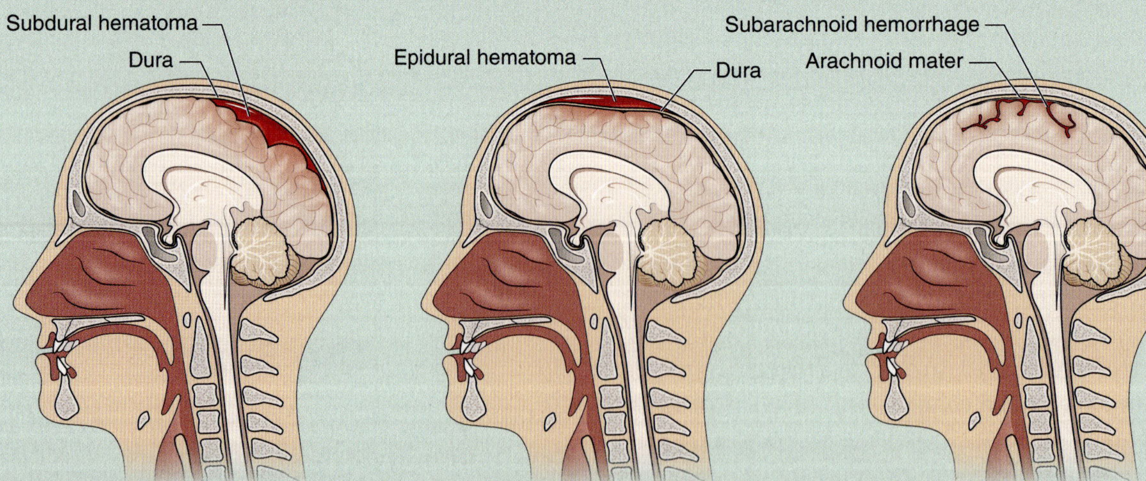

Figure 10.2 Subdural hematoma, epidural hematoma, and subarachnoid hemorrhage.

The subject of head trauma is a topic under intense study. Protocols for immediate evaluation, treatment, and return to activity have been established and are under continual revision as studies are further conducted. Head protection in activities such as football, soccer, ice hockey, field hockey, rugby, and automobile and cycling racing, and in other activities that place the head in a likely position for trauma, is constantly being studied, and devices are being created in research laboratories by equipment companies and independent researchers. Helmets protect the skull from fractures; however, they do not prevent concussions or brain trauma. Some traditional experts (e.g., coaches, athletic trainers, and equipment personnel) have recommended the elimination of helmets to reduce the use of the head as a factor in tackling, blocking, and other techniques that might involve the head being subjected to contact.

Trauma to the head or body, between either individuals or immovable objects (e.g., a floor or a wall), is a likely cause of concussion. When a force stops the skull from moving, the brain may continue in motion and strike against the inside of the skull. Depending on the direction of the force, certain areas of the brain (i.e., frontal, temporal, or occipital lobes) may be affected and can be associated with various signs and symptoms of concussion. Also, depending on the force and the immovable object, the brain may strike one portion of the skull and rebound to strike an area of the skull that is 180° opposite the original contact. A classic example of this trauma involves the frontal lobe of the brain striking the anterior internal area of the skull and then rebounding to the occipital lobe of the brain and striking the internal area of the posterior skull. Trauma to the brain, particularly the frontal lobe, may cause memory problems, and trauma to the occipital lobe may cause vision problems.

The brain itself consists of three major structures (figure 10.3): the cerebrum, the brain stem, and the cerebellum.

Cerebrum

The cerebrum (upper brain; figure 10.4) is the most developed portion of the central nervous system and can be considered the wider end of the spinal cord because it is contained in the more spacious skull area compared to the more confined space within the vertebrae of the spinal column. The right and left sides are bilaterally symmetrical (interconnected by a large fiber bundle known as the *corpus callosum*). The outer layer of brain tissue contains the gray matter known as the *cortex*. The gray matter mainly consists of nerve cell bodies and dendrites. Beneath the gray matter is the white matter, which is composed of nerve axons. The cerebrum has many folds of tissue, with the raised portions, known as *gyri*, and the depressions between the

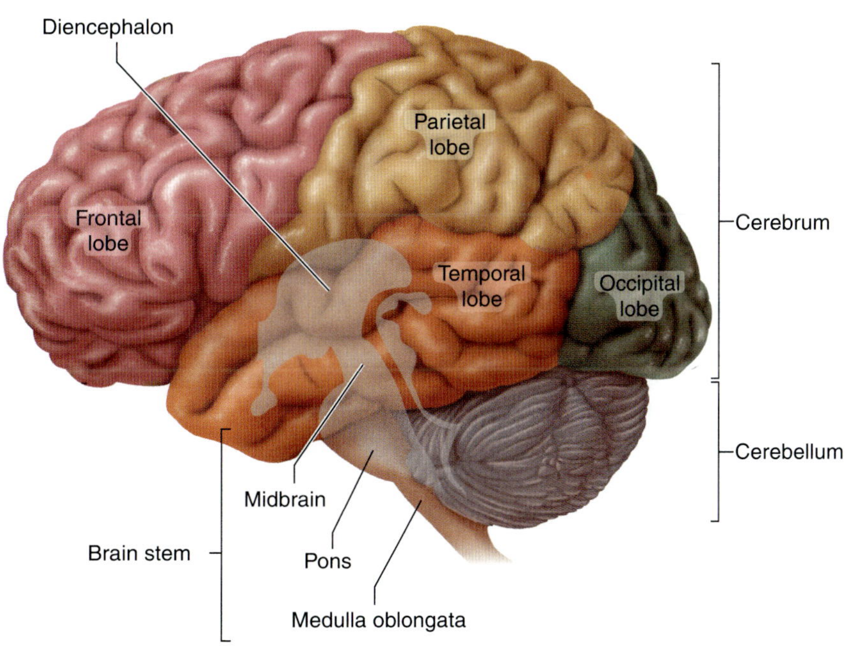

Figure 10.3 Brain divisions.

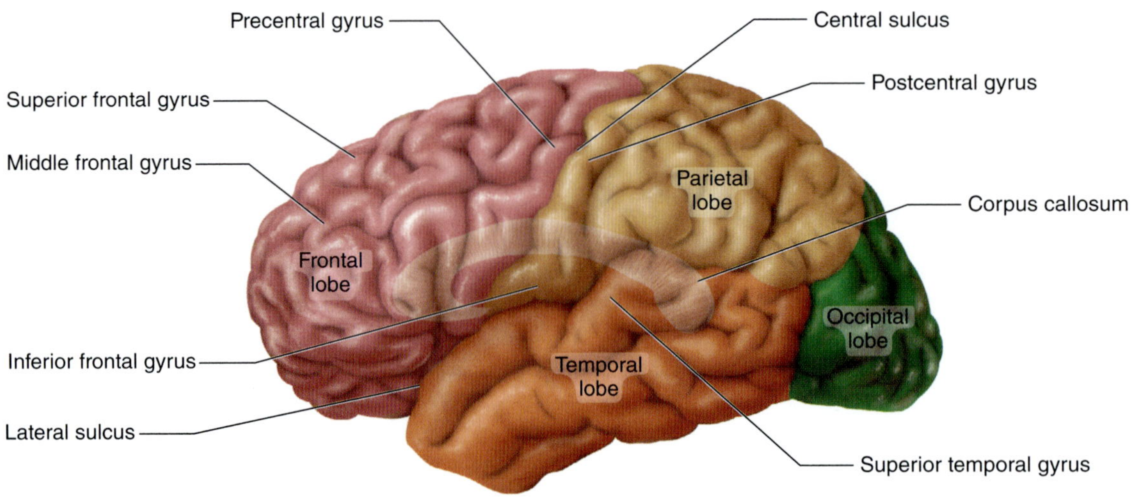

Figure 10.4 The cerebrum.

gyri, known as *sulci*. The four most prominent sulci divide the cerebrum into four distinct lobes. The median sulcus (longitudinal fissure) divides the cerebrum into its bilaterally symmetrical sides. The central sulcus (fissure of Rolando) divides the cerebrum into the frontal and parietal lobes. The lateral sulcus (fissure of Sylvius) separates the parietal and temporal lobes. The occipitoparietal sulcus separates the parietal and occipital lobes. Although there are also numerous gyri, the most prominent are the precentral gyrus (frontal lobe); postcentral gyrus (parietal lobe); superior, middle, and inferior temporal gyri; and superior, middle, and inferior frontal gyri.

The cerebrum houses the area of consciousness that makes a person aware of their surroundings. There is cross-control, meaning the left side of the cerebrum controls the right side of the body, and the right side of the cerebrum controls the left side of the body. Although trauma to either side of the motor area of the cerebrum causes major problems to the musculature on the opposite side of the body, lesser problems can also occur on the same side of the body. Voluntary activity of the musculature is initiated in the motor area of the cerebral cortex.

The cerebrum contains specific areas responsible for various functions (figure 10.5). The motor area, located in the precentral gyrus, contains the majority of the motor nerves of the cerebrum. The body sense area, located in the postcentral gyrus, also has cross-control, with the cutaneous and muscular sensations on one side of the body being recognized by the opposite side of the cerebrum. The visual center, located in the parietal lobe, contains nerve fibers via the optic nerve. The olfactory center, located in the temporal lobe, controls the ability to distinguish different odors. The taste center is thought to be located posterior to the olfactory area of the cerebrum. With the taste center so close to the olfactory center, it makes sense that taste and smell are often closely associated. The auditory center, located in the temporal lobe, contains nerve fibers that detect sound from both ears. The speech and writing centers, located in the frontal lobe, control the ability to speak and write words. The word hearing center, located in the posterior aspect of the temporal lobe, controls the ability to recognize and understand the spoken word. The word visual center, located in the posterior aspect of the parietal lobe, controls the ability to recognize written words.

The remaining large portion of the surface of the cerebrum contains the association areas, which control the ability to respond to various stimuli. A particular sound or odor will be associated with a certain response, but, depending on previous experiences, this might not be predictable from person to person. These responses bring about another ability of the cerebrum: the ability to discriminate, or discriminatory power. A high degree of discriminatory power allows people to vary in their abilities to judge tastes and

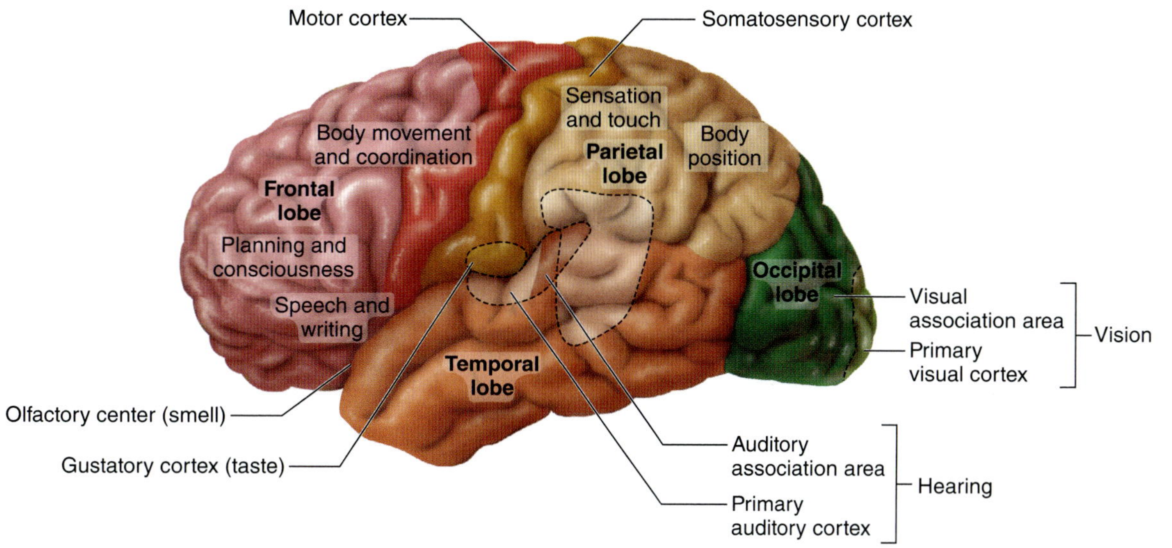

Figure 10.5 Cross section of the motor and sensory areas of the cerebrum.

smells. This power leads to the retention of various impulses (e.g., sight, sound, taste, or smell), known as *memory*. Although good memory is important, it should not be confused with intelligence. The power of reasoning involves the process of bringing together, via the associative powers of memories, old knowledge with new knowledge to identify solutions to various situations. Before one can reason, one must have memory. The combination of one's memory of facts and the power to reason results in one's degree of intelligence. The frontal lobe of the cerebrum is the area believed to be the center for various emotions, and there is no extensive

POP QUIZ

Based on the photo, what part of the brain is most subject to possible trauma?

a. frontal lobe
b. occipital lobe
c. parietal lobe
d. temporal lobe

David Jensen/Icon Sportswire via Getty Images

Trauma involving which lobe of the brain could likely affect the athlete's vision?

a. frontal lobe
b. occipital lobe
c. parietal lobe
d. temporal lobe

Trauma involving which lobe of the brain could likely affect the athlete's memory?

a. frontal lobe
b. occipital lobe
c. parietal lobe
d. temporal lobe

evidence that this area has a major role to play regarding intelligence.

Brain Stem

The brain stem contains the higher reflex centers and is actually the expanded superior end of the spinal cord. Nerves travel to and from the cerebrum through the brain stem, and cranial nerves originate here. The brain stem consists of several parts, the most important being the medulla oblongata (or myelencephalon). This structure contains the major portion of the respiratory center; the cardioinhibitory center, which decreases heart rate through the vagus nerve; the vasomotor center, which controls the size of small blood vessels; the sweat center, which stimulates the sweat glands to secrete sweat as a result of a rise in body temperature; and the origins of the 9th, 10th, 11th, and 12th cranial nerves.

As the motor tract of nerves passes from the motor aspect of the brain's cortex through the medulla oblongata, a major portion of the nerve fibers cross over to the opposite side (i.e., left cortex to right medulla oblongata) as the fibers enter the spinal cord. This process is known as *pyramidal decussation*.

The medulla oblongata contains the reflex center, where emotional activities such as fear and excitement either stimulate or inhibit visceral movements, and the center that reflexively controls activities of the organs of the alimentary tract. The medulla oblongata is where, through skeletal activity, the smooth muscle of the alimentary tract, blood vessels, and various hollow organs is innervated, which allows the smooth muscle of these structures to contract or relax.

The other parts of the brain stem include the afterbrain (metencephalon or pons), the middle brain (mesencephalon or midbrain), and the second brain (diencephalon, which includes the thalamus and hypothalamus). These areas contain the origins of the first through eighth cranial nerves as well as higher reflex centers for sight, sound, smell, and muscle tonus.

Cerebellum

The third portion of the brain is the cerebellum. It is actually part of the afterbrain (metencephalon). As opposed to the crossed control of the cerebrum, the cerebellum (consisting of two equal sides) is homolateral, meaning the right half controls the right side of the body and the left half controls the left side of the body. Muscular coordination and body equilibrium are under the control of the cerebellum. The cerebellum has an amplifying effect on the motor impulses coming from the cerebral cortex.

The body's voluntary muscles are connected to the motor area of the brain (precentral gyrus) via nerves (cells and axons) of either the direct or crossed pyramidal tracts. The vast majority of nerves cross from the brain to the opposite side of the body (crossed pyramidal tract) as they pass through the medulla oblongata. The remaining nerves that do not cross over pass down the same side of the body (direct pyramidal tract).

Peripheral Nervous System

In addition to the central nervous system components (brain and spinal cord), the nervous system has a peripheral component, made up of the cranial nerves, spinal nerves, and autonomic nervous system. The cranial nerves (12 pairs; figure 10.6) and the autonomic nervous system (figure 10.7) have their origins in the cranium and are therefore closely associated with the central nervous system. The spinal nerves (cervical, thoracic, lumbar, sacral, and coccygeal) are discussed in chapter 1 and other chapters relevant to the anatomical areas they innervate.

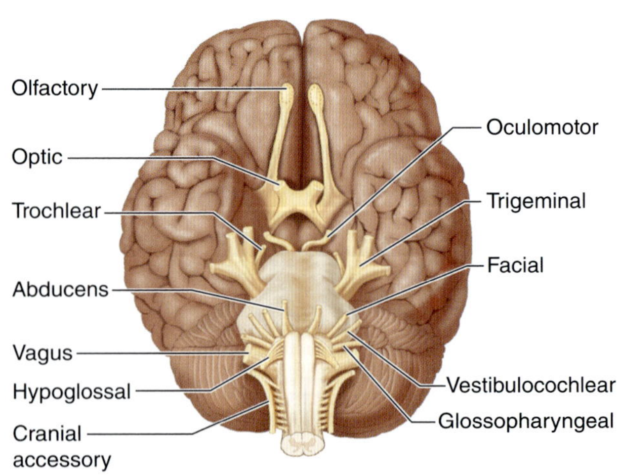

Figure 10.6 Cranial nerves.

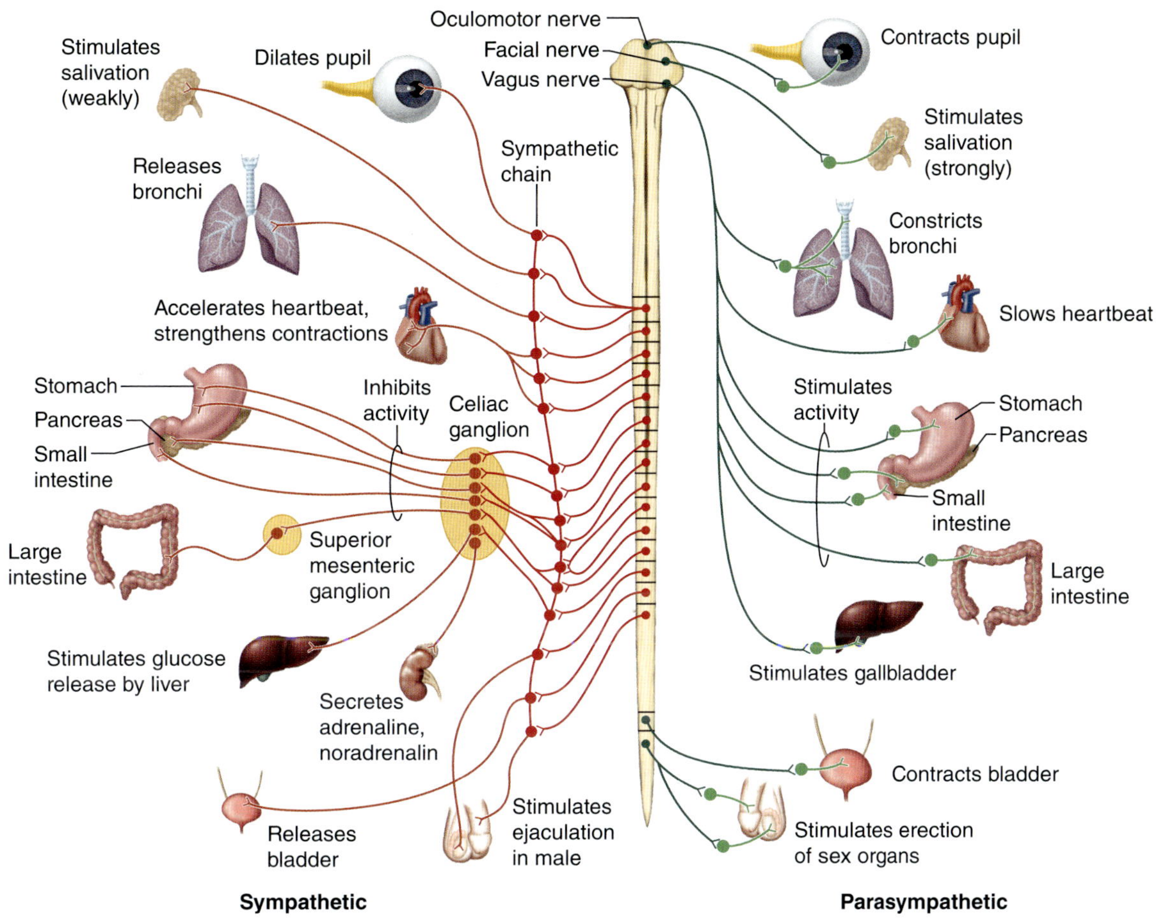

Figure 10.7 Sympathetic and parasympathetic nervous systems.

The autonomic nervous system has two specific systems within it: the sympathetic (also known as the *thoracolumbar system*) and parasympathetic (also known as the *craniosacral system*). These systems, when stimulated, reflexively control the activity of all structures of the body that are not voluntarily controlled. Nerves conducting impulses to voluntary muscles are referred to as *motor nerves*; nerves conducting impulses to the heart, smooth muscle tissue, and various glands of the body are considered part of the autonomic nervous system. The nerves consist of efferent, afferent, and mixed fibers that conduct nerve impulses both to (efferent) and from (afferent) the central nervous system.

As can be observed from their alternative names, the cells of these nerve fibers originate from various levels of the central nervous system. The nerves of the sympathetic system originate from the gray matter of the spinal cord from the first thoracic to the third lumbar level.

The 12 pairs of cranial nerves (table 10.1) contain both efferent (motor) and afferent (sensory) nerve fibers and also function in some aspects of the parasympathetic system.

The first pair of cranial nerves is the olfactory, running from the upper aspect of the nose to the temporal lobe, where the center of smell is located. The second pair of cranial nerves is the optic. Impulses are conducted between the retina and the center for vision, located in the occipital lobe. Both the olfactory and optic cranial nerves are considered sensory nerves; the oculomotor (third cranial), trochlear (fourth cranial), and abducens (sixth cranial) are motor nerves that innervate the voluntary muscles of the eye. The trigeminal (fifth cranial) nerve has sensory fibers that sense warmth, cold, pressure, and pain in

Table 10.1 Cranial Nerves

	FUNCTION	
Name	**Motor**	**Sensory**
1. Olfactory		Smell
2. Optic		Sight
3. Oculomotor	Superior, medial, and ciliary sphincter pupillae; inferior recti and inferior eyeball muscles; oblique eye muscles	
4. Trochlear	Superior oblique, eyeball muscles	
5. Trigeminal	Muscles of mastication, teeth, gums, tensor tympani, face, skin, palate, veli palatini, mylohyoid, anterior 2/3 of tongue, anterior digastric	
6. Abducens	Lateral rectus muscle of the eyeball	
7. Facial	Muscles of face, scalp, anterior 2/3 of lacrimal gland, submaxillary platysma, stylohyoid, sublingual salivary gland, stapedius, posterior digastric	Tongue's taste buds
8. Auditory		Hearing
Cochlear		Hearing
Vestibular	Equilibrium via semicircular canals	
9. Glossopharyngeal	Stylopharyngeus muscle, pharynx, parotid salivary gland, lower 1/3 of tongue	Taste buds
10. Vagus	Muscles of the soft palate, pharynx, larynx, trachea, esophagus, stomach, intestines, other abdominal viscera, and heart	Heart, lungs, trachea, pharynx, larynx, esophagus, and gastrointestinal tract
11. Accessory	Trapezius and sternocleidomastoid muscles	
12. Hypoglossal	Tongue muscles	

the teeth and skin of the face and motor fibers that innervate the muscles of mastication. The facial nerve (seventh cranial) and the glossopharyngeal nerve (ninth cranial) have sensory fibers connecting the brain with the taste buds of the tongue and the pharynx. The motor fibers of the glossopharyngeal nerve innervate the muscles of the tongue and pharynx, and the facial nerve innervates the muscles of the face. The auditory (eighth cranial) nerve is sensory in nature and has two parts: one for hearing (located in the cochlea), connecting to the temporal lobe, and one for equilibrium (located in the semicircular canals), connecting with the cerebellum. The vagus or pneumogastric (10th cranial) nerve contains both sensory and motor fibers. The sensory fibers involve the heart, lungs, trachea, pharynx, larynx, esophagus, and gastrointestinal tract. The motor fibers of the vagus nerve innervate the muscles of the soft palate, pharynx, larynx, trachea, esophagus, stomach, intestines, and other abdominal viscera, and the heart. The spinal accessory (11th cranial) and the hypoglossal (12th cranial) are motor nerves originating in the brain stem and innervating the tongue, face, and neck.

Arteries of the Head and Brain

The epidural space contains the major arteries of the brain, and the subdural space contains the major veins of the brain. The left and right common carotid arteries ascend toward the head. Figure 10.8 illustrates the right common carotid artery. Arising from the subclavian arteries are the vertebral arteries, which, in addition to supplying

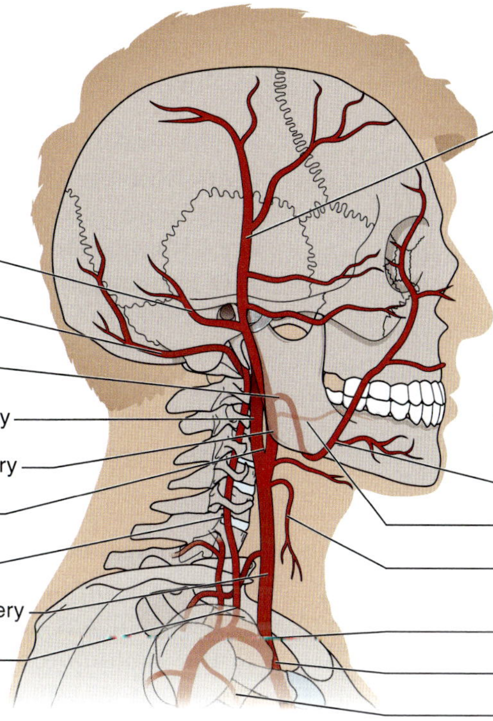

Figure 10.8 Arteries of the head, neck, and face.

blood to the spinal cord and the vertebrae and musculature of the cervical spine, supply blood to the brain. The basilar artery arises between the two vertebral arteries (and branches, known as *posterior cerebral arteries*) to supply blood to structures within the skull.

The right carotid artery, located at the angle of the jaw (mandible), arises from the brachiocephalic artery and separates into the right external and internal carotid arteries. The left carotid artery arises independently from the aortic arch and then separates into the left external and internal carotid arteries. The external carotid artery and its branches supply blood to the skin, muscles, tongue, glands, and other tissues of the upper aspect of the neck and head. The branches of the external carotid artery include the following:

- *Ascending pharyngeal artery:* supplies the dura mater, tympanic area, pharyngeal wall, soft palate, cervical glands, and prevertebral muscles and has branches that include the inferior tympanic, posterior meningeal, prevertebral, palatine, and pharyngeal.
- *External maxillary artery:* supplies the mylohyoid, masseter, and buccinator muscles; the muscles of the upper and lower lip; the pharynx, auditory tube, palatine tonsils, submaxillary and sublingual glands, and the skin of the lips and nose; and the muscles and alar cartilages of the nose. Branches include buccal, angular, ascending palatine, glandular, masseteric, submental, superior and inferior labial, lateral nasal, and tonsillar.
- *Lingual artery:* supplies the posterior aspect of the tongue, the palatine tonsils, the sublingual gland, and the following muscles: geniohyoid, mylohyoid, and genioglossus. Branches include the deep lingual, sublingual, dorsal lingual, and hyoid.
- *Occipital artery:* supplies the dura mater, transverse sinus, internal jugular, medial auricular surface, sagittal sinus, and the postvertebral and sternocleidomastoid muscles. Branches include auricular, meningeal, and mastoid.

- *Posterior auricular artery:* supplies the auricle, posterior temporal scalp, parotid gland, external acoustic meatus, vestibule, semicircular canals, mastoid cells, and the following muscles: sternocleidomastoid, occipitalis, digastric, stylohyoid, styloglossus, stylopharyngeus, and stapedius. Branches include the occipital, parotid, auricular, and stylomastoid.
- *Superior thyroid artery:* supplies the thyroid gland, larynx membrane, and thyrohyoid and laryngeal muscles. Branches include the hyoid, cricothyroid, and superior laryngeal.

At the terminal end of the external carotid artery are two important arteries: the internal maxillary and the superficial temporal arteries and their branches.

- *Internal maxillary artery:* supplies most structures of the head, excluding the brain and eyes, including the auditory tube, tympanic membrane, buccal membrane, dura mater, external acoustic meatus, maxilla, gums, teeth, lacrimal gland, eyelids, mandibular joint, maxillary sinus, sphenoidal sinus, skin of the face, hard and soft palate, palatine bone, vomer bone, parotid gland, nasal cavity, fifth cranial nerve root, and pharynx along with the following muscles: buccinator, masseter, internal and external pterygoids, tensor veli palatini, levator veli palatini, inferior oblique, and inferior rectus. Branches include the pterygoid, sphenopalatine, accessory, meningeal, anterior tympanic, buccinator, deep auricular, deep temporal, infraorbital, alveolar, pharyngeal, greater palatine, and middle meningeal.
- *Superficial temporal artery:* the posterior branch of the external carotid artery, which supplies the upper head structures, including the skin, fascia, external acoustic meatus, temporomandibular joint, lateral auricular surface, and parotid gland, along with the following muscles: frontalis, anterior and superior auricular, masseter, temporalis, and orbicularis oculi. Branches include the transverse facial, frontal, zygomatico-orbital, articular, anterior articular, middle temporal, parietal, and parotid.

The internal carotid artery is the posterior branch of the common carotid artery, which supplies most of the brain, the meninges, the eyes, the frontal sinus, and the skin of the nose. The branches of the internal carotid artery are extensive and considered in more advanced coursework in human anatomy. To illustrate how numerous these deep arterial branches are, they and the structures they supply are outlined here, but they are not illustrated in this text:

Caroticotympanic: provides connections (**anastomoses**) between other cranial arteries

Cavernous: supplies blood to the cavernous sinus; the abducens; and the trochlear, trigeminal, and oculomotor nerves

Choroidal: supplies the optic tract

Hypophyseal: supplies the hypophysis

Meningeal: supplies the dura mater

Ophthalmic: supplies the eye, lacrimal gland, forehead surface, meninges, frontal sinus, skin of the upper nose, anterior and lateral scalp, and upper and lower eyelids, along with the superior and lateral recti and orbital muscles; branches are the anterior ethmoidal, anterior meningeal, central artery of retina, dorsal nasal, frontal, lacrimal, palpebral, posterior ethmoidal, and ciliary

Posterior communicating artery: supplies the thalamus and optic chiasma (forms the lateral side of the circle of Willis)

The two terminal branches of the internal carotid artery are the anterior and middle cerebral arteries.

- *Anterior cerebral artery:* supplies numerous structures of the brain, including the lobes of the anterior aspect of the brain, through the following branches: the anteromedial basal artery; anterior, middle, and posterior medial frontal; and medial orbital
- *Middle cerebral artery:* supplies structures of the brain, including the lateral orbital surface of the frontal lobe, lateral surface of the occipital lobe, parietal lobe, temporal

lobe, inferior and middle frontal gyri, and anterior and posterior central gyri, through the following branches: temporal, ascending frontal and parietal, parietotemporal, inferior lateral frontal, and lateral orbital

The vertebral arteries also have numerous deep branches that supply structures of the brain, the meninges, the muscles of the neck, the spinal cord, and the cervical vertebrae. Again, an in-depth discussion of these structures is beyond the scope of an entry-level study of human anatomy. The deep branches are listed here along with the structures they supply, but they are not illustrated.

Anastomotic: connects occipital and deep cervical arteries

Anterior spinal: supplies the pia mater

Meningeal: connects arteries within the meninges

Muscular: supplies suboccipital muscles and deep cervical spine muscles

Posterior inferior cerebellar: supplies the medulla oblongata; the brain stem; structures within the hemisphere of the cerebellum; and the 9th, 10th, and 11th cranial nerves

Posterior spinal: supplies the pia mater

Spinal: supplies the bodies of the vertebrae and the intervertebral discs

The basilar artery is formed between the two vertebral arteries and supplies the pons; internal ear; anterior medullary velum; superior surface of the cerebellar hemispheres; visual area of the cerebral cortex; brain stem; hippocampal, fusiform, and lingual gyri; and the fourth nerve root. Branches include the anterior inferior cerebellar, superior cerebellar, pontine, internal auditory, and posterior cerebral. The terminal branches of the basilar artery are known as the *posterior cerebral arteries*.

There is a circular formation of arteries on the base of the brain known as the *circle of Willis* (figure 10.9). This circle is formed anteriorly by

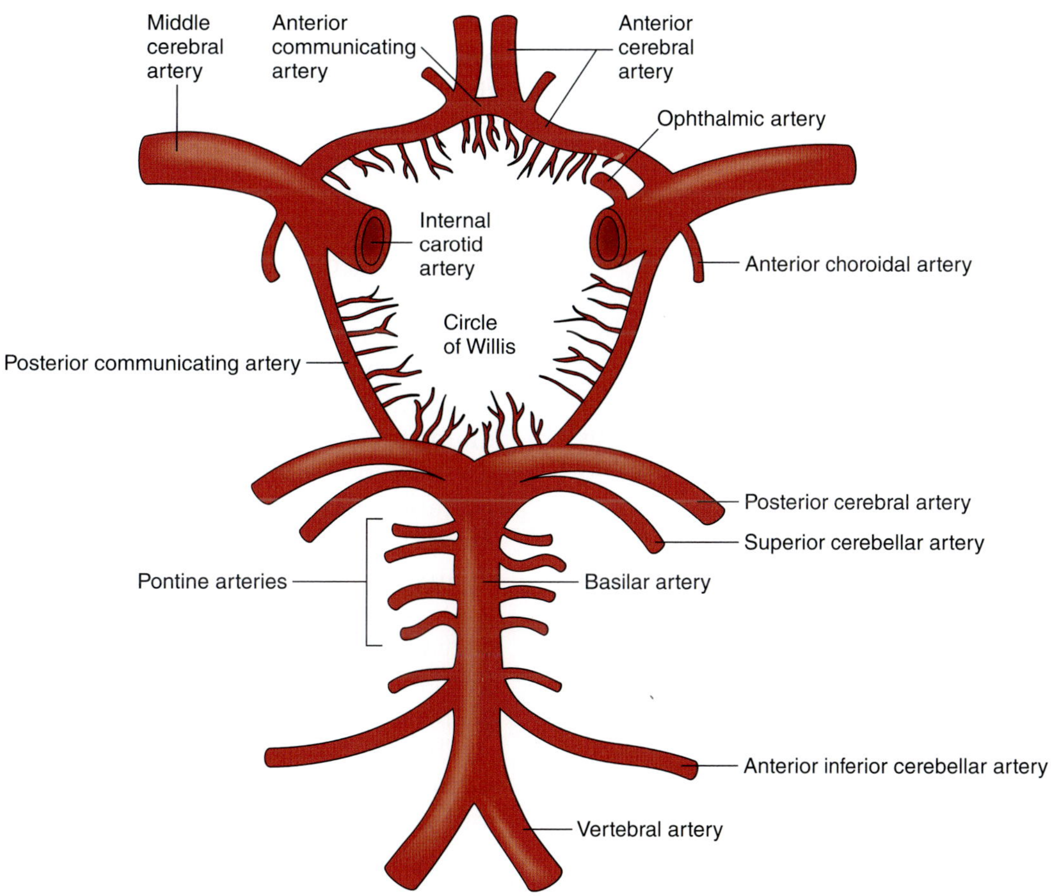

Figure 10.9 Circle of Willis.

FOCUS ON

Epidural Hematoma

As noted earlier, the arteries supplying blood to the brain are found primarily in the epidural space. A portion of the artery is attached to the dura mater, a portion lies in the epidural space, and the upper portion of the artery lies within actual grooves on the underside of the skull. When a body is in motion and the head is forced to suddenly stop, a brief period can ensue when the skull is stopped but the brain is still moving, which results in the brain striking the skull (a concussion). The motion of the brain may cause a laceration (tearing) of the artery as it moves away from its position in a groove of the skull. A common form of this trauma involves the middle cerebral artery and its branches primarily because of their anatomical position, resulting in an epidural hematoma (see figure 10.2). Because this form of hemorrhage causes rapid expansion between the brain's dura mater and skull (via arterial bleeding), vision, orientation, reasoning, and other brain functions can rapidly deteriorate and even endanger life. The epidural hematoma can be one of the most dangerous forms of head trauma, and its signs and symptoms should be known to anyone involved with activities that put participants at high risk of head trauma.

the anterior cerebral arteries and the internal carotid artery branches, connected by the anterior communicating artery, and posteriorly by two posterior cerebral arteries and the basilar branches, connected on either side to the internal carotid by the posterior communicating artery.

Veins of the Head and Brain

The blood supplied to the head and brain by arteries is returned to the heart by veins (figure 10.10). There are deep veins, superficial veins, and sinuses (which are expanded spaces in the head that receive venous blood). The deep veins typically take the names of the arteries they drain, while the superficial veins usually have their own distinct names.

The deep veins of the head are the internal jugular vein and the vertebral vein. The internal jugular vein drains blood from the face, pharynx, thyroid gland, cochlea, and tongue into the subclavian vein.

Like the arteries, the veins have branches (known as *tributaries*). The tributaries of the internal jugular vein include the cochlear vein, common facial vein, lingual vein, superior and middle thyroid vein, occipital vein, pharyngeal vein, and the inferior petrosal sinus.

The vertebral vein and its tributaries primarily drain the arteries of the cervical spine area into the innominate vein (which is formed by the joining of the subclavian and internal jugular veins). The primary tributaries include the anterior vertebral vein, which drains areas of the anterior cervical vertebrae; the deep cervical vein, which drains the structures of the posterior cervical area; and the posterior intercostal vein, which drains the first intercostal space.

Although there are only two deep veins in the head and neck, the superficial veins are numerous. The external jugular vein and its tributaries drain the muscles of the scalp, head, and neck and empty into the subclavian vein. The tributaries include the anterior jugular vein, which drains the anterior neck and lower lip surfaces; the posterior auricular vein, which drains the posterior and lateral surfaces of the head and neck; the posterior external jugular vein, which drains the posterior and lateral aspects of the neck and the occipital area of the head; and the transverse cervical vein and transverse scapular vein, which drain the area supplied by the arteries of the same name.

The superficial veins that drain the scalp include the frontal vein, which drains the medial and anterior surfaces of the scalp and joins the supraorbital vein to form the angular vein. The supraorbital vein drains the posterior and lateral

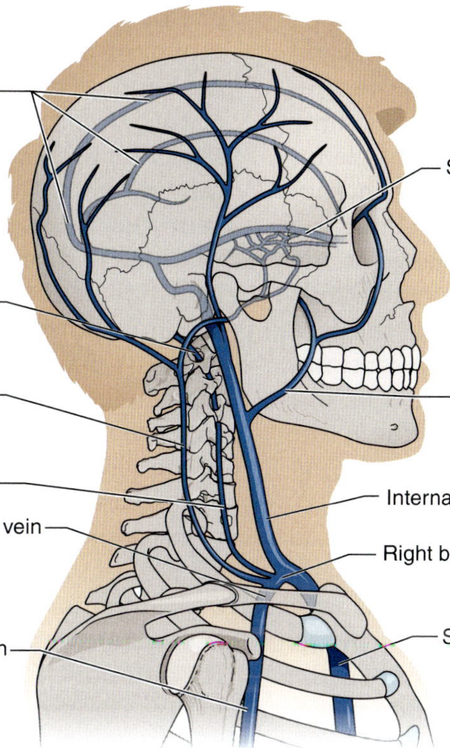

Figure 10.10 Veins of the head, neck, and face.

surfaces of the scalp. The occipital vein drains the occipital and parietal surfaces of the scalp. The posterior auricular vein drains the posterior and lateral surfaces of the head and neck. The superficial temporal vein drains the frontal and lateral surfaces of the head and includes the following tributaries: the transverse facial vein, which drains the lateral side of the face; the orbital vein, which drains the lateral areas of the eyelids; the parotid vein, which drains the parotid gland; the middle temporal vein, which drains the temporalis; the articular vein, which drains the temporomandibular joint; and the anterior auricular vein, which drains the area anterior to the ear.

The superficial veins that drain the face include the angular vein (formed by the frontal and supraorbital veins), which drains the surface of the nose and then becomes the anterior facial vein. The anterior facial vein combines with the posterior facial vein to become the common facial vein, which drains the muscles and skin of the face; its tributaries include the buccinator vein, which drains the buccinator muscle; the deep facial vein, which drains the facial muscles and bones; the superior and inferior labial veins, which drain the upper and lower lips; the superior and inferior palpebral veins, which drain the upper and lower eyelids; the masseteric vein, which drains the masseter muscle; the submaxillary vein, which drains the submaxillary glands; the palatine vein, which drains the palate; and the submental vein, which drains the area supplied by the artery of the same name. The common facial vein (anterior and posterior facial veins) and a communicating vein from the anterior jugular vein drain into the internal jugular vein. The posterior facial vein, formed by the junction of the superficial temporal and internal maxillary veins and its tributaries, has an anterior branch that joins with the anterior facial vein to form the common facial vein and a posterior branch that joins the posterior auricular vein to form the external jugular vein.

Like the arteries of the head and brain, the veins of the head and brain also have an extensive number of deep vein branches that are beyond an entry-level study of human anatomy. The following is a brief description of the veins considered in this advanced category.

The veins of the cranium are contained in the spongy material (diploe) of the cranial bones and are interrelated with the veins of the pericranium (periosteum), meninges, and sinuses of the dura mater. There are four veins responsible for draining blood from the cranium. The anterior temporal diploic vein drains the frontal bone and empties into one of the deep temporal veins and the sphenoparietal sinus. The posterior temporal diploic vein drains the posterior area of the parietal bone and empties into the transverse sinus. The frontal diploic vein drains the anterior area of the frontal bone. The occipital diploic vein (the largest diploic vein) drains the occipital bone and empties into the transverse sinus.

The veins of the meninges are both deep and superficial, draining the arteries of the dura mater and emptying into the cranial sinuses. As with the arteries, these veins are embedded in the dura mater and also lie in the grooves on the interior surface of the skull. This position makes the potential for an intracranial hemorrhage very high, particularly when associated with a fracture of the skull.

The veins of the brain drain into the cranial sinuses and are described as either cerebral or cerebellar. The cerebral veins are both superficial (external) and deep (internal). The superficial veins include the inferior cerebral veins, which drain the inferior surface of the brain into the transverse, superior, cavernous, and petrosal sinuses. The inferior cerebral veins have two tributaries: the middle cerebral vein, which drains the cerebrum into the cavernous sinus, and the ophthalmic vein, which connects the cerebral veins with the orbit and empties into the superior ophthalmic vein. The superior cerebral veins (12 plus) are located in the pia mater and the subarachnoid space; they drain the superior and lateral areas of the cerebrum and empty into the superior sagittal sinus. The basal vein is a combination of the anterior cerebral, deep middle cerebral, and inferior striate veins. It drains the midbrain, the hippocampal gyrus, the inferior portion of the lateral ventricle, and the interpeduncular fossa. The anterior cerebral vein drains the corpus callosum and cingulate gyrus and empties into the basal vein. The deep middle cerebral vein drains the insula and opercula and empties into the basal vein. The inferior striate vein drains the corpus striatum and empties into the basal vein.

Two deep cerebral veins drain the internal aspects of the cerebrum and empty into the great cerebral vein. Receiving blood from the basal vein, the choroid vein and the vena terminalis combine to form the great cerebral vein.

The veins of the cerebellum are grouped into superior and inferior veins. The inferior cerebellar veins drain the inferior area of the cerebellum and empty into the dura mater sinuses. The superior cerebellar veins drain the superior area of the cerebellum and empty into the transverse and superior petrosal sinuses. The deep cerebellar veins drain the cerebellum and empty into the superficial veins.

The veins of the medulla oblongata drain into a superficial plexus of veins that drain into the anterior and posterior median and radicular veins. The veins of the pons empty into the superficial plexus.

The veins of the ear include those of the external acoustic meatus and external ear, which drain into the posterior facial and posterior auricular veins. The veins of the tympanic membrane empty into the posterior facial vein and the superior petrosal sinus. The labyrinth of the ear is drained by the internal auditory veins into the inferior petrosal sinus and vestibular vein.

The veins of the nose drain the walls of the nasal cavity and form the sphenopalatine vein, which empties into the pterygoid plexus. The superior labial and the lateral nasal veins are tributaries of the anterior facial vein.

The veins of the orbit are branches of the ophthalmic artery and the frontal and supraorbital veins. These veins combine to form the superior and inferior ophthalmic veins and drain into the cavernous sinus.

The sinuses of the dura mater are widened venous areas lying between the periosteal and meningeal layers of the dura mater. The sinuses receive blood from the veins of the brain and the meningeal and diploic veins. The cranial sinuses are divided into paired lateral sinuses and unpaired median sinuses.

The paired sinuses include the superficial plexus on either side of the sphenoid bone; as the superficial plexus extends toward the temporal bone, it divides into the superior and inferior petrosal sinuses. The inferior petrosal sinus is located at the petrous area of the temporal bone and empties into the internal jugular vein. Tributaries of the inferior petrosal sinus include the veins of the internal ear and the inferior cerebellar veins. The superior petrosal sinus extends from the apex of the petrous area of the temporal bone at the posterior end of the cavernous sinus to the transverse sinus at the superior edge of the petrous area of the temporal bone. The occipital sinus is attached to the dura mater at the posterolateral borders of the foramen magnum. The occipital sinus has tributaries from the inferior cerebellar veins and empties into the transverse sinus. The cavernous sinus lies on each side of the sphenoid bone and has tributary veins, including the inferior cerebral and ophthalmic veins and veins from the sphenoparietal sinuses. The sphenoparietal sinuses are found in the dura mater under the sphenoid bones. This is where the veins from the dura mater empty into the cavernous sinus. Tributaries of the sphenoparietal sinuses include the diploic, tympanic, and superior and inferior cerebellar veins. The transverse sinuses extend from the external occipital protuberance to the internal jugular vein. The transverse sinuses have tributary veins, including the inferior cerebral and cerebellar veins, the diploic veins, and the superior petrosal sinuses.

The unpaired sinuses include the basilar plexus in the dura mater over the basilar area of the occipital bone and extending to the cavernous sinus along the edge of the foramen magnum. It connects the two inferior petrosal sinuses. The circular sinus encompasses the two cavernous sinuses and the connecting anterior and posterior intercavernous sinuses. The inferior sagittal sinus is found in the posterior margins of the cerebri and connects to the great cerebral vein. Tributaries of the inferior sagittal sinus include the veins from the cerebri and the medial aspects of the hemispheres. The occipital sinus is formed by the inferior sagittal sinus and the great cerebral vein and empties into the transverse sinus. Tributaries of the occipital sinus include veins from the cerebri, the occipital lobes, and the superior cerebellum. The superior sagittal sinus is in the median groove of the inner area of the cranium along the cerebri and extends from the foramen cecum to the internal occipital protuberance. Its tributaries are the superior cerebral veins; it empties into the transverse sinus.

Nerves of the Thorax and Trunk

As presented in the chapters on the upper and lower extremities (see chapters 6 and 14, respectively), the nerves innervating the thorax and trunk originate from plexuses (networks) of nerves from the spinal cord. In particular, one plexus that innervates the lungs is vital for respiration and the maintenance of life. Additionally, the major nerves that serve the entire lower extremity originate at the lower end of the spinal column (lumbosacral plexus). This section of the chapter concentrates on the nerves that supply the muscles of the thorax, trunk, heart, and lungs. These structures are identified in the summary table found at the end of this section for the head, spinal column, and thorax.

The nerves that innervate the head and cervical spine arise from the cervical plexus (figure 10.11). The sensory nerves of the plexus are superficial, whereas the motor nerves are the deep branches. The deep branches innervate the muscles of the scalp and face, and also the cervical spine muscles, such as the sternocleidomastoid, the trapezius, and the levator scapulae. Branches from a cranial nerve, the spinal accessory, also communicate with C2, C3, and C4 and innervate the sternocleidomastoid and trapezius muscles. Additional deep branches of the cervical plexus innervate the rectus capitis anterior and lateralis muscles, the longus capitis and colli muscles, the prevertebral muscles, the levator scapulae, the scalenus medius muscle, the sternocleidomastoid, and the trapezius. The most important nerve of the cervical plexus for respiration is the phrenic nerve (C3, C4, and C5), which innervates the diaphragm, which is essential for breathing. Other nerves of the brachial plexus that innervate the muscles of the cervical spine and the lateral aspect of the thorax are described in chapter 6.

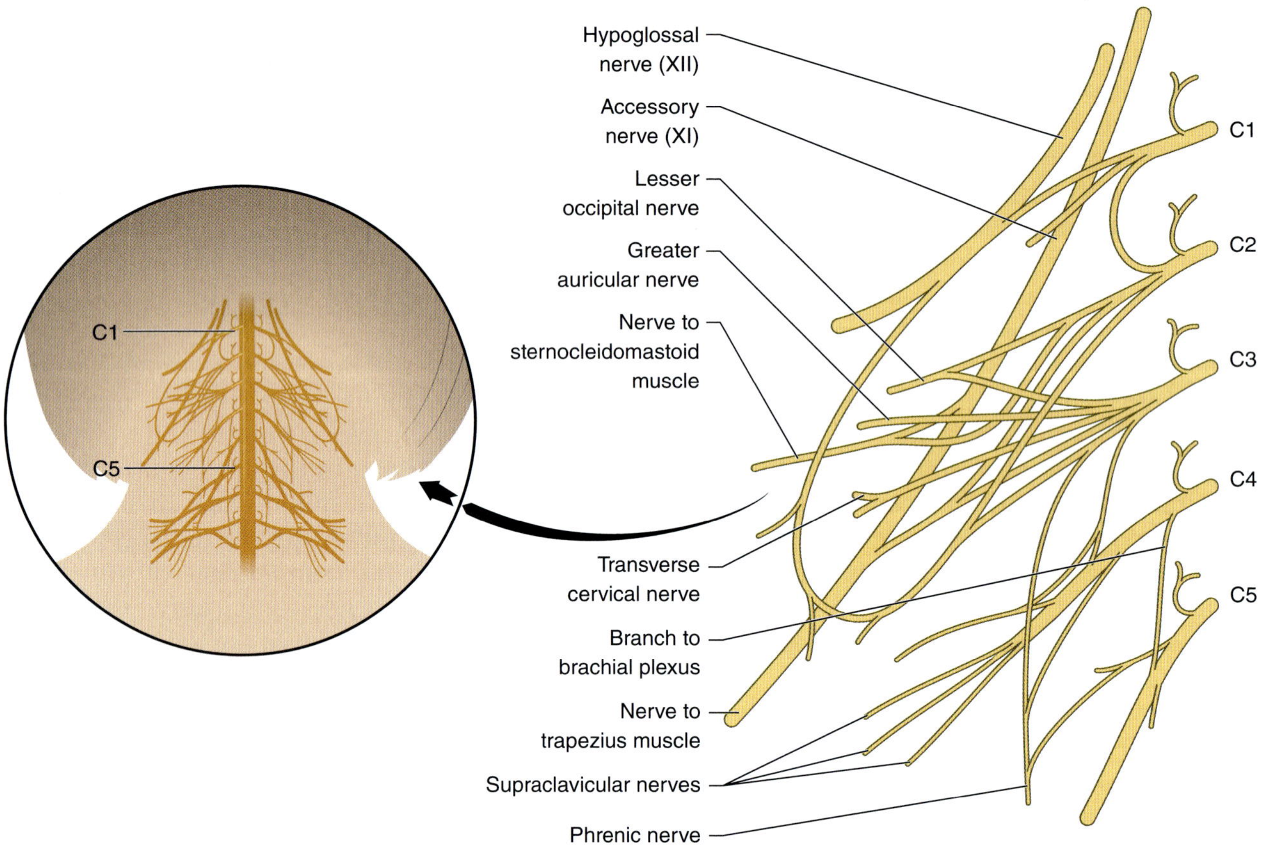

Figure 10.11 The cervical plexus from C1 through C5 nerve roots.

There are 12 thoracic nerves, often described as 11 intercostal nerves and 1 subcostal nerve, that arise from the thoracic section (T1-T12) of the spinal column (figure 10.12). The first thoracic intercostal nerve innervates the levatores costarum muscles, the intercostal muscles, and the superior portion of the serratus posterior muscles. The second thoracic intercostal nerve innervates the intercostal muscles, the levatores costarum muscles, and the subcostal muscles. The third thoracic intercostal nerve innervates the levatores costarum, intercostal, and subcostal muscles. The fourth thoracic intercostal nerve innervates the levatores costarum, intercostal, subcostal, and transversus thoracis muscles. The fifth and sixth thoracic intercostal nerves innervate the intercostal, subcostal, and transversus thoracis muscles. The seventh and eighth thoracic intercostal nerves innervate the intercostal muscles. The 9th, 10th, and 11th thoracic

FOCUS ON

Cervical Spine Injury

When a person suffers a cervical spine injury, it is extremely important to assess the person's vital signs, including respiration. Trauma to the phrenic nerve can disrupt breathing, necessitating the first-aid technique known as rescue breathing. Disruption of breathing combined with cessation of the pulse calls for the initiation of cardiopulmonary resuscitation (CPR). Obviously, when someone uses a CPR technique (chest compressions only) that does not include rescue breathing, any consideration of phrenic nerve trauma that may disrupt breathing is not a factor to consider.

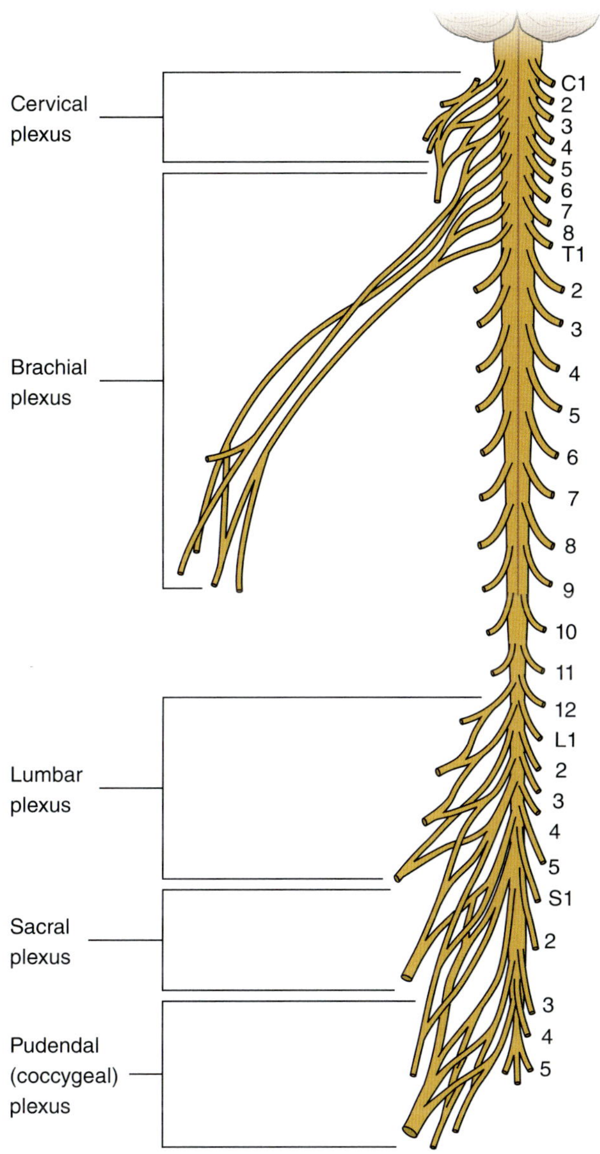

Figure 10.12 The spinal nerves and major plexuses.

intercostal nerves innervate the intercostals, the inferior portion of the serratus posterior, and the abdominal muscles. The 12th thoracic subcostal nerve innervates the abdominal muscles.

The lumbosacral plexus is formed by the anterior rami of the nerves of the lumbar plexus, the sacral plexus, and the pudendal (coccygeal) plexus (see figure 10.12). Most of these nerves innervate the structures of the lower extremity—including the hips, buttocks, groin, and organs of the pelvic region—and are therefore discussed in chapter 14.

Arteries of the Thorax and Trunk

The ascending aorta arises from the heart and has right and left coronary arteries, which have branches to the heart itself. The area between the ascending and descending aorta is known as the *aortic arch* (figure 10.13), which first gives rise to the brachiocephalic artery, which supplies blood to the right arm, the thorax, and the right side of the head and neck. The first branches from the brachiocephalic artery are the right common carotid and right subclavian arteries. The right common carotid artery, just below the jaw (mandible), divides into the external and internal carotid arteries. The left carotid artery arises independently from the aortic arch and then separates into the left external and internal carotid arteries.

Hands On

To obtain a carotid artery pulse, place your index and middle fingers on your Adam's apple, and move them laterally between this structure and the anterior muscles of your neck (figure 10.14). Apply gentle pressure with your fingertips. Do you feel a pulse? What vessel are you palpating to get that pulse? What muscle are you palpating to find this pulse?

When performing CPR, the rescuer must take care in applying pressure on the victim's neck to determine a pulse; the pressure must be applied to the same side of the neck on the victim that the rescuer is on, rather than on the side opposite the rescuer. Improperly applied, the rescuer could possibly pull structures across the airway.

The external carotid artery and its branches supply the muscles of the head and upper portion of the neck. The internal carotid artery and its branches supply the brain and other structures of the head. Additionally, the vertebral arteries, arising from the subclavian arteries, supply the brain, upper portions of the spinal cord, cervical vertebrae, and deep muscles of the neck. An artery between the two vertebral arteries (known as the *basilar artery*) and its branches (the posterior cerebral arteries) supply structures within the skull.

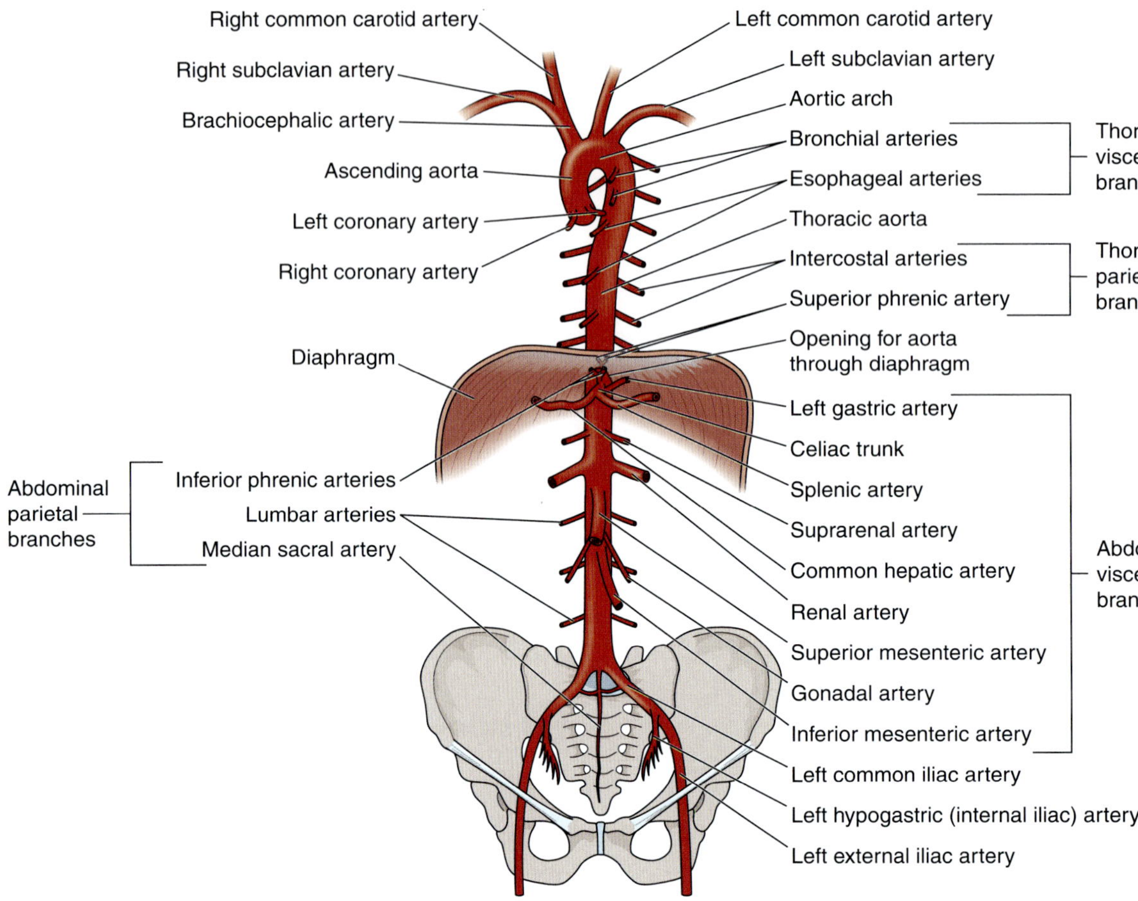

Figure 10.13 Major arteries of the trunk, thorax, and abdomen.

The descending aorta, initially designated as the thoracic aorta, has both thoracic and abdominal parietal and thoracic and abdominal visceral branches (see figure 10.13). The thoracic parietal branches (intercostal, subcostal, and superior phrenic) supply muscles such as the intercostals, the vertebral column muscles, and the diaphragm. The thoracic visceral branches (bronchial and esophageal) supply the structures within the thoracic cavity. Below the diaphragm, the descending aorta becomes known as the *abdominal aorta* (see figure 10.13). The abdominal parietal branches (inferior phrenic, lumbar, and median sacral) supply the diaphragm and the musculature of the abdominal, lumbar, and sacroiliac areas. The abdominal visceral branches (gastric, celiac, splenic, suprarenal, renal, common hepatic, superior and inferior mesenteric, and gonadal) supply the organs of the abdominal cavity, including the liver, spleen, stomach, suprarenals, kidneys, pancreas, intestines, and reproductive structures.

At approximately the L4 level of the vertebral column, the abdominal aorta divides into the right and left common iliac arteries (see figure 10.13). The arteries in the lower abdominal cavity and pelvic region are divisions of the common iliac artery: the external iliac and the hypogastric (internal iliac) arteries. The external iliac artery supplies the abdominal muscles, the iliacus, the psoas muscles, the sartorius, and the tensor fasciae latae of the lower extremity. At the level of the inguinal ligament, the external iliac artery becomes known as the *femoral artery*, which is discussed further in chapter 14 on the nerves and blood vessels of the lower extremity. The hypogastric (internal) iliac artery has branches that supply muscles of the pelvic girdle and the hip joint.

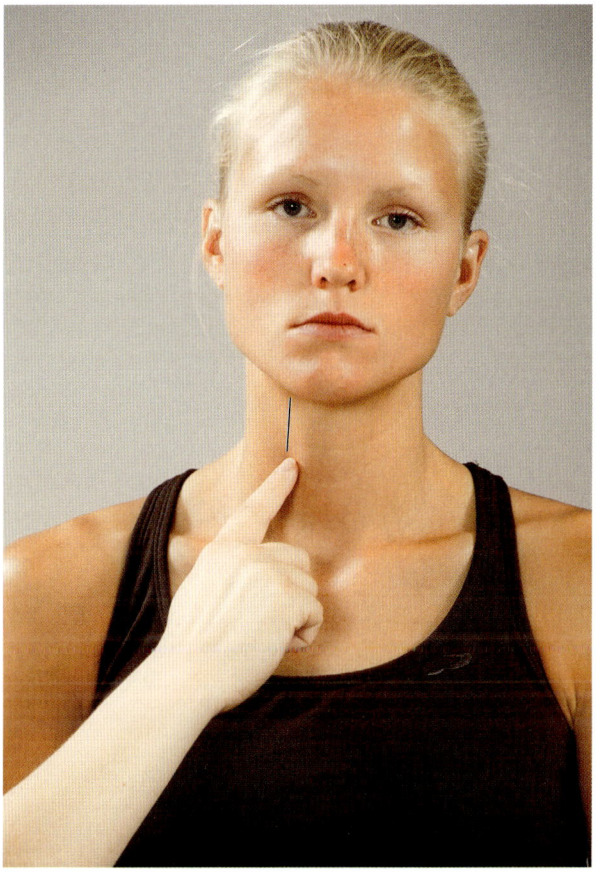

Figure 10.14 Locating the carotid artery.

Veins of the Thorax and Trunk

With a few exceptions, the veins of the thorax and trunk (figure 10.15) parallel the arteries they drain and have the same or similar names. The superior vena cava is the major vein that drains the head, neck, shoulders, upper extremity, and parts of the thorax and abdomen into the heart. The brachiocephalic vein receives blood from the subclavian veins and drains into the superior vena cava. The veins draining the head and neck include the internal jugular, the vertebral, and the external jugular veins and their branches. The internal jugular drains the face and structures of the throat into the subclavian vein. The vertebral vein drains the structures of the skull, the posterior neck muscles, and the first intercostal area into the brachiocephalic vein. The external jugular and its more superficial branches drain the superficial muscles of the head and neck into the subclavian vein.

The inferior vena cava, which empties into the heart, is formed by the juncture of the left and right common iliac veins. It drains the lower extremities, pelvis, and abdominal region. The common iliac vein is formed by the junction of the external and internal iliac veins. The external iliac vein, which is a continuation of the femoral vein (see chapter 14), drains the lower extremity. The hypogastric (internal) iliac vein and its branches drain structures of the lower abdominal and pelvic regions.

Nerves, Arteries, and Veins of the Heart and Lungs

The two major organs housed within the thorax are the heart and lungs. As with all other anatomical structures, the heart and lungs need nerves and blood vessels to accomplish their functions.

Heart Nerves

The nerves of the heart are cardiac branches of the vagus nerve fibers arising from trunks of the sympathetic nervous system. When stimulated, the sinoatrial node (SA node) of the heart, found in the area of the right atrium near the superior vena cava, sends the impulse to the right and left atria myocardium. Specialized tissue (myocardial cells) of the atrioventricular bundle (AV node; located in the lower aspect of the interatrial septum) receives the impulse after it has passed through the atrium (figure 10.16). The impulse continues through the bundle of His (atrioventricular bundle), which divides into left and right branches (becoming known as *myofibers of conduction*, or *Purkinje fibers*) that enter the muscular walls (myocardium) of the ventricles and the papillary muscles. The result is atrial contraction rapidly followed by ventricular contraction.

Respiratory Nerves

The respiratory center is a group of cell bodies located on each side of the medulla oblongata. Arising from these centers are two sets of nerves (figure 10.17): (1) the phrenic nerves, arising from the cervical plexus and leading to the diaphragm

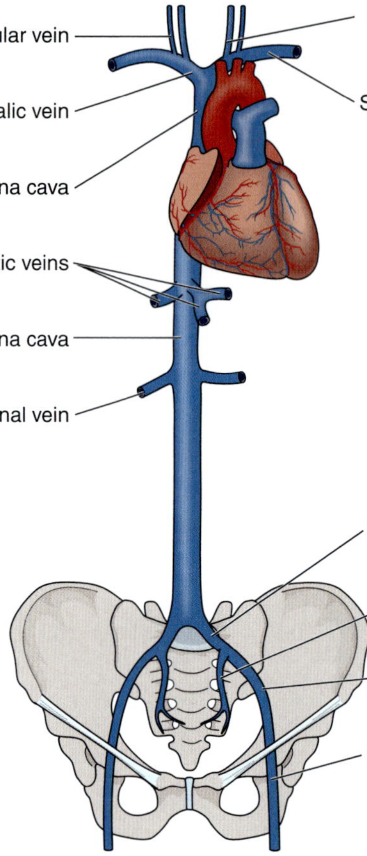

Figure 10.15 Major veins of the trunk, thorax, and abdomen.

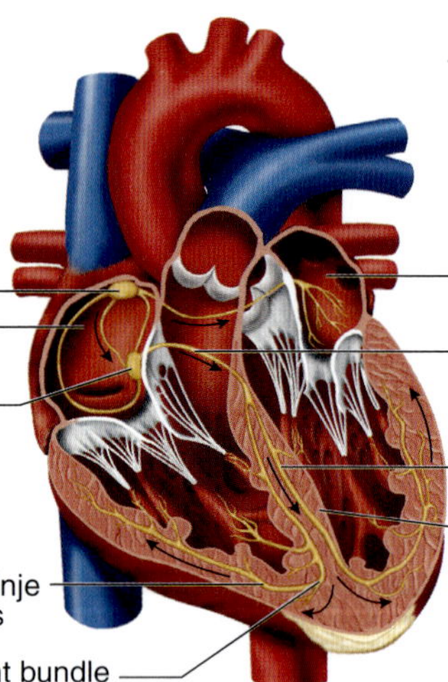

Figure 10.16 Nerves of the heart.

and (2) the intercostal nerves, which innervate the intercostal muscles. When the respiratory center is stimulated by carbon dioxide, it sends an impulse over the phrenic nerves, causing the diaphragm to contract (pull downward) and increasing space in the thoracic cavity. At the same time, the intercostal nerves cause the intercostal muscles to contract, lifting the ribs and increasing the space within the thoracic cavity. This change in capacity of the thoracic cavity creates a change in atmospheric pressure, causing air to rush in and distend the lungs (inspiration). When the lungs have been distended to a certain point, sensory nerves running from the air sacs, via the vagus nerve, send an impulse to the respiratory center to inhibit it. This stops the center's impulses to the phrenic and intercostal nerves, causing the diaphragm and intercostal muscles to relax and resulting in a reduction in the size of the thoracic cavity, forcing air out of the lungs (expiration).

Heart Arteries

Two of the largest arteries of the heart are the pulmonary artery, coming from the right ven-

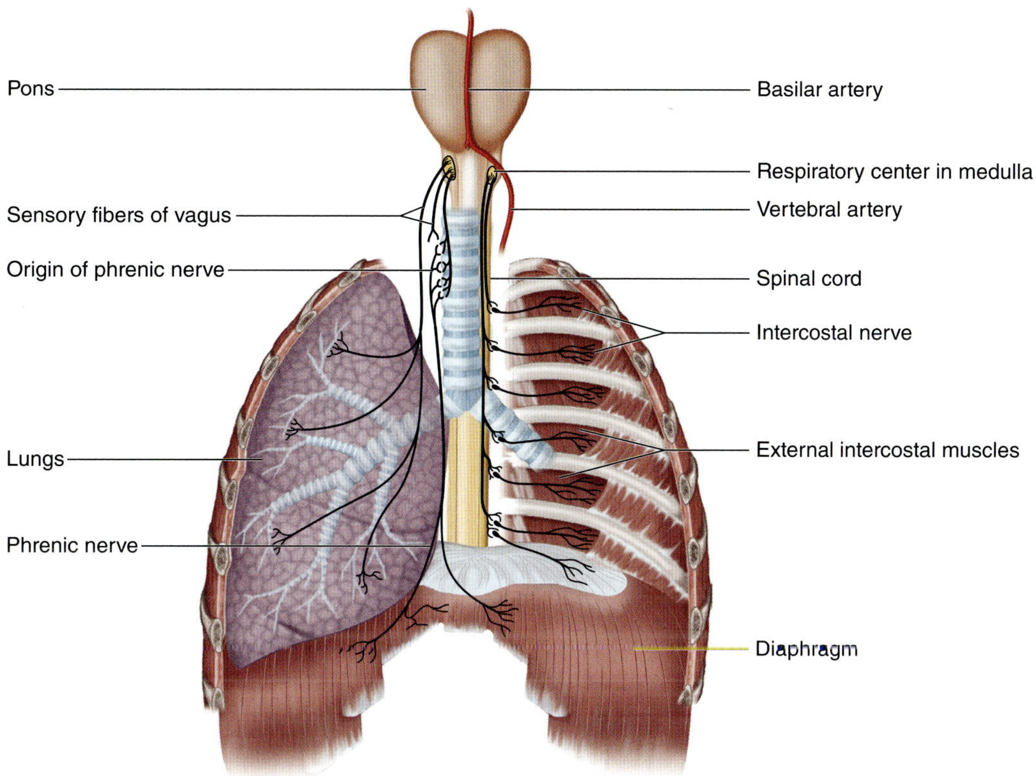

Figure 10.17 Nerves of the respiratory system.

tricle, and the aorta, coming from the left ventricle (figure 10.18). These arteries have valves at their ventricular ends to prevent any backflow of blood. The pulmonary artery has a three-flap valve known as the *pulmonary semilunar valve*, and the aorta has a similar valve known as the *aortic semilunar valve*.

The right and left coronary arteries come from the aorta and are located on the outer surface of the heart, supplying blood flow to the muscular walls of the heart (myocardium). The branches of the right coronary artery include the posterior (dorsal) interventricular and marginal arteries, which supply the anterior surface of the right ventricle; the aortic and pulmonary branches, which supply the aorta and pulmonary arteries; the interventricular, which supplies both ventricles; the right atrial, which supplies the right atrium surface; and the right marginal, which supplies the inferior surfaces of both ventricles. The branches of the left coronary artery include the aortic and pulmonary branches, which supply the aorta and pulmonary arteries; the circumflex, which supplies the left atrium and ventricles; the anterior (ventral) interventricular, which supplies both ventricles; and the left atrial, which supplies the left atrium. While not the only area of concern regarding the heart, it should be noted that the right and left coronary arteries (and their branches) often are the sites of blockage that diminish or cease blood flow to areas of the heart tissue (muscle) and thus cause a "heart attack." Frequently, one of the major causes for this condition is the accumulation of low-density lipoprotein (LDL) cholesterol within these vessels, which reduces or stops blood flow to the heart muscle. Poor diet, lack of exercise, being overweight, and smoking are some of the habits that can contribute to problems of the heart.

Heart Veins

Blood from the body drains to the heart via two major veins: The superior vena cava (and its tributaries) drains the upper extremities, head,

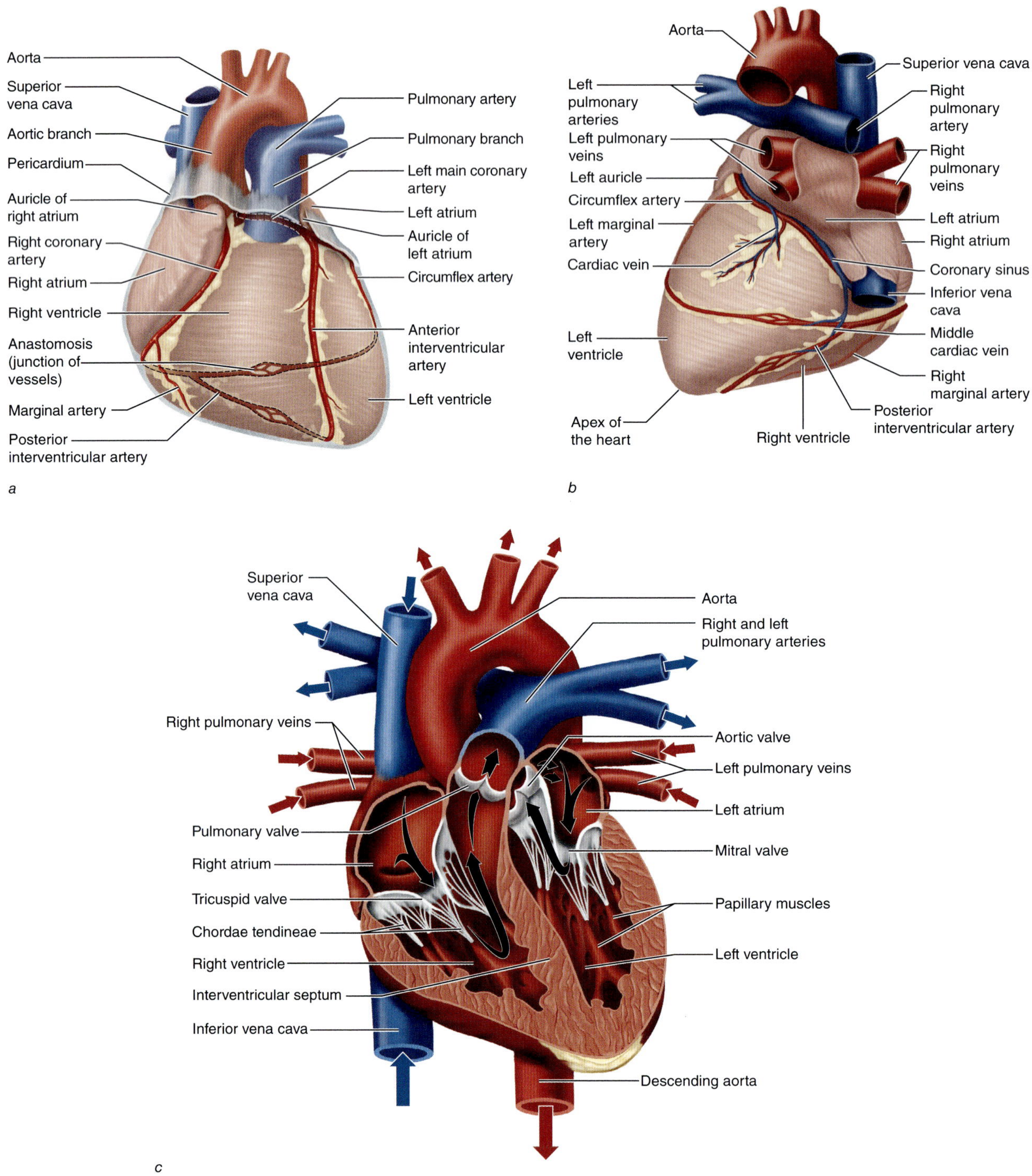

Figure 10.18 Heart blood vessels: *(a)* anterior heart. *(b)* posterior heart, and *(c)* cross section of the heart.

neck, shoulders, thorax, and a portion of the abdominal wall into the heart's right atrium. The inferior vena cava (and its tributaries) drains the lower extremities, pelvis, abdominal viscera, and a portion of the abdominal wall into the right atrium. The left atrium contains the opening for the pulmonary veins, which bring the blood from the lungs to the heart.

The right and left coronary veins drain into the coronary sinus, which empties into the right atrium. Tributaries of the coronary sinus include the great cardiac vein, which drains the left atrium and both ventricles into the coronary sinus. The great cardiac vein also has a tributary: the left marginal vein, which drains the left margin of the heart. Other coronary sinus tributaries include the inferior cardiac vein of the left ventricle, which drains the inferior surface of the left ventricle; the middle cardiac vein, which drains both ventricles and empties into the coronary sinus; the oblique vein of the left atrium, which drains the left atrium into the coronary sinus; and the small cardiac vein, which drains the right atrium and right ventricle into the coronary sinus (see figure 10.18).

Respiratory Arteries and Veins

Blood flow to and from the lung tissues is accomplished through branches of the bronchial arteries and bronchial veins. More in-depth discussions of oxygen and carbon dioxide levels and changes in atmospheric pressures are found in coursework and texts in human physiology.

Arterial blood supplies the body cells with oxygen and is therefore well oxygenated. Venous blood is less oxygenated as it returns to the lungs. The blood in the pulmonary veins returning from the lungs to the heart is highly oxygenated, however (figure 10.19). In other words, the pulmonary veins are the only veins in the body that carry oxygen-rich blood.

A few veins of the heart do not drain into the coronary sinus. These veins include the anterior

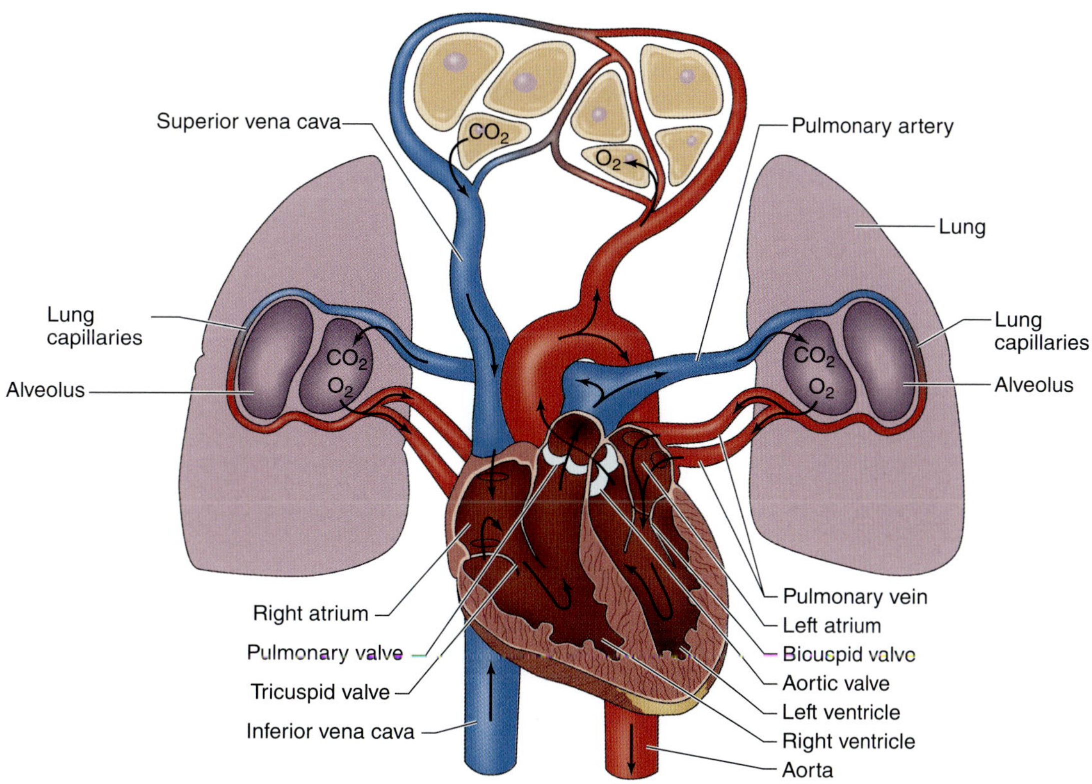

Figure 10.19 Air exchange in the heart and lungs.

cardiac veins, which arise from the wall of the right ventricle and empty into the right atrium, and the venae cordis minimae, which are small veins in the heart walls that drain into the atria.

In the capillaries of the pulmonary vessels in the walls of the air sacs in the lungs is where the respiratory exchange of oxygen and carbon dioxide takes place (figure 10.20).

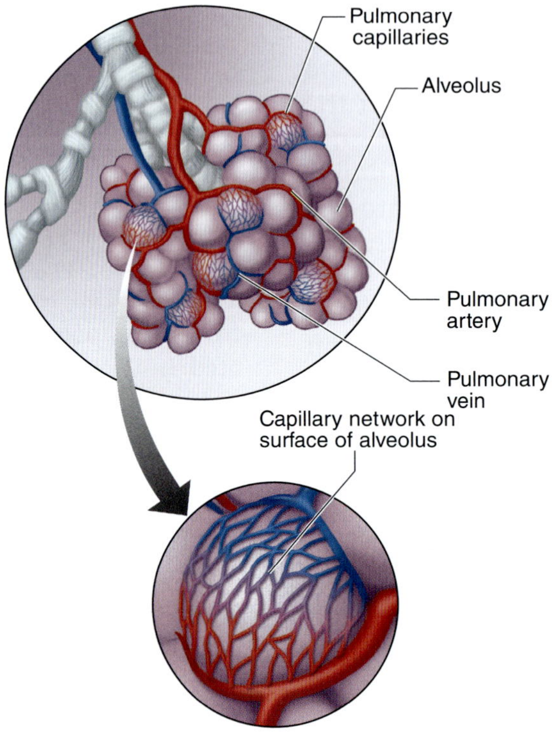

Figure 10.20 Air sac artery and vein arrangement.

Visit the web resource to complete the activities for this chapter.

Articulations of the Head, Spinal Column, Thorax, and Pelvis

Joint	Type	Bones	Ligaments	Movement
HEAD				
Temporomandibular	Diarthrodial	Mandible and temporal bone	• Capsule • Temporomandibular • Articular disc • Support from the stylomandibular and sphenomandibular	Elevation, depression, protrusion, medial and lateral mandibular motion
SPINAL COLUMN				
Atlantooccipital	Condyloid	Occipital and atlas	• Capsules • Anterior atlantooccipital • Posterior atlantooccipital • Lateral atlantooccipital	Flexion, extension, and lateral flexion
Atlantoaxial	Pivot	Atlas and axis	• Capsules • Anterior atlantoaxial • Posterior atlantoaxial • Transverse	Rotation
Intervertebral (bodies)	Amphiarthrodial (connected by cartilage)	Vertebral bodies	• Anterior longitudinal (occipital to sacrum) • Posterior longitudinal (occipital to coccyx) • Intervertebral discs	Slight gliding (limited flexion, extension, rotation)
Intervertebral (arches)	Arthrodial (plane/gliding)	Vertebral articular processes and arches	• Capsules • Ligamentum flavum (axis to sacrum) • Ligamentum nuchae (occipital to C7) • Supraspinous (thoracic to lumbar) • Interspinous (lumbar) • Intertransverse (lumbar)	Gliding

(continued)

Articulations of the Head, Spinal Column, Thorax, and Pelvis *(continued)*

Joint	Type	Bones	Ligaments	Movement
THORAX				
Costovertebral	Arthrodial (plane/ gliding)	Vertebral bodies and heads of ribs	• Capsule • Costovertebral radiate • Interarticular	Gliding
Costotransverse	Arthrodial (plane/ gliding)	Vertebral transverse processes and the necks and tubercles of ribs	• Capsule • Anterior costotransverse • Posterior costotransverse • Ligament of the neck • Ligament of the tubercle	Gliding
Sternocostal	Synarthrodial (fixed, joined by cartilage)	Manubrium of sternum and first rib	Articular disc	None
Sternocostal	Arthrodial (plane/ gliding)	Sternum and ribs (2-7)	• Costosternal capsule • Costosternal radiate • Interarticular • Costoxiphoid	Gliding
Costochondral	Synarthrodial (fixed)	Ribs and costocartilage	• Capsule • Periosteum	None
Interchondral	Synarthrodial (fixed)	Ribs (6-8) and costocartilage	• Capsule • Interchondral	None
PELVIS				
Sacroiliac	Amphiarthrodial (connected by cartilage)	Sacrum and ilium	• Anterior sacroiliac • Posterior sacroiliac (three bands: long, short, and interosseous) • Iliolumbar	Very limited; slight in pregnancy
Pubic symphysis	Amphiarthrodial (connected by cartilage)	Pubic bones	• Interpubic disc • Anterior pubic • Posterior pubic • Superior pubic • Inferior pubic (arcuate)	Slight in pregnancy
NONARTICULAR LIGAMENTS OF THE PELVIS				
		Sacrum, coccyx, and ilium	Sacrotuberous Sacrospinous	Stabilize the sacrum and coccyx
		Anterior superior iliac spine to the pubic tubercle	Inguinal ligament	A landmark that separates the anterior abdominal wall from the thigh (the groin area)

Muscles, Nerves, and Blood Supply of the Head, Thorax, and Spinal Column

Muscle	Origin	Insertion	Action	Nerve	Blood supply
CRANIAL (EPICRANIUS)					
Frontalis	Galea aponeurotica	Procerus and orbicularis oculi	Elevates eyebrows and pulls scalp forward	Facial	Ophthalmic and superficial branch of external carotid
Occipitalis	Lateral 2/3 of superior nuchal lines	Galea aponeurotica	Pulls scalp backward	Facial	Occipital artery
CRANIAL (EXTERNAL EAR)					
Anterior auricularis	Temporal fascia	Ear cartilage	Anterior ear movement	Facial	Superficial temporal branch
Posterior auricularis	Mastoid process	Cranial ear surface	Backward ear movement	Facial	Posterior auricular
Superior auricularis	Galea aponeurotica	Medial ear surface	Upward ear	Facial	Superficial temporal
CRANIAL (MIDDLE EAR)					
Stapedius	Pyramidal eminence of temporal bone	Posterior aspect of neck of stapes	Backward movement of head of stapes	Facial (stapedius branch)	Posterior auricular artery
Tensor tympani	Roof of auditory tube and great wing of sphenoid	Anterior aspect of handle of malleus	Tenses tympanic membrane	Otic ganglion	Internal maxillary branch
FACIAL MUSCLES (EYELID)					
Corrugator supercilii	Medial aspect of superciliary arch	Orbicularis oculi	Pulls eyebrow down and wrinkles forehead	Facial	Superficial external branch of external carotid and ophthalmic
Levator palpebrae superioris	Small wing of sphenoid	Superior tarsal plate of eyelid	Elevates upper eyelid	Oculomotor	Superficial temporal branch of external carotid and ophthalmic branch of internal carotid
Orbicularis oculi	Sphincter muscle in and around eyelid	Sphincter muscle in and around eyelid	Closes eyelid, compresses lacrimal sac, and wrinkles forehead	Facial	Labial branch of external axillary

(continued)

Muscles, Nerves, and Blood Supply of the Head, Thorax, and Spinal Column *(continued)*

Muscle	Origin	Insertion	Action	Nerve	Blood supply
FACIAL MUSCLES (NOSE)					
Depressor septi nasalis	Incisive fossa of maxilla	Septum	Pulls nostrils down, constricts nostrils, and depresses tip of nose	Facial	Lateral nasal branch of external maxillary branch of external carotid
Dilator naris anterior	Greater alar cartilage	Margin of nostril skin	Dilates external nostrils	Facial	Lateral nasal branch of external maxillary
Dilator naris posterior	Nasal notch of maxilla	Margin of nostril skin	Dilates external nostrils	Facial	Lateral nasal branch of external maxillary
Nasalis alar	Greater alar cartilage	Tissue of tip of nose	Pulls nostrils down and constricts nostrils	Facial	Lateral nasal branch of external maxillary branch of external carotid
Nasalis transverse	Maxilla	Membrane over nose	Pulls nostrils up and laterally	Facial	Lateral nasal branch of external maxillary branch of external carotid
Procerus	Tissue stretched over nose and nasal bone	Skin over root of nose and fibers attached to frontalis	Pulls skin over central portion of forehead downward and wrinkles skin at root of nose	Infraorbital branch of facial	Angular branch of external maxillary branch and nasal branch of ophthalmic
FACIAL MUSCLES (MOUTH)					
Buccinator	Buccinator crest of mandible and molar aspect of outer alveolar process of maxilla	Fibers to the orbicularis oris in both upper and lower lip	Pulls corners of mouth laterally, pulls lips against teeth, flattens cheeks, and assists mastication	Buccal branch of facial	Buccinator branch of internal maxillary
Caninus	Canine fossa of maxilla	Orbicularis oris	Elevates and medially pulls corners of mouth	Zygomatic branch of facial	Branches of external and internal maxillaries
Mentalis	Incisor fossa of mandible	Skin of chin	Draws up skin of chin, protrudes lower lip	Mandibular branch of facial	External maxillary

Muscle	Origin	Insertion	Action	Nerve	Blood supply
FACIAL MUSCLES (MOUTH)					
Orbicularis oris	Muscles of mouth area	Lip skin and mucosa	Sphincter closing mouth	Facial	Labial branch of external maxillary
Quadratus labii inferior	Lateral surface of mandible and medial to triangularis	Orbicularis oris and lower lip	Protrudes and pulls up lower lip	Mandibular branch of facial	Mental branch of external maxillary
Quadratus labii superior	Three parts: caput muscles				
Risorius	Parotid masseteric fascia	Skin of angle of mouth	Pulls angle of mouth laterally	Buccal branch of facial	Branches of external and internal maxillaries
Triangularis	Oblique line of mandible	Orbicularis oris and skin of lower lip	Pulls corners of mouth both downward and laterally	Buccal branch of facial	External maxillary
Zygomaticus	Temporal process of zygomatic bone	Orbicularis oris and skin of upper lip	Elevates and pulls corners of mouth laterally	Zygomatic branch of facial	Branches of external and internal maxillaries and ophthalmic
FACIAL MUSCLES (MASTICATION)					
Internal pterygoids	Palatine bone and maxillary tuberosity	Medial aspect of ramus and angle of mandible	Protraction, elevation, and lateral movement of mandible	Mandibular branch of fifth nerve	Pterygoid branch of internal maxillary
External pterygoids	Great wing and lateral pterygoid plate of the sphenoid	Neck of mandible	Protraction, closing, and lateral movement of mandible	Mandibular branch of fifth nerve	Pterygoid branch of internal maxillary
Masseter	Inferior border of zygomatic arch	Lateral aspect of ramus and angle of mandible	Elevation and protraction of mandible	Mandibular branch of fifth nerve	Internal maxillary branch of external carotid
Temporalis	Temporal fossa	Coronoid process and ramus of mandible	Elevation and retraction of mandible	Mandibular branch of fifth nerve	Temporal branch of external carotid

(continued)

Muscles, Nerves, and Blood Supply of the Head, Thorax, and Spinal Column *(continued)*

Muscle	Origin	Insertion	Action	Nerve	Blood supply
EYEBALL MUSCLES (EXTRINSIC)					
Superior oblique	Superior aspect of optic foramen	Sclera of lateral surface of eyeball	Moves eyeball medially and inferiorly	Trochlea	Ophthalmic
Inferior oblique	Orbital aspect of maxilla	Sclera of lateral surface of eyeball	Moves eyeball laterally and superiorly	Oculomotor	Ophthalmic and infraorbital branch of internal maxillary
Orbitalis	Spans inferior orbital fissure	Spans inferior orbital fissure	Protrudes eye	Sympathetic	Ophthalmic branch of internal carotid
Rectus superior	Fibrous ring around optic foramen	Sclera of eyeball	Medial and superior movement of eyeball	Oculomotor	Ophthalmic
Rectus medialis	Fibrous ring around optic foramen	Sclera of eyeball	Medial movement of eyeball	Oculomotor	Ophthalmic and infraorbital branch of maxillary
Rectus lateralis	Fibrous ring around optic foramen	Sclera of eyeball	Lateral movement of eyeball	Abducens	Lacrimal branch of ophthalmic
Rectus inferior	Fibrous ring around optic foramen	Sclera of eyeball	Medial and inferior movement of eyeball	Oculomotor	Ophthalmic and infraorbital branch of internal maxillary
EYEBALL MUSCLES (INTRINSIC)					
Ciliary	Posterior aspect of sclera	Ciliary process	Accommodation	Short ciliary nerve of oculomotor	Ophthalmic branch of internal carotid
Dilator pupillae	Involuntary and radiating fibers converging toward center of eyeball and blending with circular fibers	Involuntary and radiating fibers converging toward center of eyeball and blending with circular fibers	Dilates pupil	Sympathetic fibers from superior cervical ganglion	Ophthalmic branch of internal carotid
Sphincter pupillae	Involuntary and circular fibers surrounding pupil margins	Involuntary and circular fibers surrounding pupil margins	Constricts pupil	Oculomotor (third cranial) through ciliary ganglion	Ophthalmic branch of internal maxillary

Muscle	Origin	Insertion	Action	Nerve	Blood supply
TONGUE MUSCLES (EXTRINSIC)					
Chondroglossus	Lesser cornu of hyoid bone	Side of tongue	Tongue depression	Hypoglossal (12th cranial)	Lingual artery of external carotid
Genioglossus	Upper genial tubercle of the mandible	Tip of tongue and hyoid bone	Depression, protrusion (posterior fibers), and retraction (anterior fibers) of tongue	Hypoglossal (12th cranial)	Lingual artery of external carotid
Hyoglossus	Body and greater cornu of hyoid bone	Side of tongue	Tongue depression	Hypoglossal (12th cranial)	Lingual artery of external carotid
Styloglossus	Anterior border of styloid process and stylohyoid ligament	Side and inferior surface of tongue	Elevation and retraction of tongue	Hypoglossal (12th cranial)	Deep lingual branch of lingual and posterior auricular artery
TONGUE MUSCLES (INTRINSIC)					
Superior longitudinalis	Base of tongue	Tip of tongue	Controls shape of tongue	Hypoglossal (12th cranial)	Deep lingual branch of lingual
Inferior longitudinalis	Inferior portion of tongue	Tip of tongue	Controls shape of tongue	Hypoglossal (12th cranial)	Deep lingual branch of lingual
Transversus linguae	Median septum of tongue	Dorsum of tongue	Controls shape of tongue	Hypoglossal (12th cranial)	Deep lingual branch of lingual
Verticalis linguae	Dorsum of tongue	Extrinsic tongue muscles	Controls shape of tongue	Hypoglossal (12th cranial)	Deep lingual branch of lingual
SOFT PALATE MUSCLES					
Palatoglossus	Dorsum and side of tongue	Soft palate	Pulls tongue up and back, depresses sides of soft palate	Glossopharyngeus	Lingual branch of external carotid
Uvulae	Posterior nasal spine and soft palate aponeurosis	Uvula	Elevates uvula	Pharyngeal plexus	Palatine branch of internal maxillary
Levator veli palatini	Cartilaginous aspect of auditory tube and petrous part of temporal bone	Soft palate aponeurosis	Elevates soft palate, narrows pharyngeal opening, and expands isthmus of auditory tube	Vagus	Branch of internal maxillary

(continued)

Muscles, Nerves, and Blood Supply of the Head, Thorax, and Spinal Column *(continued)*

Muscle	Origin	Insertion	Action	Nerve	Blood supply
SOFT PALATE MUSCLES					
Tensor veli palatini	Scaphoid fossa and angular spine of sphenoid bone and lateral area of cartilaginous aspect of auditory tube	Posterior border and aponeurosis of hard palate	Tenses soft palate and opens auditory tube during swallowing	Mandibular branch of trigeminal (fifth cranial)	Internal maxillary branch of external carotid
Palatopharyngeus	Posterior border of thyroid cartilage and aponeurosis of pharynx	Soft palate	Depresses soft palate and elevates pharynx and larynx	Vagus (10th cranial)	Palatine branch of internal maxillary
Salpingopharyngeus	Lower aspect of auditory tube cartilage	Inferior margin of medial lamina of auditory tube	Lifts pharynx and larynx	Vagus (10th cranial)	Branch of internal maxillary
PHARYNX MUSCLES					
Stylopharyngeus	Medial aspect of styloid process root	Superior and posterior border of thyroid cartilage and wall of pharynx	Talking and swallowing	Glossopharyngeal (ninth cranial)	External maxillary and lingual artery
Superior pharyngeal constrictor	Posterior border of medial pterygoid lamina, pterygomandibular raphe, and mylohyoid line of mandible		Swallowing and talking	Glossopharyngeal	Ascending pharyngeal nerve and external maxillary
Inferior pharyngeal constrictor	Oblique line of thyroid cartilage and cricoid cartilage	Posterior wall of pharynx	Swallowing and talking	Glossopharyngeal	Superior and inferior thyroid artery
Middle pharyngeal constrictor	Stylohyoid ligament and both cornu of hyoid bone	Median raphe of pharynx	Swallowing and talking	Glossopharyngeal	Ascending pharyngeal nerve and external maxillary
LARYNX MUSCLES (EXTRINSIC/INFRAHYOID)					
Levator glandulae thyroideae	Hyoid bone	Isthmus/ pyramid of thyroid gland	Raises thyroid gland	Branches from hypoglossal	Superior thyroid
Sternohyoid	Posterior aspect of manubrium and posterior sternoclavicular ligament	Medial aspect of hyoid bone	Depression of hyoid bone	First three cervical nerves via ansa hypoglossal	Superior thyroid

Muscle	Origin	Insertion	Action	Nerve	Blood supply
LARYNX MUSCLES (EXTRINSIC/INFRAHYOID)					
Sternothyroid	Posterior aspect of manubrium and first costal cartilage	Oblique line of thyroid cartilage	Depression of hyoid bone and thyroid cartilage	First three cervical nerves via ansa hypoglossal	Superior thyroid
Thyrohyoid	Oblique line of thyroid cartilage	Body and great cornu of hyoid bone	Depression of hyoid bone and elevation of thyroid cartilage	First two cervical nerve branches through hypoglossal	Superior thyroid
Omohyoid (anterior belly)	Tendon of posterior omohyoid belly	Lateral inferior border of hyoid bone	Depression of hyoid	First three cervical nerves via ansa hypoglossal	Superior thyroid and superficial cervical
Omohyoid (posterior belly)	Superior border of scapula and superior transverse scapular ligament	Anterior omohyoid belly	Depression of hyoid	First three cervical nerves via ansa hypoglossal	Superior thyroid and superficial cervical
LARYNX MUSCLES (EXTRINSIC/SUPRAHYOID)					
Geniohyoid	Inferior genial tubercle of mandible	Anterior area of body of hyoid bone	Depression of tongue, anterior fibers retract tongue, posterior fibers protrude tongue	Hypoglossal (12th cranial)	Branch of lingual
Mylohyoid	Mylohyoid ridge of mandible	Posterior border of hyoid bone	Elevation of hyoid bone	Trigeminal	Lingual
Stylohyoid	Posterior border of styloid process	Body of hyoid bone	Elevation and posterior movement of hyoid bone	Facial	External maxillary and lingual artery
Digastric (anterior belly)	Body of hyoid bone and intermediate tendon	Digastric fossa on inferior border of mandible	Opens mouth, elevates and moves hyoid bone forward	Trigeminal	External maxillary branch of external carotid
Digastric (posterior belly)	Mastoid notch of temporal bone	Body of hyoid bone and intermediate tendon	Elevates and moves hyoid bone backward	Branch of facial	Posterior auricular branch of external carotid

(continued)

Muscles, Nerves, and Blood Supply of the Head, Thorax, and Spinal Column *(continued)*

Muscle	Origin	Insertion	Action	Nerve	Blood supply
LARYNX MUSCLES (INTRINSIC)					
Aryepiglotticus	Apex of arytenoid cartilage	Thyroepiglottic ligament and lateral edge of epiglottic cartilage	Pulls down epiglottis	Superior laryngeal	Superior laryngeal branch of superior thyroid
Arytenoid oblique	Muscular process of arytenoid cartilage	Apex of opposite arytenoid cartilage and angle of thyroid	Constricts laryngeal aperture and vestibule of larynx	Anterior branch of inferior laryngeal nerve	Superior laryngeal branch of superior thyroid branch of external carotid
Arytenoid transverse	Posterior concave area of one arytenoid cartilage	Posterior concave area of other arytenoid cartilage	Narrows glottis	Posterior branch of inferior laryngeal nerve	Superior laryngeal branch of superior thyroid branch of external carotid
Posterior cricoarytenoid	Posterior surface of cricoid lamina	Muscular process of arytenoid cartilage	Tenses vocal ligaments and widens glottis	Posterior branch of inferior laryngeal nerve	Inferior laryngeal branch of inferior thyroid
Lateral cricoarytenoid	Upper margin and outer surface of cricoid arch	Anterior surface of muscular process of arytenoid cartilage	Stretches and approximates vocal ligaments	Anterior branch of inferior laryngeal nerve	Superior thyroid branch of external carotid
Cricothyroid	Arch of cricoid cartilage	Caudal margin of thyroid cartilage	Elevates cricoid cartilage and depresses lamina and arytenoid cartilage	External branch of inferior laryngeal nerve	Branch of superior thyroid
Thyroarytenoid	Lamina of thyroid cartilage	Margin of arytenoid cartilage	Relaxes vocal ligaments	Anterior branch of inferior laryngeal nerve	Superior laryngeal branch of superior branch of external carotid
Thyroepiglotticus	Inner surface of thyroid lamina	Quad membrane and lateral border of epiglottis	Closes epiglottis	Anterior branch of inferior laryngeal nerve	Superior laryngeal branch of superior branch of external carotid

Muscle	Origin	Insertion	Action	Nerve	Blood supply
LARYNX MUSCLES (INTRINSIC)					
Ventricularis	Lateral edge of arytenoid cartilage	Lateral margin of epiglottic cartilage	Pulls epiglottis downward	Internal branch of superior laryngeal nerve	Superior laryngeal branch of superior thyroid branch of external carotid
Vocalis	Angle of thyroid lamina	Vocal process and oblong fovea of arytenoid cartilage	Relaxes vocal ligaments	Anterior branch of inferior laryngeal nerve	Superior laryngeal branch of superior thyroid branch of external carotid
THORAX					
Diaphragm	Sternum (xiphoid), last seven ribs, first four lumbar vertebrae	Central tendon	Pulls central tendon down	Phrenic	Superior and inferior phrenic
Internal intercostals	Upper portion of costal groove	Upper edge of rib cartilage below	Depression of rib	T2-T6	Intercostals
External intercostals	Lower edge of rib cartilage	Upper edge of rib below	Elevation of rib	T2-T6	Intercostals
Levatores costarum	Transverse process of C7 and T1-T11	Angle of rib below	Elevation of rib	T2-T6	Intercostals
Subcostals	At angle on inner aspect of rib	Inner aspect of second and third ribs below	Elevation of rib	T2-T6	Intercostals
Serratus posterior superior	Spinous processes of C7 and T1-T3	Angles of ribs (2-5)	Elevation of ribs	First to third intercostal	Intercostal and transverse cervical
Serratus posterior inferior	Spinous processes of T11-T12 and L1-L3	Inferior edge of ribs (9-12)	Depression of ribs	10th and 11th intercostal	Intercostal and transverse cervical
Transversus thoracis	Distal 1/3 of posterior aspect of sternum and cartilage of ribs (4-7)	Posterior aspect of costal cartilage of ribs (2-6)	Depression of ribs, aid in expiration	Second to sixth thoracic	Intercostals

(continued)

Muscles, Nerves, and Blood Supply of the Head, Thorax, and Spinal Column *(continued)*

Muscle	Origin	Insertion	Action	Nerve	Blood supply
SPINAL COLUMN, CERVICAL (ANTERIOR)					
Sternocleidomastoid	Middle 1/3 of clavicle and sternum	Mastoid process	Flexion, lateral flexion of cervical spine, and contralateral rotation of head	Spinal accessory, second and third cervical	Transverse cervical
Prevertebrals					
Rectus capitis anterior	Lateral aspect of atlas	Occipital bone	Flexion and lateral flexion of cervical spine	C1 and C2	Pharyngeal and inferior thyroid
Rectus capitis lateralis	Transverse process of atlas	Inferior aspect of occipital bone	Flexion and lateral flexion of cervical spine	C1 and C2	Pharyngeal and occipital
Longus colli					
Vertical	Body of C5-C7 and T1-T3	Bodies of C2-C5	Flexion of cervical spine	C2-C8	Inferior and superior thyroid
Superior oblique	Anterior aspect of C3-C5 transverse processes	Anterior atlas	Flexion of cervical spine	C2-C8	Inferior and superior thyroid
Inferior oblique	T1-T3 bodies	C5-C6 transverse processes	Flexion of cervical spine	C2-C8	Inferior and superior thyroid
Scaleni					
Scalenus anterior	C3-C7 transverse processes	First rib	Lateral flexion of cervical spine, elevation of 1first rib, and ipsilateral rotation of head	C4-C8	Transverse cervical andinferior thyroid
Scalenus medius	C2-C7 transverse processes	First rib	Lateral flexion of cervical spine, elevation of first rib, and ipsilateral rotation of head	C4-C8	Cervical and inferior thyroid
Scalenus posterior	C4-C6 transverse processes	Second rib	Elevation of second rib and ipsilateral rotation of head	C4-C8	Cervical and inferior thyroid

Muscle	Origin	Insertion	Action	Nerve	Blood supply
SPINAL COLUMN, CERVICAL (POSTERIOR)					
Rectus capitis posterior major	Spinous process of axis (C2)	Occipital bone	Extension, lateral flexion of cervical spine, and ipsilateral rotation of the head	C1	Vertebral
Rectus capitis posterior minor	Posterior atlas (C1)	Occipital bone	Extension, lateral flexion of cervical spine, and ipsilateral rotation of the head	C1	Vertebral
Obliquus capitis superior	Transverse process of atlas (C1)	Occipital bone	Extension of cervical spine	C1	Vertebral and occipital
Obliquus capitis inferior	Spinous process of axis (C2)	Transverse process of the (C1)	Ipsilateral rotation of the head	C1	Vertebral
Splenius capitis	C7-T3 spinous processes and nuchal ligament	Mastoid process	Extension, lateral flexion of cervical spine, and ipsilateral rotation of the head	C2-C4	Occipital and transverse cervical
Splenius cervicis	T4-T6 spinous processes	C1-C3 transverse processes	Extension, lateral flexion of cervical spine, and ipsilateral rotation of the head	C2-C4	Occipital and transverse cervical
SPINAL COLUMN, THORACIC OR LUMBAR (ANTERIOR)					
Rectus abdominis	Pubic symphysis	Pubic bone, cartilage or ribs (5-7), and xiphoid process	Flexion of trunk, compresses abdominal viscera, and tenses abdominal wall	Last six intercostals, iliohypogastric, ilioinguinal, and genitofemoral	Inferior epigastric branch of external iliac
Internal oblique	Lateral inguinal ligament, middle of iliac crest	Pubis, ribs (9-12), and linea alba	Trunk flexion, lateral flexion, and ipsilateral rotation of spine	Last six intercostals	Lower intercostals and epigastric

(continued)

Muscles, Nerves, and Blood Supply of the Head, Thorax, and Spinal Column *(continued)*

Muscle	Origin	Insertion	Action	Nerve	Blood supply
SPINAL COLUMN, THORACIC OR LUMBAR (ANTERIOR)					
External oblique	Inferior edge of last eight ribs, sternum	Pubis, linea alba, and iliac crest	Trunk flexion, lateral flexion, and contralateral rotation of spine	Last three intercostals, iliohypogastric, and ilioinguinal	Inferior epigastric branch of external iliac
Transversus abdominis	Inner aspect of last six ribs, internal aspect of middle 1/2 of iliac crest, lateral 1/3 of inguinal ligament	Linea alba, pubic crest	Compresses abdominal viscera, tenses abdominal wall	Last seven intercostals, iliohypogastric	Lower intercostals and inferior epigastric branch of external iliac
SPINAL COLUMN, THORACIC OR LUMBAR (POSTERIOR)					
Spinals					
Spinalis capitis	C7 and T1 spinous processes	Occipital bone	Extension of cervical spine	T6-T9	Occipital
Spinalis cervicis	Ligamentum nuchae, C7 and T1-T2 spinous processes	C2-C5 spinous processes	Extension and lateral flexion of cervical spine	T6-T9	Vertebral and transverse cervical
Spinalis dorsi (thoracis)	T11-L2 spinous processes	T2-T8 spinous processes	Extension and lateral flexion of vertebral column	T6-T9	Intercostals
Semispinals					
Semispinalis capitis	C4-C7 articular processes and T1-T6 transverse processes	Occipital bone	Extension, lateral flexion, and contralateral rotation of cervical spine	C1-C5	Occipital, transverse cervical, vertebral, and deep cervical
Semispinalis cervicis	T1-T6 transverse processes	C2-C5 spinous processes	Extension, lateral flexion, and contralateral rotation of cervical spine	T3-T6	Deep cervical, occipital, and vertebral
Semispinalis dorsi (thoracis)	T6-T10 transverse processes	C6-C7 and T1-T4 spinous processes	Extension of cervical spine	T3-T6	Intercostals

Muscle	Origin	Insertion	Action	Nerve	Blood supply
SPINAL COLUMN, THORACIC OR LUMBAR (POSTERIOR)					
Iliocostals					
Iliocostalis cervicis	Angle of ribs (3-6)	C4-C6 transverse processes	Extension and lateral flexion of spine	Cervical	Thyrocervical
Iliocostalis dorsi (thoracis)	Angle of last six ribs	Upper edge of first six ribs and C7 transverse process	Extension and lateral flexion of spine	Thoracic	Intercostals
Iliocostalis lumborum	Posterior aspect of iliac crest	Angle of last seven ribs	Extension and lateral flexion of spine	Lumbar	Lumbar
Longissimi					
Longissimus capitis	C4-C7 articular processes and T1-T5 transverse processes	Mastoid process	Extension, lateral flexion, and ipsilateral rotation of cervical spine	C1-L5	Transverse cervical, external carotid, and vertebral
Longissimus cervicis	T1-T5 transverse processes	C2-C6 transverse processes	Extension, lateral flexion, and ipsilateral rotation of cervical spine	C1-L5	Transverse cervical, external carotid, and vertebral
Longissimus dorsi (thoracis)	L1-L5 transverse processes and lumbodorsal fascia	Ribs (3-12) and T1-T12 transverse processes	Extension and lateral flexion of cervical spine; rib depression	C1-L5	Intercostals and lumbar
Quadratus lumborum	Posterior aspect of iliac crest	Lowest edge of 12th rib and L1-L4 transverse processes	Lateral flexion of lumbar spine, depression of 12th rib, and unilaterally raises pelvis	L1-L4	Lumbar and lower intercostal
Multifidus	C4-C7 articular processes, T1-T12 and L1-L5 transverse processes, posterior superior iliac spine, and posterior sacrum	Spinous processes of the vertebra (2-4) above the origin	Extension and contralateral rotation of spine	C1-S5	Lumbar

PART IV

Lower Extremity

CHAPTER 11

The Hip and Thigh

Any discussion of the hip must include the bones of the pelvis and thigh and the muscles of the thigh, as well as the ligaments and muscles of the hip. Several muscles that cross the hip joint also cross the knee joint, and discussion of these muscles in this chapter is restricted to their function at the hip joint. Their function at the knee joint is discussed in the next chapter.

Bones of the Hip Joint and Thigh

The hip joint is the articulation between the pelvis and the femur, or thigh bone. The pelvis is discussed in chapter 8. The hip joint is the body's most weight-bearing joint. It allows us to stand and walk in an upright position. The important structure of the pelvis regarding the hip joint is the anatomical area labeled the acetabulum (figure 11.1; also see figure 8.32). Remember that the acetabulum is formed by the articulation of the three bones making up the innominate bone (the ilium, the ischium, and the pubic bone). The hip joint is classified as a ball-and-socket, or triaxial, joint. The acetabulum is considered the socket of the joint, and the ball of the hip joint is the structure known as the *head of the femur* (figure 11.2). Although the hip joint and the shoulder joint are often compared because of their similarities as ball-and-socket, or triaxial, joints, the hip joint is a much more stable joint than the shoulder joint because of the depth of the acetabulum compared with the very shallow glenoid fossa of the scapula. It is through the hip joint that the entire weight of the trunk and upper extremities is transferred to the lower extremities.

Dislocations, fractures, and other damage in and around the hip that result from falls, excessive motion beyond normal ranges of motion, and direct trauma are common. In recent years, we have seen advanced medical techniques that include complete hip joint replacement. Prior to surgical replacement, depending on the cause of a hip joint condition, initial attempts might

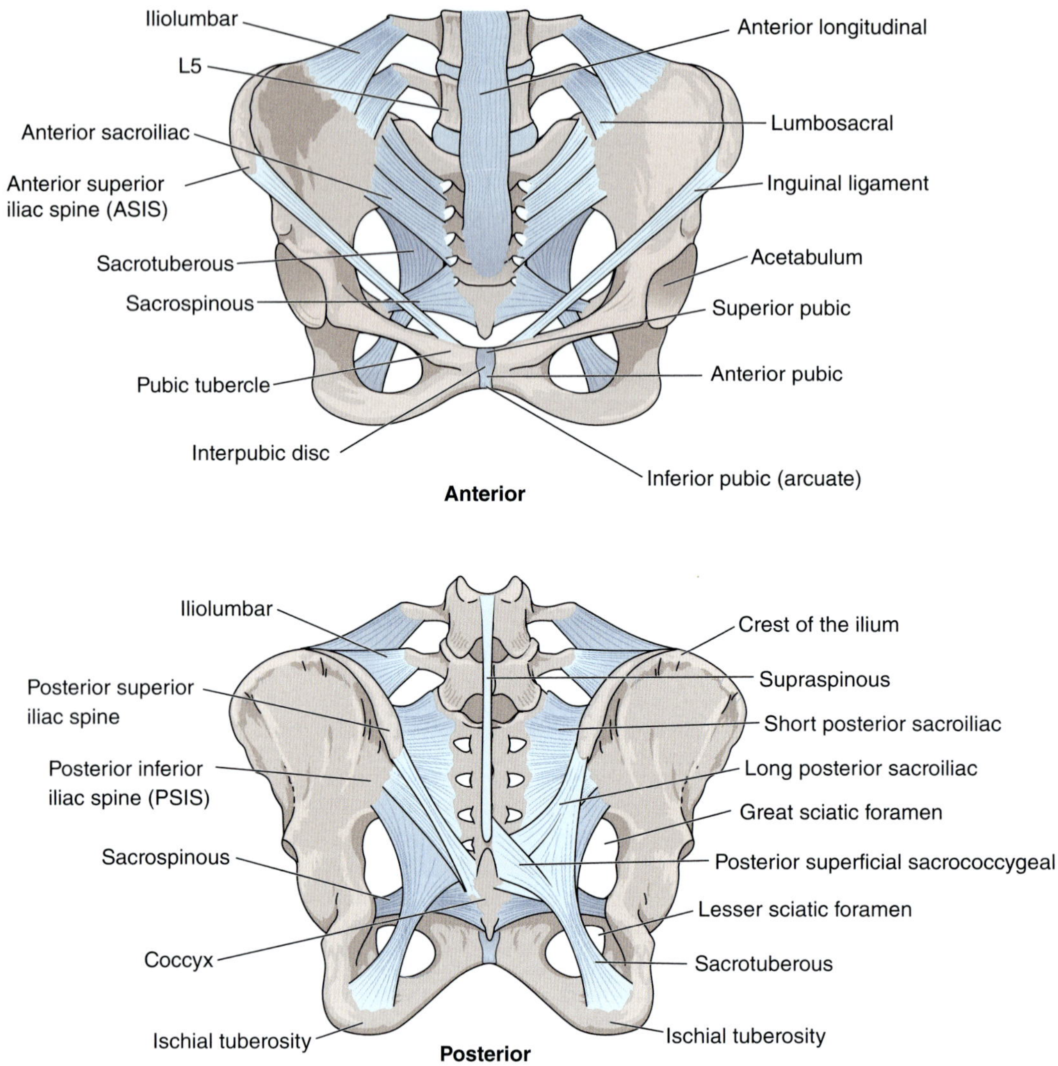

Figure 11.1 The pelvis.

include rest, physical therapy, medication, and the use of properly fitted crutches or a cane. As a last resort, if total hip joint replacement is chosen, extreme care must be taken by everyone involved to guard against infection. Weeks or even months of physical therapy will follow to lead to full restoration of the joint.

The femur, the longest and largest bone in the body, is discussed in this chapter only as it pertains to the hip joint (at the proximal end). Further discussion of the distal end of the femur is presented in chapter 12 on the knee.

The ball, or head, of the femur is separated from the shaft of the bone by the neck of the femur. At the proximal end of the shaft of the femur, just distal to the neck, are two large projections of bone known as the *greater and lesser trochanters*. These structures serve as points of insertion for the major muscles responsible for hip joint actions.

Hands On

Place your hand over your hip joint. Rotate one leg toward the other leg (internally rotate your hip joint), and feel the greater trochanter as it moves against your skin (figure 11.3).

The lesser trochanter is distal, medial, and somewhat posterior to the greater trochanter and more readily observed in a posterior view of the femur. The ridge of bone running between the lesser and greater trochanters is known as

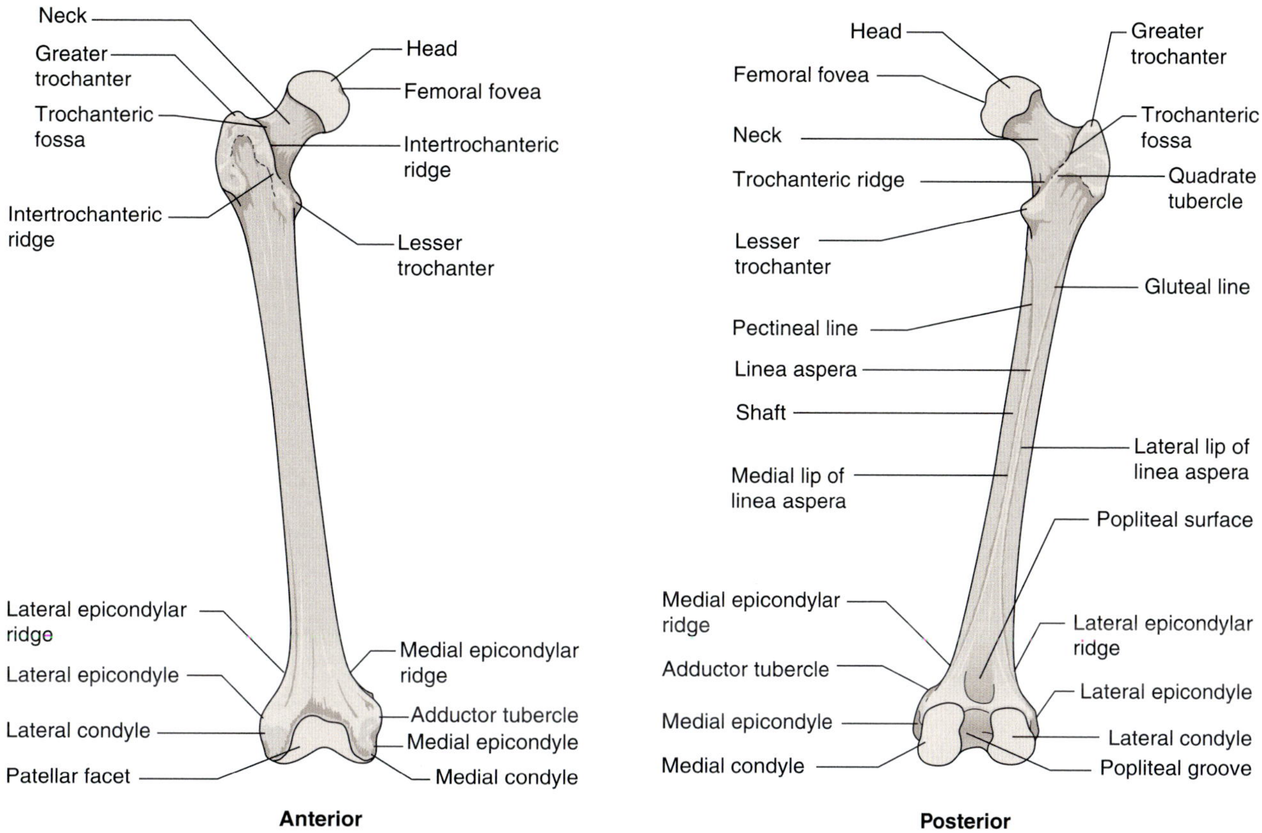

Figure 11.2 The femur.

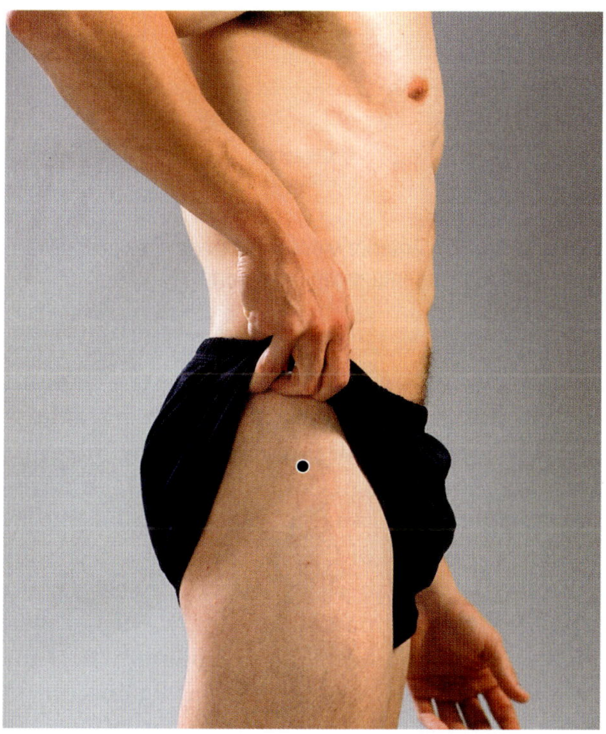

Figure 11.3 Locating the greater trochanter of the femur.

the *intertrochanteric ridge* (see figure 11.2). A large ridge or crest of bone running down the posterior aspect of the shaft of the femur is known as the *linea aspera*, Latin for "rough line" (see figure 11.2). The linea aspera serves as a source of muscular attachment for numerous muscles of both the hip and knee.

Ligaments of the Hip Joint

Although the hip joint (figure 11.4) derives most of its stability from the bony formation between the head of the femur and the acetabulum of the pelvis, it is also reinforced by several strong ligaments. There are seven ligaments of the hip joint. As with all synovial joints, there is a capsular ligament running from the edge of the acetabulum of the pelvis to the neck of the femur. The acetabular labrum, like the shoulder joint's glenoid labrum, is a fibrocartilaginous rim on the outer edge of the acetabulum that helps deepen its socket. Three ligaments that strengthen the capsular ligament are easily identified because

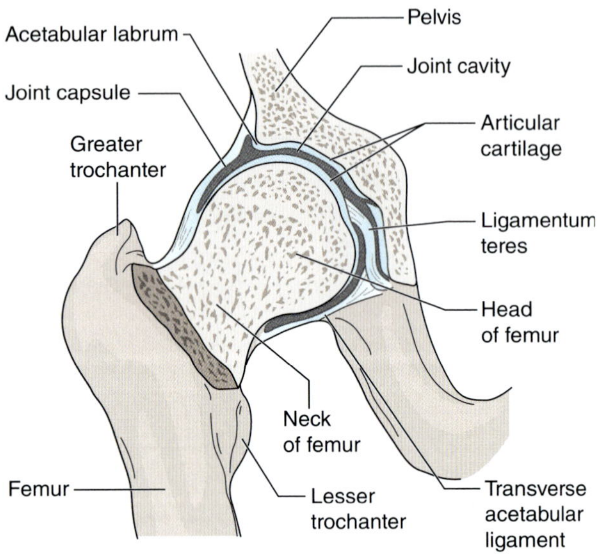

Figure 11.4 Longitudinal section of the hip joint.

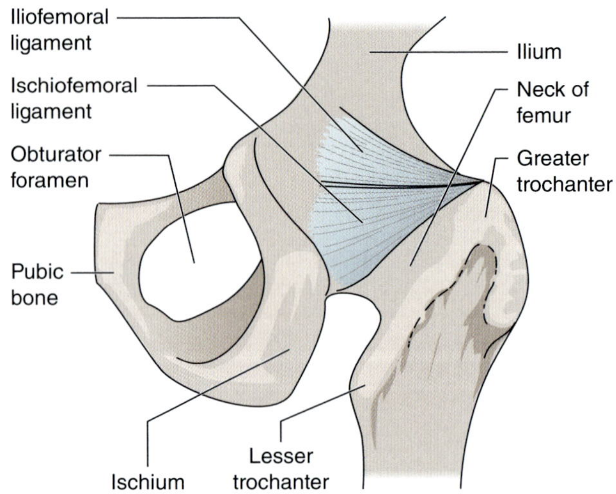

Figure 11.6 Posterior view of the ischiofemoral ligament.

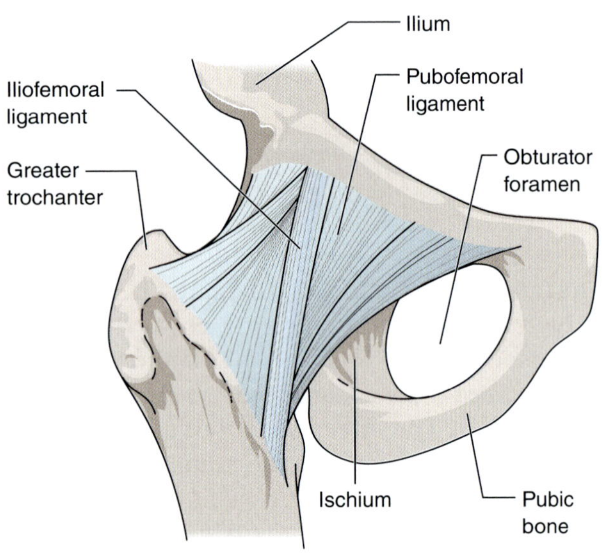

Figure 11.5 Anterior view of the iliofemoral (or Y) and pubofemoral ligaments.

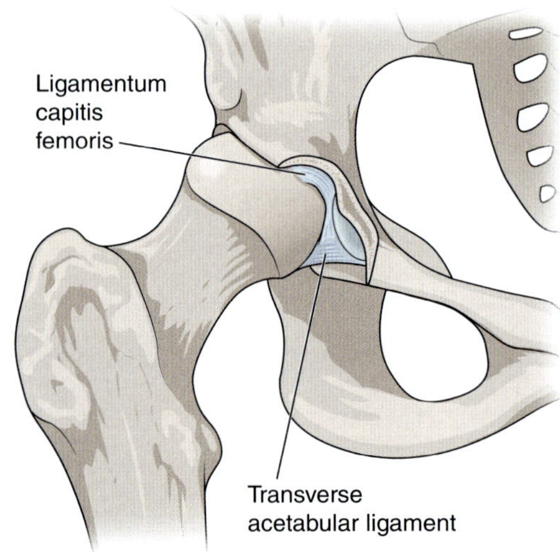

Figure 11.7 Anterior view of the transverse acetabular ligament and ligamentum capitis femoris.

their names represent the bones they tie together: the iliofemoral, the ischiofemoral, and the pubofemoral ligaments (figures 11.5 and 11.6). The iliofemoral ligament looks like an upside-down Y and, therefore, is frequently referred to as the *Y ligament*. Because of the anatomical relationships between the ilium and the femur and between the pubic bone and the femur, the iliofemoral and pubofemoral ligaments cross the anterior aspect of the hip joint and attach to the neck of the femur. Likewise, because the ischium is posterior to the femur, the ischiofemoral ligament crosses the posterior aspect of the hip joint and attaches to the neck of the femur. The transverse acetabular ligament spans a small notch in the lower portion of the acetabulum and creates a foramen through which blood vessels to and from the hip joint pass (figure 11.7). The last ligament, the ligamentum teres (see figure 11.7), runs between the transverse acetabular ligament and a hole in the superior edge of the femoral head known as the *femoral fovea* (see figure 11.2). This ligament attaches the head of the femur to the acetabulum and plays the least significant role of any of the ligaments of the hip for joint stability; however, the ligamentum teres also contains an artery that provides a small amount of blood supply to the head of the femur.

Fundamental Movements of the Hip Joint

The hip joint is classified as a triaxial joint, having movement in all three planes about all three axes: flexion and extension in the sagittal plane about a frontal horizontal axis, abduction and adduction in the frontal plane about a sagittal horizontal axis, and internal and external rotation in the horizontal plane about a vertical axis. The musculature that creates those six fundamental movements includes flexors, extensors, abductors, adductors, and rotators. They are presented in groups based on their position: anterior, posterior, lateral, or medial to the hip joint.

Hands On

As you stand in the anatomical position, flex your hip. What group of muscles primarily flexed your hip (anterior, posterior, lateral, or medial)? Extend your hip joint to return to the anatomical position. Again, what muscle group (anterior, posterior, lateral, or medial) was primarily involved in extending your hip? Now, abduct and then adduct your hip joint. Determine which muscle groups are primarily responsible for these actions of the hip joint. To determine which muscle groups are primarily responsible for internal and external rotation of the hip joint, lie on your back, and internally and externally rotate your hip joint while palpating the muscles surrounding your hip joint. Now that you have completed the hip exercises, we can look at the muscles that cross the hip joint.

Muscles of the Hip Joint and Upper Leg (Thigh)

Muscles that cross the hip joint are grouped by position as anterior, posterior, medial, or lateral to the joint.

Anterior Muscles

From the previous exercise, you know that the muscles that perform flexion of your hip are located anterior to the hip joint. Interestingly enough, some of these muscles also cross the knee joint. In this chapter, however, only their function at the hip joint is considered. The anterior muscles are listed here (figure 11.8).

- *Iliopsoas:* The iliopsoas muscle is a combination of the iliacus and psoas major muscles. These muscles, along with the psoas minor, are commonly referred to as the *true groin muscles* (or hip flexor muscle group). The term *groin* refers to the anterior aspect of the hip joint. Their main function is to flex the hip joint, but if the femur is stabilized, these muscles will also cause flexion of the lumbar spine when they contract. An example of this would be the classic sit-up exercise with the legs held down by a partner or a strap around a table so that the femur is stabilized. The medial muscles of the hip joint (the adductors)

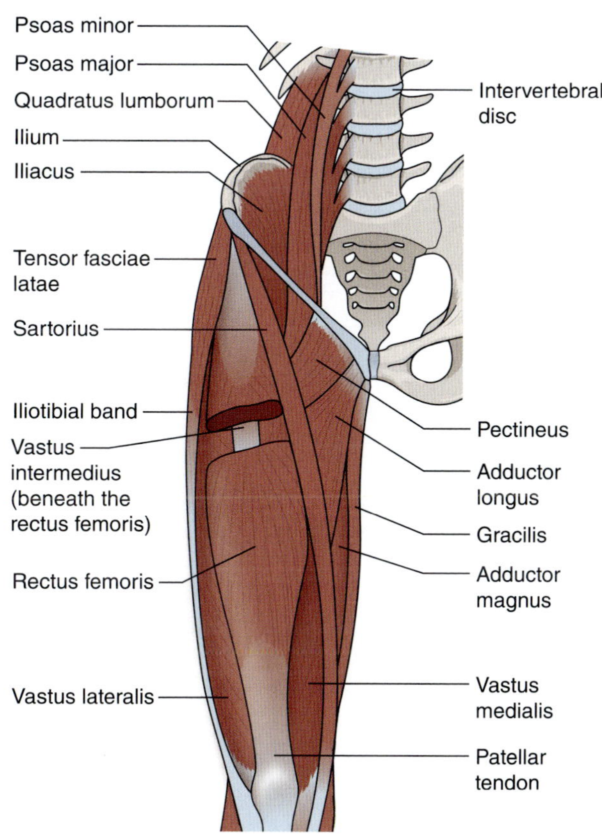

Figure 11.8 Anterior muscles of the hip.

are often referred to as the *common groin muscles* (or hip adductor muscle group) and are discussed later in this chapter.

- *Psoas major:* This muscle originates on the transverse processes of all five lumbar vertebrae and the bodies and intervertebral discs of the 12th thoracic vertebra and all five lumbar vertebrae. The psoas major inserts on the lesser trochanter of the femur. Along with the iliacus muscle, this muscle flexes the hip joint, assists with adduction and external rotation of the hip joint, and assists with flexion and rotation of the lumbar spine (figure 11.9).

- *Iliacus:* This muscle originates from the iliac fossa of the pelvis to the tendon of the psoas major and inserts just distal to the lesser trochanter of the femur. Along with the psoas major muscle, this muscle flexes the hip joint; assists with adduction and external rotation of the hip joint; and, when the femur is stabilized, assists with flexion of the lumbar spine (see figure 11.9).

- *Psoas minor:* This muscle is not present on one or both sides in more than 50% of the population. The muscle originates from the 12th thoracic and 1st lumbar vertebrae and the intervertebral disc between these two vertebrae and inserts on the pubic bone. Therefore, the psoas minor is not considered a muscle of the hip joint. Its function is to assist the psoas major with flexion of the lumbar spine (see figure 11.9).

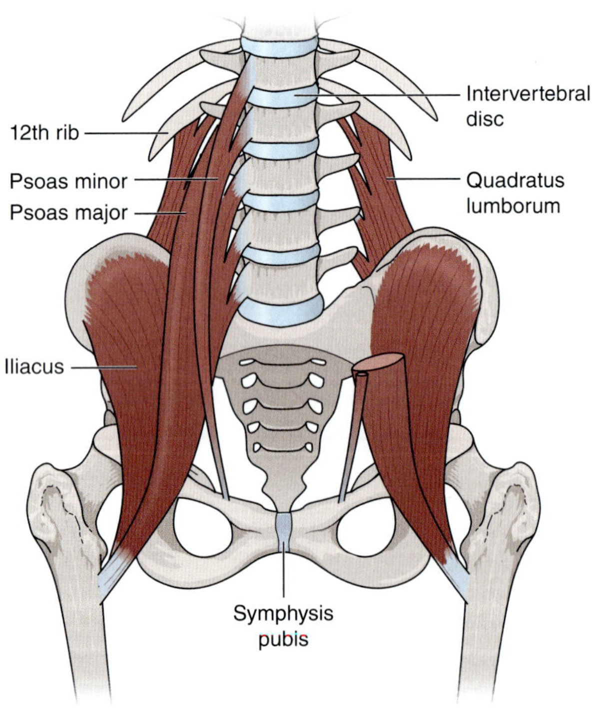

Figure 11.9 The psoas major, psoas minor, and iliacus.

- *Sartorius:* This muscle, the longest muscle of the body, originates from the anterior superior spine of the ilium and inserts on the medial aspect of the proximal end of the tibia, a bone of the lower leg (see chapter 12, figure 12.5), just inferior to the medial condyle (see figure 11.8). Note that this muscle crosses both the hip and knee joints. The functions of the sartorius at the hip joint are flexion, abduction, and external rotation.

Hands On

Sitting on the floor with your ankles crossed and knees apart (the tailor's position), you should be able to palpate your sartorius ("tailor's muscle") just beneath the anterior superior iliac spine (figure 11.10). Both of the terms *sartorius* (from a Latin word for *tailor*) and *tailor's position* refer to a tailor's practice in bygone days of sitting for hours to do his work in a cross-legged position, which often caused soreness in this muscle after a long day of sewing. Put yourself in this position on the floor or on a table, and think about the positions of the origin and the insertion of the sartorius muscle. Why was the tailor experiencing soreness in this muscle after a long day of sewing?

- *Rectus femoris:* This muscle is part of a powerful group of four muscles known as the *quadriceps femoris*. Three of these muscles, the vastus lateralis, the vastus intermedius, and the vastus medialis, cross only the knee joint, but the rectus femoris crosses both the knee and hip joints. The muscle originates on the anterior inferior iliac spine and inserts on the tibial tuberosity. The rectus femoris flexes the hip joint and acts as a very weak abductor because of its angle of pull.

Hands On

Lie on a table with your hips in extension and knees flexed over the end of the table, and extend your

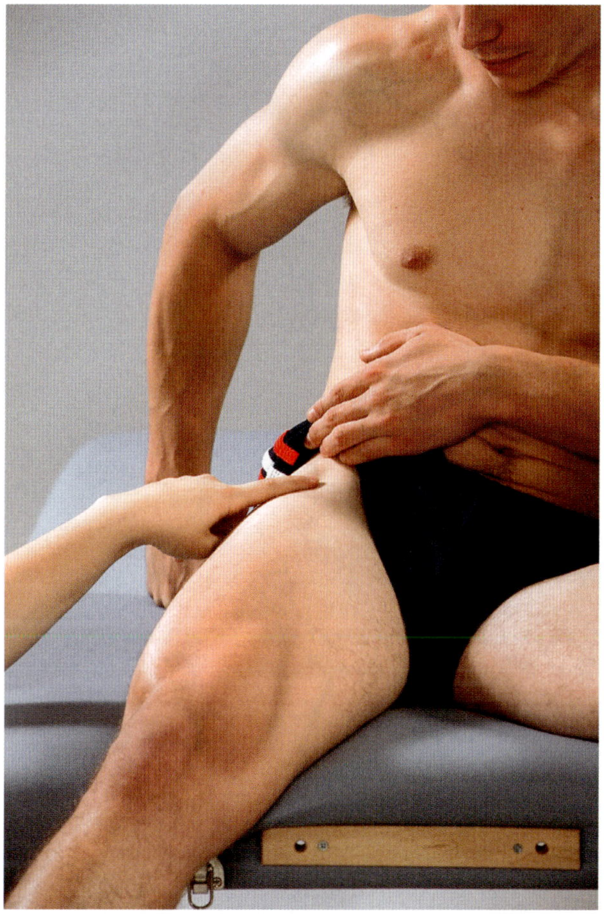

Figure 11.10 Locating the sartorius.

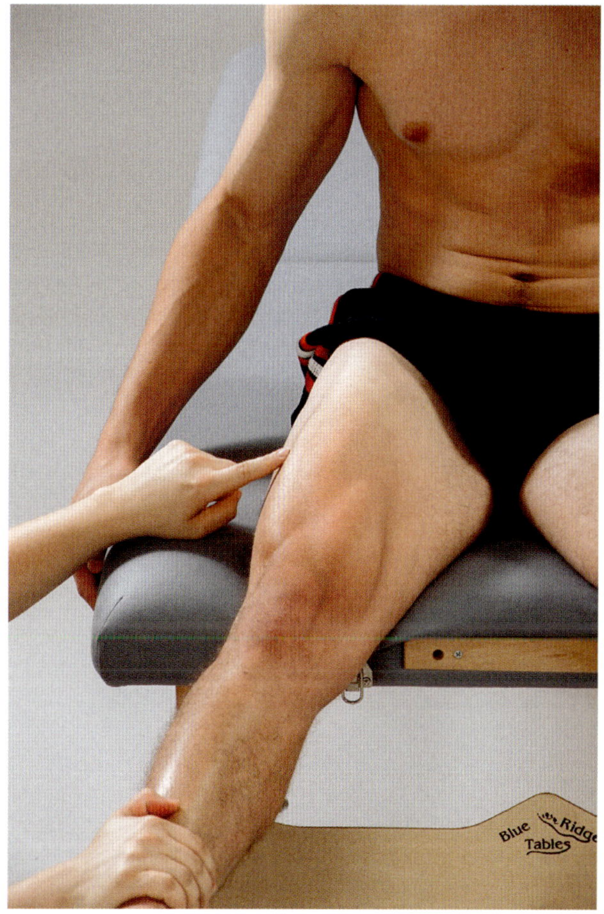

Figure 11.11 Locating the rectus femoris.

knee against a partner's manual resistance. Note the prominence of muscular tissue running the length of the anterior thigh area (figure 11.11). This prominence is the rectus femoris contracting against your resistance.

- *Tensor fasciae latae:* This muscle originates on the external rim of the crest of the ilium and combines with the gluteus maximus muscle to form another structure, the iliotibial band or tract, which inserts in the area of the lateral condyle of the tibia (figures 11.8, 11.12, and 12.5). The tensor fasciae latae flexes and abducts the hip joint. Even though the term latae might lead one to believe this is a lateral muscle of the hip joint, it is actually considered an anterior hip joint muscle, although a very lateral anterior muscle. More about this muscle and its involvement with the iliotibial band is presented later in this chapter and in chapter 12 on the knee joint.
- *Pectineus:* This muscle originates between the anterior superior aspect of the pubic bone and inserts on the femur just distal to the lesser trochanter (see figure 11.8). Besides crossing the hip anteriorly, making it a hip flexor, it is a strong adductor and external rotator of the hip because of its angle of pull.

Posterior Muscles

Ten posterior muscles cross the hip joint (figure 11.13), including a group of three (the hamstrings) and a group commonly referred to as the *six deep external rotators*. The hamstrings are discussed only as they function at the hip joint. These powerful and often strained muscles

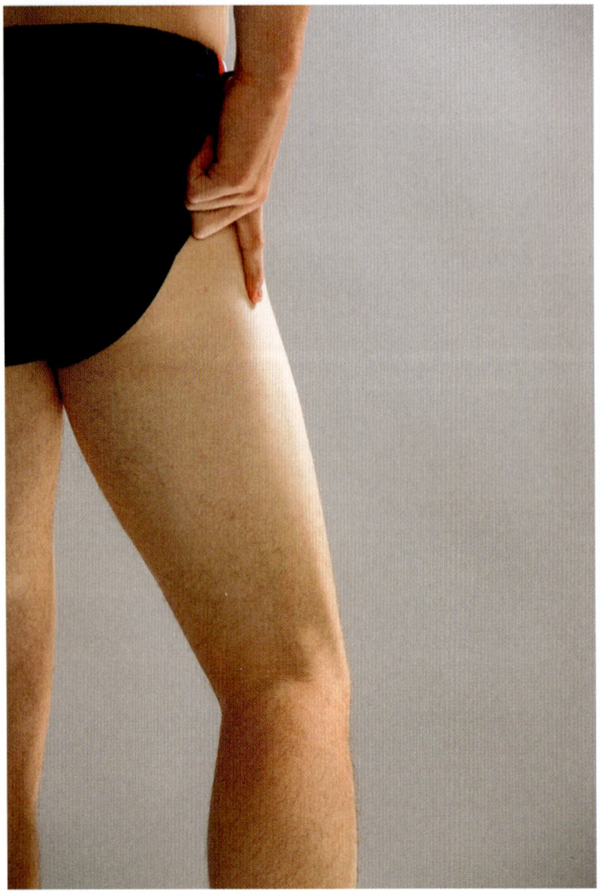

Figure 11.12 Demonstrating the tensor fasciae latae and iliotibial band.

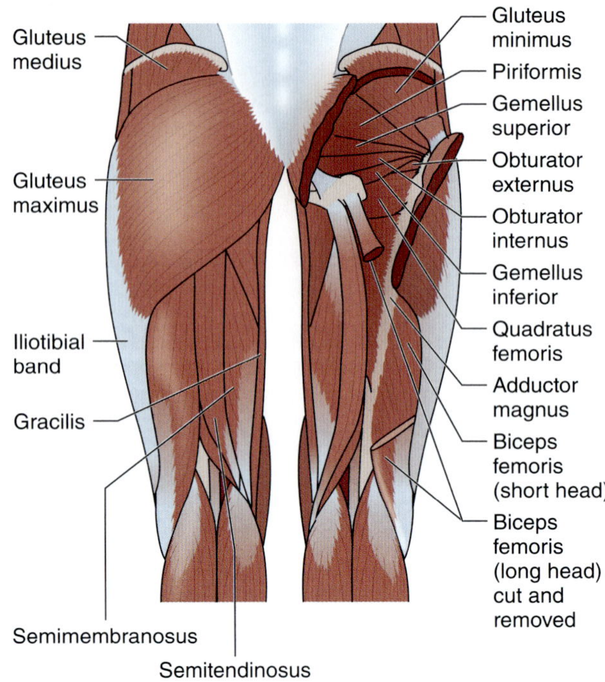

Figure 11.13 Posterior muscles of the hip.

also cross the knee joint and are discussed again in chapter 12.

• *Biceps femoris:* Although all three hamstring muscles (biceps femoris, semitendinosus, and semimembranosus) originate on the ischial tuberosity and have the same functions—extension of the hip joint and flexion of the knee joint—the biceps femoris is unique in that it has two heads: The long head originates on the ischial tuberosity with the semitendinosus and semimembranosus, and the short head originates from the linea aspera of the femur and therefore does not cross the hip joint. Both heads combine into the belly of the muscle, which then inserts on the head of the fibula, a bone of the lower leg, and the lateral condyle of the tibia (see figure 12.5). The biceps femoris (long head) functions specifically at the hip joint to extend the joint and assist in adduction and external rotation of the joint.

Hands On

The tendon of insertion of the biceps femoris can be observed on the posterior lateral aspect of the knee joint when you lie in the position illustrated in figure 11.14.

• *Semitendinosus:* Another muscle of the hamstring group, the semitendinosus, also originates on the ischial tuberosity and inserts on the medial aspect of the tibia just below the medial condyle (see figure 12.5).

Hands On

The tendon of insertion of the semitendinosus has a very prominent, cord-like structure (figure 11.15). Have your partner lie in a prone (facedown) position and flex their knee joint, and then locate this tendon on the posterior medial aspect of the knee joint. The muscle extends and assists with internal rotation and adduction of the hip joint.

• *Semimembranosus:* The final muscle of the hamstring group is the semimembranosus, which, like the semitendinosus, originates on the ischial tuberosity and inserts on the medial

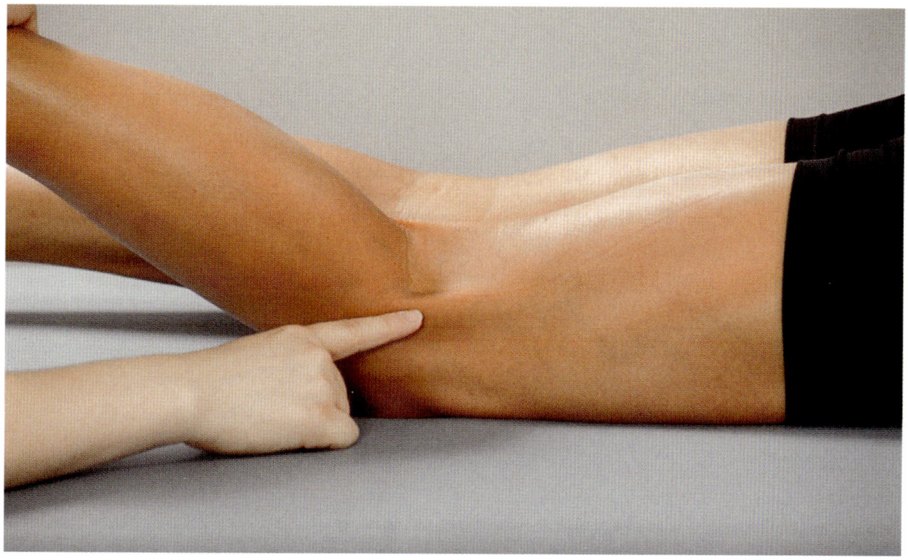

Figure 11.14 Locating the biceps femoris.

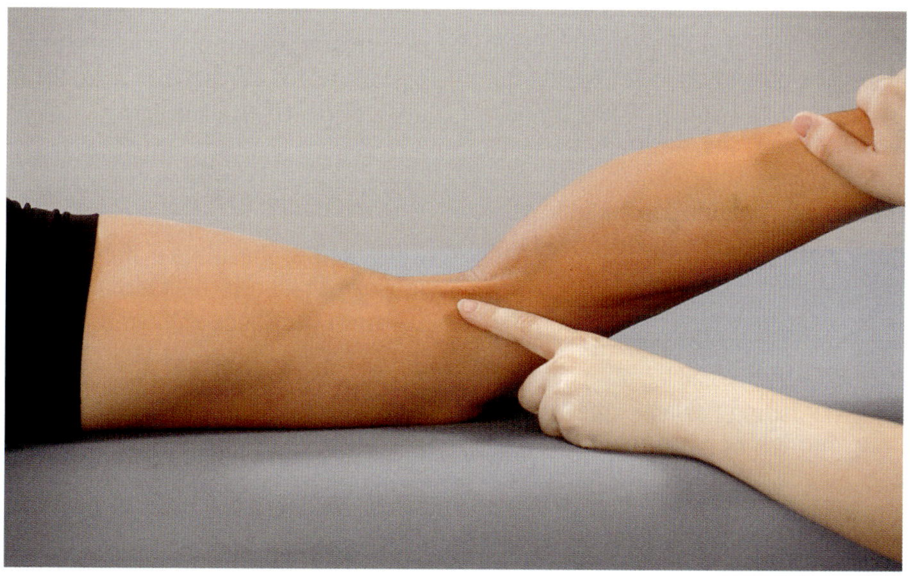

Figure 11.15 Locating the semitendinosus.

condyle of the tibia (see figure 12.5). This muscle extends the hip joint and assists with internal rotation and adduction of the hip joint.

Hands On

The semimembranosus tendon of insertion is a very broad structure just medial to the semitendinosus tendon of insertion (figure 11.16). Ask your partner to assume the same position as when you located the semitendinosus, and then locate the semimembranosus.

Hands On

A simple test for hamstring flexibility can easily be performed on yourself. Sit on the end of a table with your legs (from the knee to the foot) hanging over the edge of the table. Sit relaxed and extend one knee,

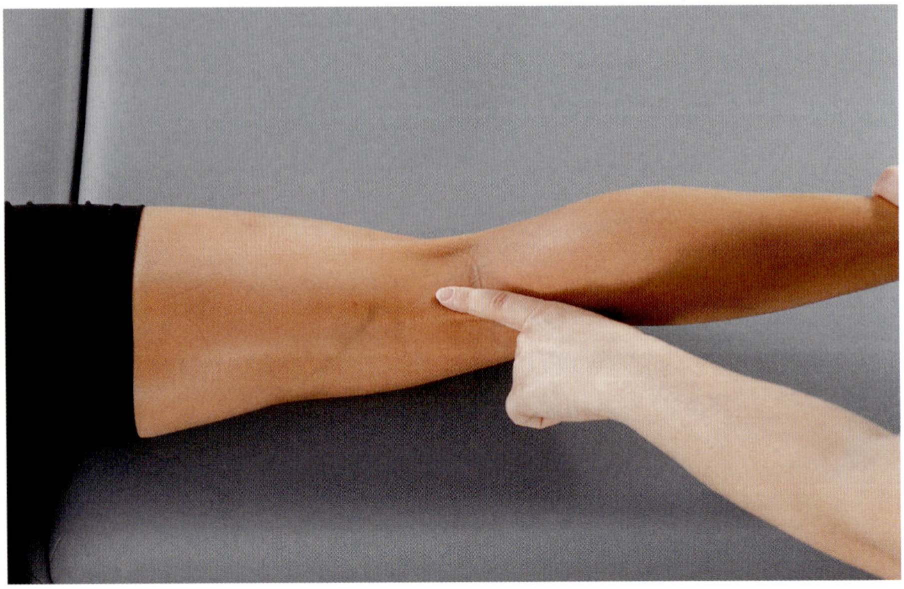

Figure 11.16 Locating the semimembranosus.

and then the other. Were you able to fully extend each knee? Most people will be able to do this. Now, sit up straight and arch your back. Did you have any trouble extending your knees? If so, the flexibility of your hamstring muscles is questionable. Why? When you sat up straight, as opposed to sitting comfortably, you rotated your pelvis. When you did that, you moved the origins of the hamstring muscles farther away from their insertions. Tilting the pelvis and extending the knee puts the hamstrings under full stress and requires excellent flexibility. Picture a hurdler as the lead leg goes over the hurdle. Without good hamstring flexibility, the hurdler will either knock over hurdles or have to jump higher to clear the hurdles. Either way, it will take more time to get from start to finish.

- *Gluteus maximus:* This rather large muscle is often identified as the buttocks muscle. It originates from the posterior aspects of the ilium, sacrum, and coccyx bones and inserts by blending with the tensor fasciae latae muscle to form the iliotibial band (see figure 11.13). The muscle extends and externally rotates the hip joint. Because one portion of the belly of the muscle crosses above the hip joint and another portion below the hip joint, the muscle is involved in both abduction and adduction of the hip. Look carefully at the hip joint in figure 11.13 and determine which portion of the gluteus maximus assists with abduction and which assists with adduction of the hip joint.

Hands On

Deep below the gluteal muscles is a group of six muscles that perform external rotation of the hip joint. Lying on your back with your legs straight, have your partner stabilize one of your feet in the anatomically neutral position. Attempt to move your foot away from the other foot by externally rotating your entire lower extremity. Apply hand pressure to the belly of the gluteus maximus muscle of the rotating leg. Did you feel the muscles contract deep within the joint? What muscles were contracting? Let us look for these answers in the discussion of the six deep external rotator muscles of the hip joint (see figure 11.13), which are presented anatomically from superior to inferior.

- *Piriformis:* This muscle originates on the upper portion of the sacrum, crosses the greater sciatic notch, and inserts on the upper surface of the greater trochanter. If contracted in spasm (sudden, involuntary contraction) as the result of strain or physical trauma, this muscle can impinge on the sciatic nerve as it passes through the sciatic notch of the pelvis, causing pain to

FOCUS ON

Hamstring Strain

The term *hamstring strain* is often used and misunderstood. There are three muscles (biceps femoris, semitendinosus, and semimembranosus) that could be strained as a result of hip or knee hyperextension or, more likely, a combination of both hip flexion and knee extension. Although determining which of the three muscles (or combination of muscles) is strained is of little importance to the general public, someone with a good knowledge of human anatomy should be able to understand that *hamstring strain* is a somewhat nonspecific term. Flexibility is a major factor in preventing hamstring strains because many athletic activities require the hip joint to flex at the same time the knee joint is extending, which puts this muscle group under tension. Note in the photo the hurdler's lead leg as it goes over a hurdle. Observe the hip position and the same leg knee position. Flexion of the hip and extension of the knee are required, and the flexibility of the hamstring muscles is essential for a successful hurdler.

radiate down the lower extremity along the course of the sciatic nerve and its branches.

- *Gemellus superior:* Originating from the spine of the ischium and inserting on the medial aspect of the greater trochanter, the gemellus superior externally rotates the hip joint.
- *Internal obturator (obturator internus):* Originating from the inner surface of the obturator foramen of the pelvis and inserting on the medial aspect of the greater trochanter, the internal obturator externally rotates the hip joint.
- *Gemellus inferior:* Originating from the ischial tuberosity and inserting on the medial aspect of the greater trochanter, the gemellus inferior externally rotates the hip joint.
- *External obturator (obturator externus):* Originating from the outer surface of the pubic bone and ischium around the obturator foramen and inserting on the trochanteric fossa of the femur, the external obturator externally rotates the hip joint.
- *Quadratus femoris:* The quadratus femoris originates on the ischial tuberosity, inserts on the intertrochanteric ridge of the femur, and externally rotates the hip joint.

Medial Muscles

Three of the four adductor muscles of the hip joint give their function away by their names. The fourth adductor, the gracilis, is the only one of the four that also crosses the knee joint. These four muscles (figure 11.17), as mentioned earlier,

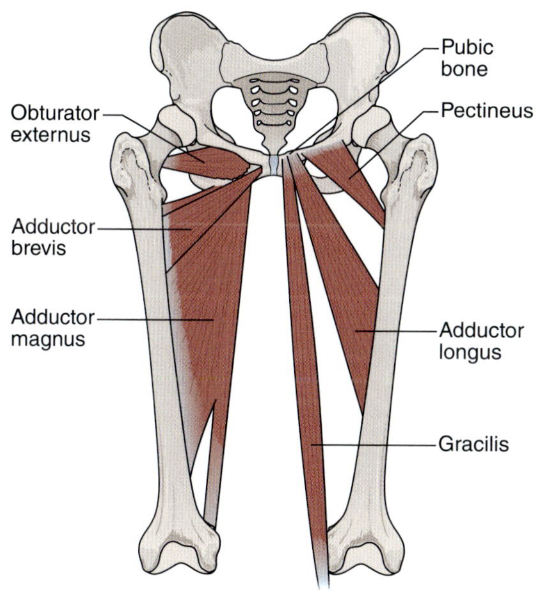

Figure 11.17 Adductors of the hip.

POP QUIZ

Observe the hip and thigh positions for the gymnast. Which muscles are being stretched in the front leg? Which muscles are being stretched in the back leg?

are known as the *common groin muscles* (or hip adductor muscle group). Compared with the true groin muscles (the iliopsoas muscle group), these groin muscles are more commonly affected when one experiences a groin strain.

- *Adductor longus:* This muscle originates from the body of the pubic bone and inserts on the middle third of the linea aspera of the femur. Its major function is adduction of the hip, and it assists with flexion and external rotation.
- *Adductor brevis:* This muscle originates from the body and inferior surface of the pubic bone and inserts on the proximal third of the linea aspera of the femur. Its major function is adduction of the hip, and it assists with external rotation and flexion.
- *Adductor magnus:* This muscle (see figures 11.17 and 11.18) originates on the inferior surface of the pubic bone, the ischium, and the ischial tuberosity and inserts on the entire length of the linea aspera and the adductor tubercle of the femur. The adductor magnus has two distinct portions: the anterior portion, which adducts, flexes, and externally rotates the hip joint, and the posterior portion, which adducts, extends, and internally rotates the hip joint.
- *Gracilis:* This muscle originates on the lower portions of the anterior aspect of the pubic symphysis and the pubic bone and inserts on the tibia just distal to the medial condyle. At the hip, the gracilis adducts and flexes the joint.

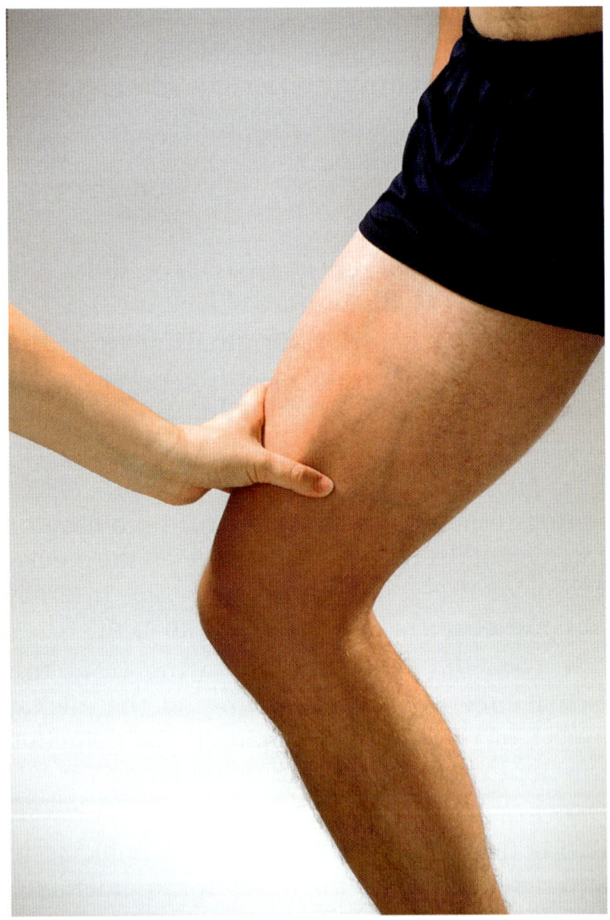

Figure 11.18 Locating the adductor magnus.

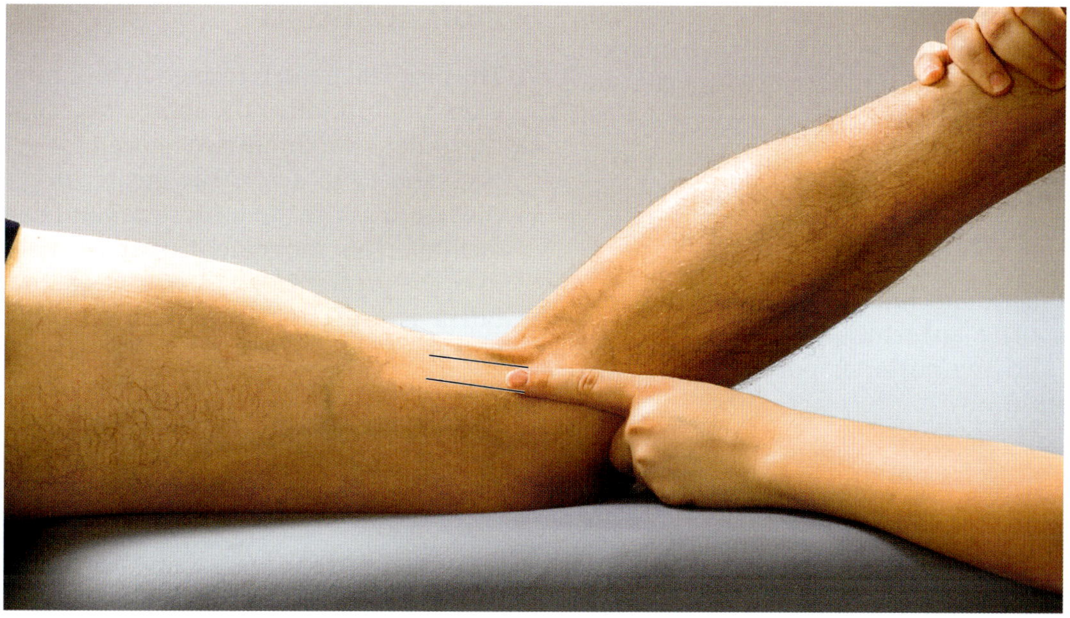

Figure 11.19 Locating the gracilis.

Hands On

Sitting in the lotus, or butterfly, position (on the floor, with the soles of your feet together, your knees flexed, and your hips externally rotated), you can easily palpate your adductor longus and gracilis muscles in the groin area. The gracilis is the more lateral of the two tendons. Additionally, you can palpate the gracilis by pressing with your finger on the area just medial to the semimembranosus tendon of insertion on the medial aspect of the knee joint (figure 11.19).

Lateral Muscles

There are only two anatomically lateral muscles of the hip joint, not including the tensor fasciae latae, which was identified as an anterior, not lateral, muscle of the hip joint. Even though the gluteus maximus shares a name with the lateral muscles of the hip joint, it is a posterior muscle. The lateral muscles of the hip joint are the gluteus medius and the gluteus minimus (see figure 11.13).

- *Gluteus medius:* The gluteus medius originates from the middle of the external aspect of the ilium to the iliac crest and inserts on the posterior lateral aspect of the greater trochanter (see figures 11.13 and 11.20). The major function

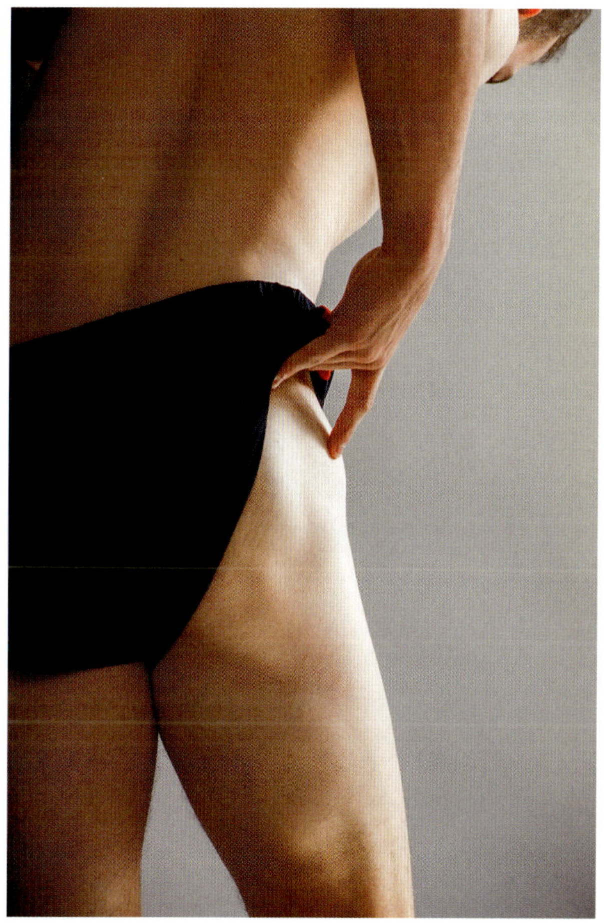

Figure 11.20 Locating the gluteus medius.

of this muscle is hip abduction. The anterior portion assists with flexion and internal rotation, and the posterior portion assists with extension and external rotation.

- *Gluteus minimus:* Originating in the same area on the ilium as the gluteus medius, the gluteus minimus inserts on the anterior surface of the greater trochanter. The major functions of this muscle are hip abduction and internal rotation. The anterior portion assists with flexion, and the posterior portion assists with extension.

Iliotibial Band

The iliotibial band is formed by two muscles that cross the hip joint: the tensor fasciae latae and the gluteus maximus (see figure 11.13). It passes back and forth across the greater trochanter of the femur as the hip joint flexes and extends. To facilitate this movement of the iliotibial band over the greater trochanter, a large trochanteric bursa is located between the iliotibial band and the greater trochanter. Excessive flexion or extension or direct trauma to this area can cause inflammation of the bursa (trochanteric bursitis).

FOCUS ON

Groin Strain

Often, we hear about athletes suffering from a groin strain. What muscle group is affected in this injury? This chapter presents two groups of groin muscles: the iliopsoas muscle group (iliacus and psoas major muscles), often referred to as the *true groin muscles*, and the adductor muscle group (adductor longus, adductor brevis, adductor magnus, and gracilis), referred to as the *common groin muscles*. The true groin muscles (the iliopsoas group) are typically strained as a result of hyperextension of the hip joint. Injury to the common groin muscles (the adductor group) is usually the result of overstretching these muscles; the adductor group is more commonly strained in athletic endeavors, hence the term *common groin strain*.

Additionally, the term *sports hernia* is used to describe a groin strain. The most common form of sports hernia is defined as a tearing away of the oblique muscles' tendons from their attachment to the pubic bone. Another, but lesser form, can be described as a tearing away from the pubic bone of the medial muscles of the thigh. Think about someone with a leg planted on a playing surface, making a twisting motion with the pelvis rotating. This is a common description for what is known as a *sports hernia*. Physicians correctly use the term *athletic pubalgia*.

Visit the web resource to complete the activities for this chapter.

CHAPTER 12

The Knee

The knee joint, one of the largest joints in the body, is a uniaxial synovial joint and is often referred to as a *hinge joint*. In reality, the knee joint is *not* a true hinge joint but rather a modified hinge joint. A true hinge joint, like a door hinge, opens and closes about a single constant axis. In the knee joint, the tibia (distal bone) glides around the distal end of the femur (proximal bone), and, although the movement remains in one plane (the sagittal plane), it occurs about an ever-changing axis. With each degree of movement in the sagittal plane, the frontal horizontal axis changes. Hence, the term *modified hinge joint* is more appropriate than just *hinge joint* for the knee joint. Although the knee joint appears to be structurally sound, it was not built to withstand many of the stresses placed on it by athletic activities. As we examine the normal anatomy of the joint, we also draw attention to the common results of abnormal stresses.

Bones of the Knee

The femur (figure 12.1) is presented in chapter 11 on the hip joint, but its role at the knee is discussed in this chapter. At the distal end of the shaft of the femur, the bone flares out, forming medial and lateral epicondylar ridges, similar to those in the humerus just proximal to the elbow joint. Just distal to these ridges are the medial and lateral epicondyles (figure 12.2). Two other large prominences of bone, easily palpated just proximal to the joint line (an imaginary line between the femur and the tibia), are the medial and lateral condyles (figure 12.3). These two condyles articulate with the tibia to transfer the body weight from the femur to the lower leg. The medial condyle is slightly more distal than the lateral condyle. This information will be important when movements of the knee joint are discussed.

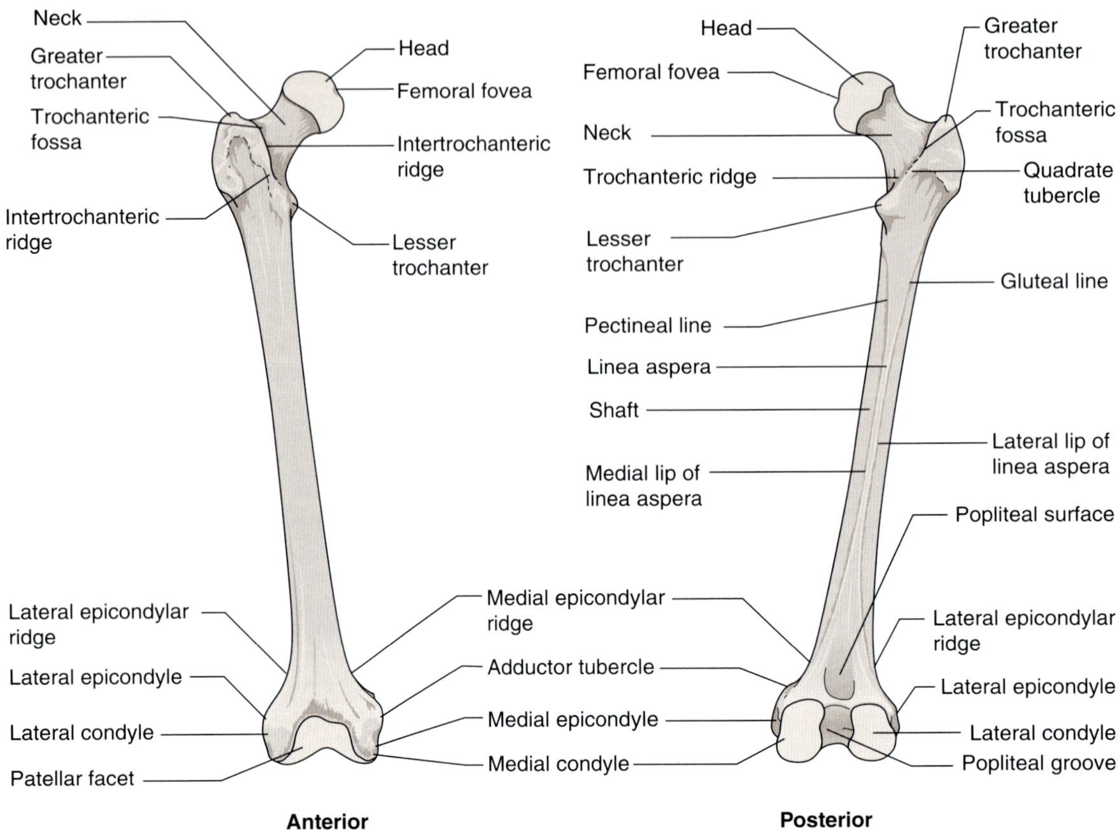

Figure 12.1 The femur.

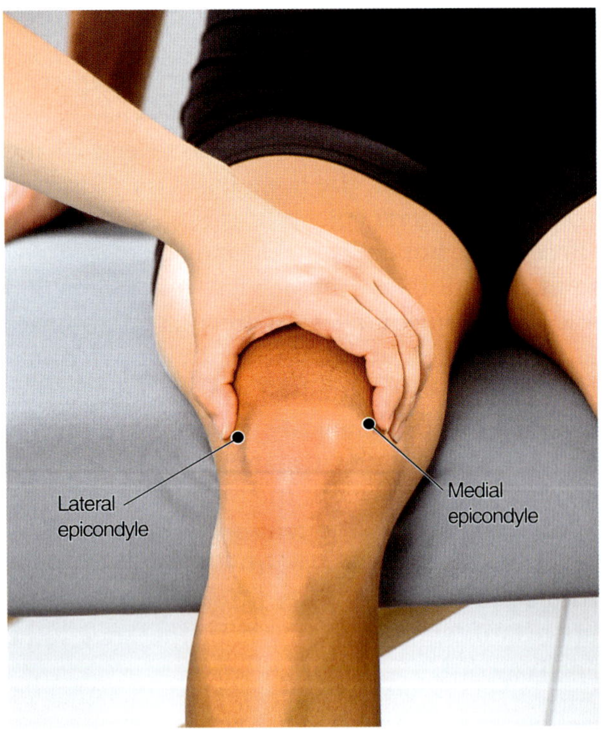

Figure 12.2 Locating the medial and lateral femoral epicondyles.

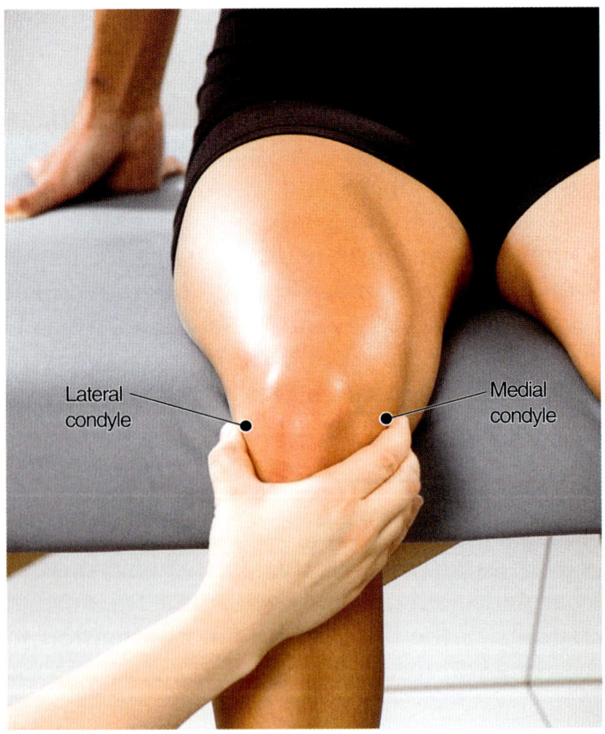

Figure 12.3 Locating the medial and lateral femoral condyles.

Hands On

After locating the lateral and medial femoral condyles (see figure 12.3), move your fingers distal to (below) the condyles. You should feel a space between the distal ends of the femur and the proximal ends of the tibia. This space is commonly known as the *joint line*. Under your fingers on the lateral and medial joint lines lie the anterior horns of the lateral and medial menisci.

Just proximal to the medial epicondyle is a small prominence known as the *adductor tubercle* (figure 12.4). On the anterior surface of the distal end of the femur, between the lateral and medial condyles, is a smooth surface covered with articular cartilage known as the *patellar facet*. This surface articulates with the posterior aspect of the patella (kneecap, discussed later), forming the patellofemoral joint. On the posterior side of the distal end of the femur, between the lateral and medial condyles, is the popliteal groove. The area just proximal to this groove is known as the *popliteal surface*. The word *popliteal* refers to the posterior knee joint and is applied to bones, ligaments, muscles, nerves, or blood vessels of the posterior knee. The space between the femoral condyles is referred to as the *intercondylar*, or *femoral*, *notch*.

The two bones of the lower leg are the fibula and tibia (figure 12.5). In this chapter on the knee joint, only the proximal ends of these two bones are described. Their distal ends are considered in chapter 13.

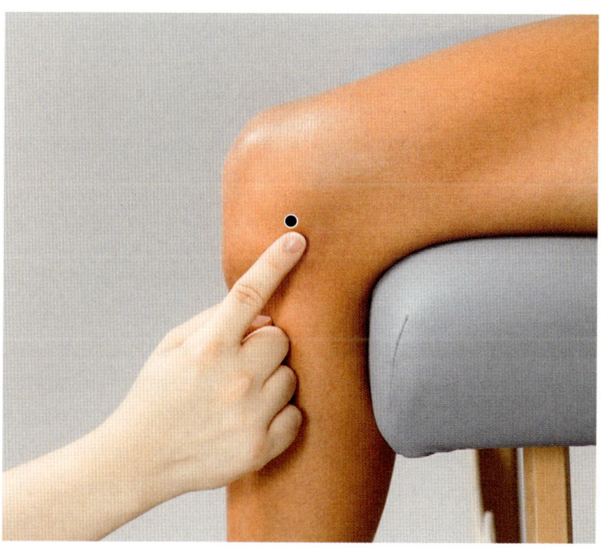

Figure 12.4 Locating the adductor tubercle.

The fibula, the smaller lateral bone, is essentially a non-weight-bearing bone that serves as a major source of soft-tissue attachment at both the proximal and distal ends. Three prominent structures at the proximal end of the fibula are the apex, the head, and the neck. The tibia, the larger medial bone of the lower leg, articulates with the femur and bears the weight of the body from the femur to the foot. At the very proximal end of the shaft of the tibia are the lateral and medial condyles, and, between them, on the proximal anterior surface of the tibia, lies a very large prominence known as the *tibial tuberosity*.

Hands On

You can easily palpate the tibial tuberosity by feeling for the large bump on the anterior surface of your lower leg just below your knee (figure 12.6).

A smaller prominence, located on the anterior aspect of the lateral condyle of the tibia, is Gerdy's tubercle (figure 12.7), where the iliotibial band inserts. On the posterior aspect of the proximal tibia, just distal to the space between the lateral and medial condyles, is the area known as the *popliteal surface*. Just beneath this surface is a small diagonal line called the *popliteal line*, also known as the *linea musculi solei* (see figure 12.5). The superior view of the tibia reveals several structures. The two large surfaces, the medial and lateral condylar surfaces, are where the condyles of the femur articulate with the tibia to form the knee joint (figure 12.8). The large prominence of bone between the two condylar surfaces is the intercondylar eminence, and the smaller prominences alongside the intercondylar eminence are the medial and lateral intercondylar tubercles (see figure 12.5).

The final bone of the knee joint is vital for proper movement of the joint. The patella (kneecap) is the largest sesamoid (free-floating; see chapter 1) bone in the body (figure 12.9). It is not directly attached to other bones to form a joint but rather is embedded in the tendon of insertion of the quadriceps femoris, which is part of the anterior knee muscle group (discussed later in the section on muscles). This muscle group inserts on the patella through a broad fibrous sheath, which

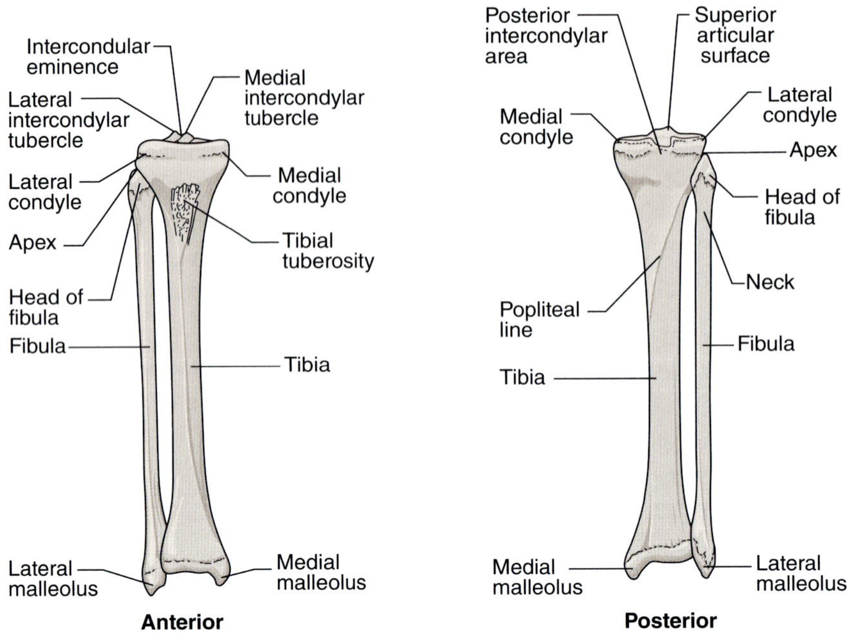

Figure 12.5 The tibia and fibula.

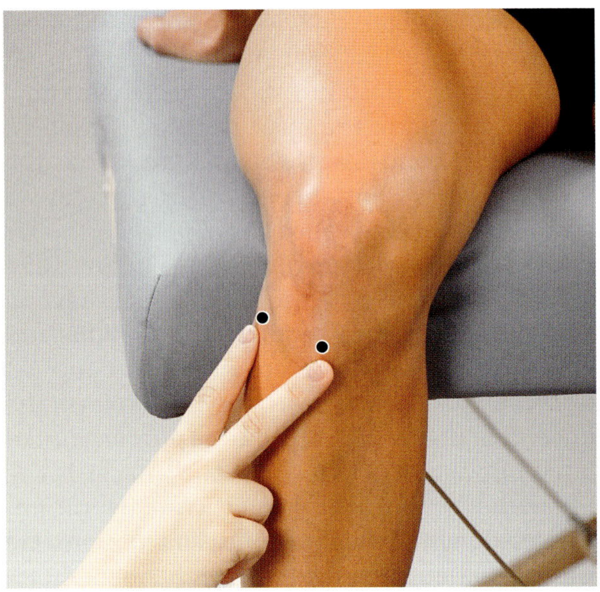

Figure 12.6 Locating the tibial tuberosity and head of the fibula.

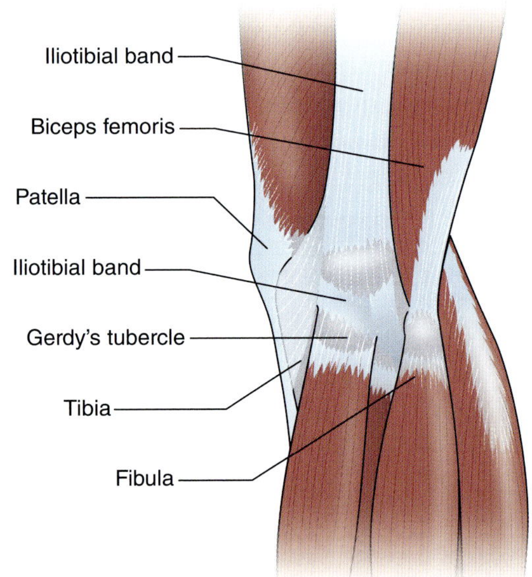

Figure 12.7 Gerdy's tubercle and the iliotibial band.

in turn attaches to the tibial tuberosity through the patellar ligament and tendon. The patella not only protects the structures beneath it but also changes the angle of pull of the quadriceps femoris to create a greater rotary force (in flexion and extension) of the knee joint as compared to the stabilizing force of the quadriceps femoris that pulls the tibia into the femur. The proximal end of the patella is known as the *base*, and the distal end is known as the *apex*. The posterior surface has lateral and medial articular surfaces that are covered with articular cartilage and articulate with the medial and lateral condyles of the femur to form the patellofemoral articulation.

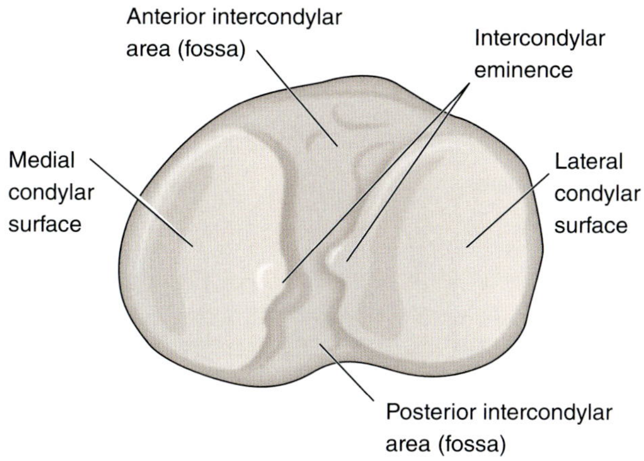

Figure 12.8 Superior view of the tibia.

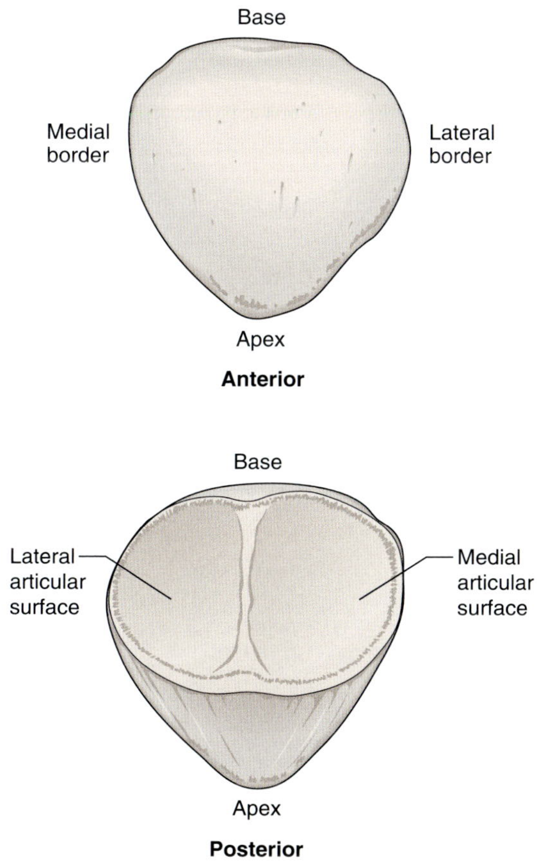

Figure 12.9 The patella.

Ligaments of the Knee

Like all synovial joints, the knee joint has a capsular ligament. This capsular ligament is unlike others because it consists of portions of other ligaments and fibrous expansions of other structures that cross the knee joint and become part of the capsule (figures 12.10 and 12.11). The components of the capsule include portions of the medial and lateral collateral ligaments and fibrous expansions of the quadriceps femoris, the iliotibial band, the vastus muscles, the sartorius muscle, and the semimembranosus muscle. Probably the best illustration of the pure capsular ligament is seen at the posterior aspect (popliteal space) of the knee joint.

The medial collateral ligament (MCL) and lateral collateral ligament (LCL) of the knee provide stability to either side of the joint, essentially preventing abduction and adduction of the joint, which makes it a uniaxial joint that flexes and extends in the sagittal plane. The MCL runs from the medial condyle of the femur to the medial condyle of the tibia, with some deep fibers attaching to the medial meniscus (discussed later in this chapter). The LCL runs from the lateral condyle of the femur to the head of the fibula. Note that the LCL, unlike the MCL, does *not* have fibers attaching to the lateral meniscus.

The term *genu* is Latin for knee. Three terms often mentioned regarding conditions of the knee are *genu varum*, *genu valgum*, and *genu recurvatum*. All three of these conditions might cause ligamentous stress. Genu varum is more commonly known as "bow legged." Which of the collateral ligaments of the knee joint might this condition stress? Genu valgum is more commonly known as "knock knees." Which of the collateral ligaments of the knee joint might this condition stress? Genu recurvatum refers to a *hyperextended knee*. Which cruciate ligament of the knee joint is likely to be stressed if the joint is hyperextended? Also, to be discussed later in this chapter, under normal conditions, is a slightly hyperextended knee more likely to occur in a male's or a female's knee?

Hands On

As you sit in a chair, place the ankle of one leg on top of the other leg's knee joint. Palpate the lateral condyles of both the femur and tibia of the leg crossed on top of the other leg. Find the joint space between the condyles. Move your finger anterior and posterior through the joint line until you feel a cord-like structure. This structure is the LCL. Now, consider the role of the MCL and LCL in the mechanics of the knee. Remember that the knee normally only flexes and extends. What ligament

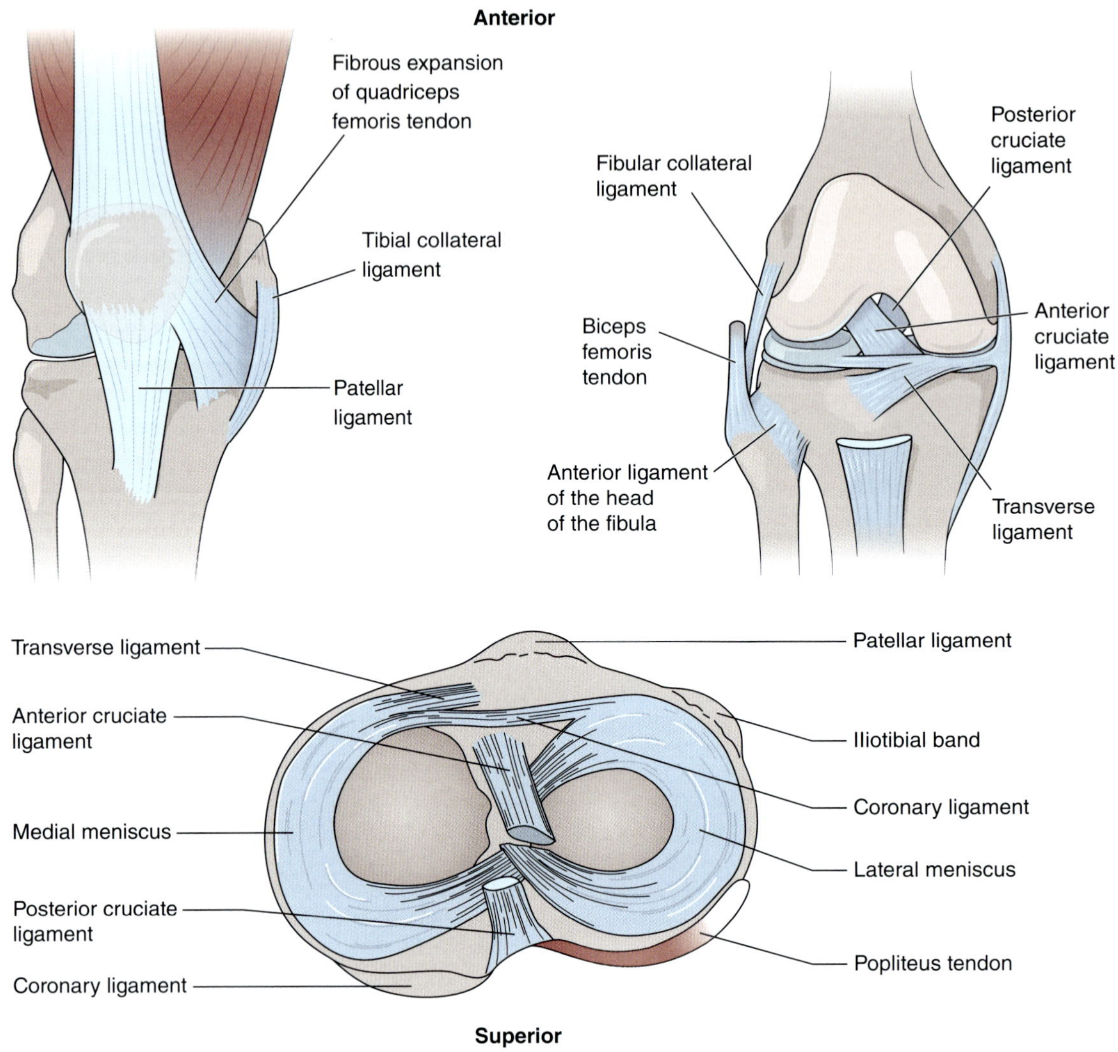

Figure 12.10 Ligaments of the knee.

would come under stress if a force were applied to the lateral side of the knee joint (often referred to as a *valgus force*), forcing the joint into abduction (i.e., increasing the space between the tibia and the femur on the medial side of the joint)? (See figure 12.12*a* for the answer.) What ligament would come under stress if a force were applied to the medial side of the knee joint (a varus force), forcing the joint into adduction (increasing the space between the tibia and the femur on the lateral side of the joint)? (See figure 12.12*b* for the answer.) These stresses placed on the MCL and LCL ligaments are commonly the mechanism of sprains of the knee joint.

In the middle of the knee joint are two ligaments known as the *anterior cruciate ligament* (ACL) and the *posterior cruciate ligament* (PCL; see figures 12.10 and 12.11). The term *cruciate* means

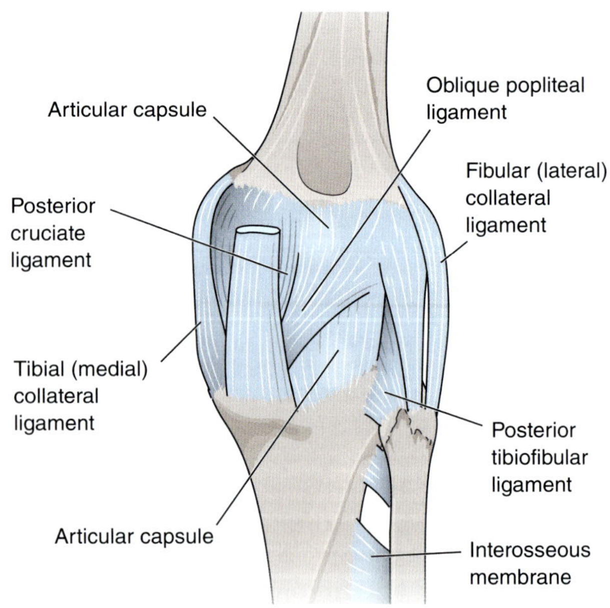

Figure 12.11 Posterior view of the ligaments of the knee.

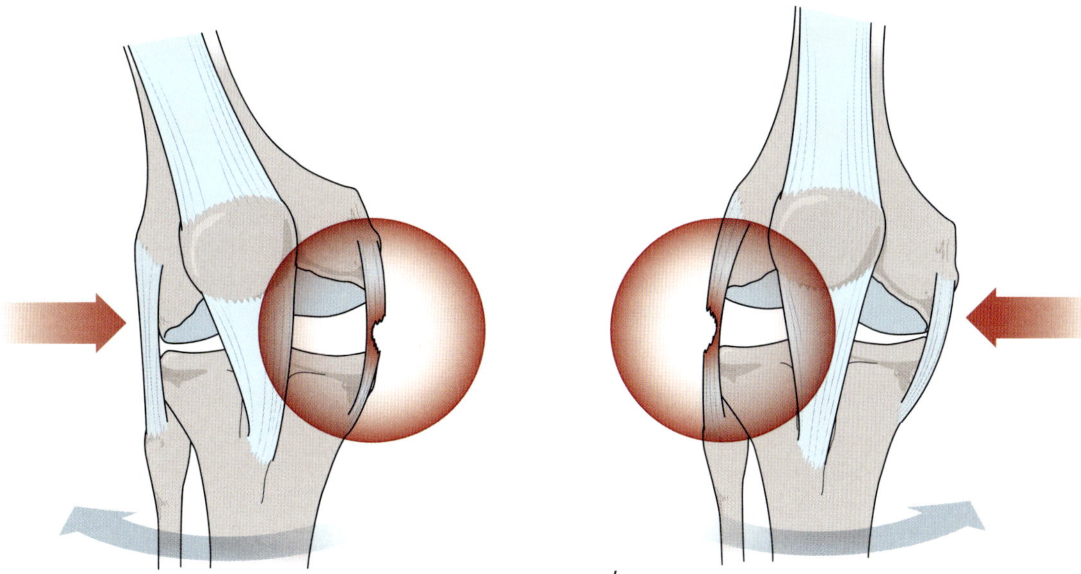

Figure 12.12 *(a)* A lateral force to the right knee places a valgus stress on the medial collateral ligament. *(b)* A medial force to the right knee places a varus stress on the lateral collateral ligament.

POP QUIZ

Based on the direction of the force being applied to the left knee joint in this photo, what collateral ligament is likely to be under stress?

Cameron Spencer/Getty Images

FOCUS ON

Anterior Cruciate Ligament

The anterior cruciate ligament (ACL) has received much attention recently in athletics. Many sports activities apply external and internal forces to the knee joint structures, including the ACL. There has also been great interest in the difference between female and male athletes in the number of ACL injuries. The theories regarding the larger number of female ACL injuries include lower-extremity malalignment (wider female pelvis and increased Q angle), ligament laxity, quadriceps to hamstring strength ratio imbalance, smaller female intercondylar (femoral) notch and femoral condyle size, the effects of estrogen on ligament tissue, and maturation rates and the effect on landing positions (straight-knee landing and one-step stop landing). Other factors considered are the intensity of training, playing surfaces, the type of activity, coaching techniques, direct trauma causing knee hyperextension, and enough additional theories to fill an entire textbook. Simply said, physicians, biomechanists, physical therapists, athletic trainers, and other interested parties continue to search for the reasons why ACL injuries remain prevalent in athletes and physically active individuals.

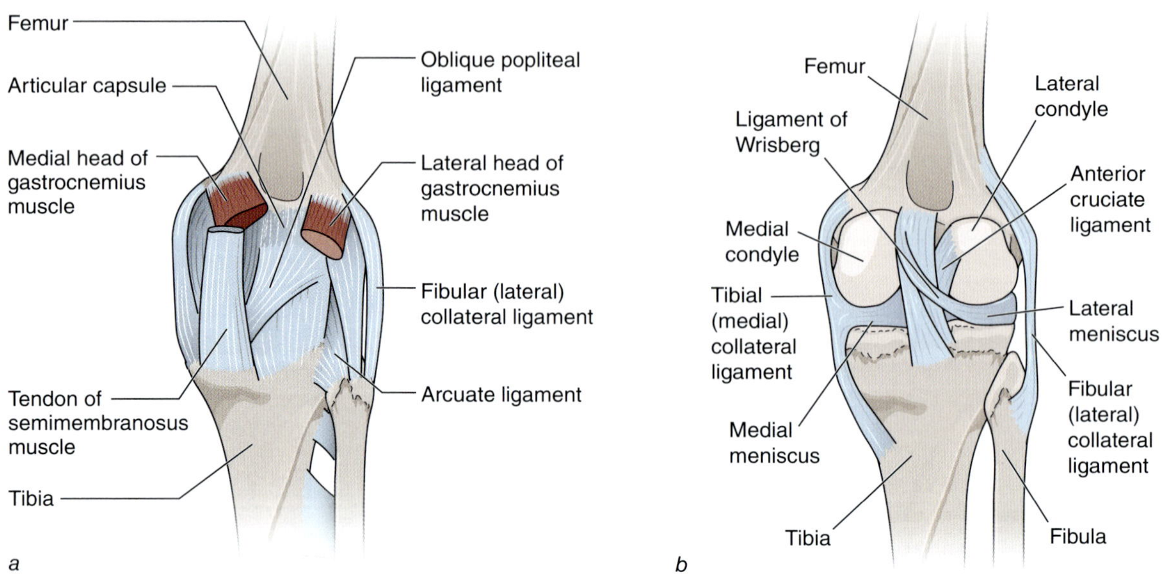

Figure 12.13 Posterior view of *(a)* the oblique popliteal and the arcuate ligaments; and *(b)* the ligament of Wrisberg.

"cross," and these two ligaments actually cross each other as they pass through the middle of the knee joint. The ACL runs from just anterior to the intercondylar eminence of the tibia to the posterior medial surface of the lateral condyle of the femur. The primary function of the ACL is to prevent anterior displacement of the tibia off the distal end of the femur. This ligament, along with the resistance of the posterior muscles crossing the knee joint, prevents the normal knee from hyperextending. The PCL runs from just posterior to the intercondylar eminence of the tibia to the anterior portion of the medial surface of the medial condyle of the femur. The primary function of the PCL is to prevent posterior displacement of the tibia off the distal end of the femur. Excessive squatting may place the knee joint into such a degree of flexion that this ligament comes under stress.

Three ligaments of the knee joint are found exclusively on the posterior aspect of the joint: the oblique popliteal ligament, the arcuate ligament, and the ligament of Wrisberg (figure 12.13). The oblique popliteal ligament runs from the posterior aspect of the lateral condyle of the femur to the posterior edge of the medial condyle of the tibia. The arcuate ligament runs from the posterior aspect of the lateral condyle of the femur to the posterior surface of the capsular ligament. The ligament of Wrisberg runs between the posterior horn of the lateral meniscus (see the section on menisci later in this chapter) and the posterior aspect of the medial condyle of the femur.

On the anterior side of the knee, running between the apex of the patella and the tibial tuberosity, is the patellar ligament (see figures 12.10 and 12.14). Because the patella is embedded in

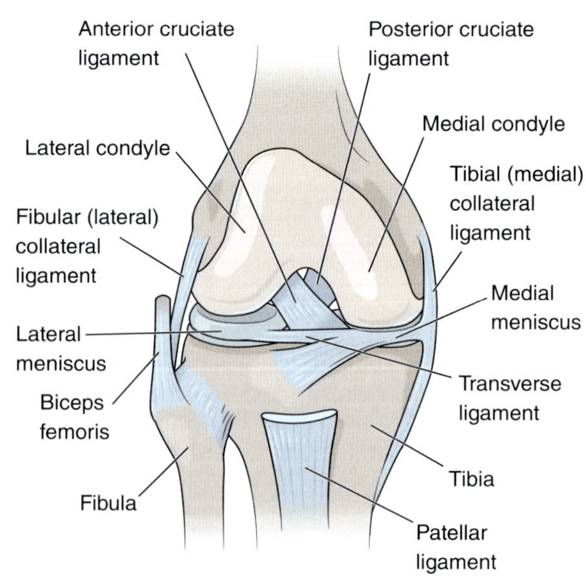

Figure 12.14 Anterior view of the ligaments of the knee.

the patellar tendon, some anatomists consider this structure to be an extension of the quadriceps femoris tendon of insertion. Others, however, label this structure the patellar ligament because it ties bone to bone (patella to tibia). In this text, the structure is known as the *patellar ligament*. Even though some anatomists hold that the quadriceps tendon inserts on the tibial tuberosity, the simple approach is that the quadriceps muscles attach to the patella and the patella attaches to the tibial tuberosity via the patellar ligament.

Hands On

The placement of the patella creates an angle of pull of the quadriceps tendon to create a force greater than was initially applied to facilitate extension of the knee joint. Draw two sketches of a person on a table showing the femur on the table with the tibia hanging over the end of the table at a 90° angle to the femur. Now, in one sketch, place a patella just above the tibia. Draw a line over the femur and the patella, and end it on the tibia in the area of the tibial tuberosity. Do the same with the other sketch, but instead omit the patella. Which sketch illustrates the greater ability of the quadriceps to extend the knee joint? Which sketch illustrates the greater ability to stabilize the knee joint? (Remember the angle of pull.)

Two additional ligaments of the knee joint are unique in that they do not tie bone to bone, which is the normal function of ligaments. The coronary ligament, which is actually a portion of the capsular ligament, is responsible for connecting the outer edges of the menisci to the proximal end of the tibia (see figure 12.10). The transverse ligament runs between the anterior horns of the medial and lateral menisci (see figure 12.14). This ligament prevents the anterior horn of each meniscus from moving forward when the knee joint moves into extension and the condylar surfaces of both the femur and the tibia exert pressure on the menisci.

FOCUS ON

Disruption of the Meniscus

The menisci, more medial than the lateral, often come under stress through athletic participation. Stress to the deep fibers of the MCL that attach to the medial meniscus could disrupt the meniscus. Excessive rotation of the femur on a fixed tibia (rigidly planted or pinned on the ground or playing surface) could cause stress to the menisci. Two of the most common disruptions of the menisci are called *parrot-beak* and *bucket-handle tears* (figure 12.15). The more common bucket-handle tear actually causes the middle portion of the meniscus (between the anterior and posterior horns) to split, causing the outer portion of the meniscus to look like the handle of a bucket as the inner portion separates from the main body of the meniscus. Other types of tears of the menisci also occur. Equipment design, playing surfaces, footwear, and specific exercises are areas of interest for decreasing the incidence of this serious condition. Medicine has advanced by leaps and bounds regarding menisci. Years ago, rather lengthy incisions were performed to completely remove a disruptive meniscus. It was thought that the meniscus was not vascularized sufficiently to promote healing. Today, with microscopic surgery providing minimal invasion, in many cases, disrupted menisci are considered repairable, and, with intense rehabilitation techniques available, meniscus problems are no longer considered to be career ending.

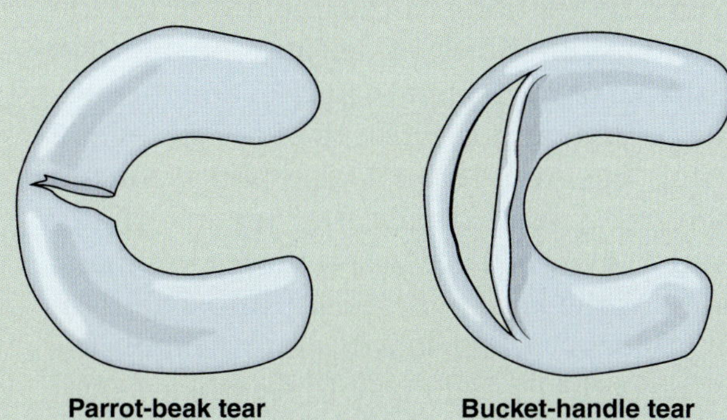

Figure 12.15 Two common tears of the meniscus of the knee.

Menisci of the Knee

Two semilunar (crescent-shaped) fibrocartilaginous structures sit on the proximal end of the tibia, on the medial and lateral condylar surfaces. These structures are known as the *lateral meniscus*, which is nearly circular in shape, and the *semicircular medial meniscus*. These structures help deepen the condylar surfaces of the tibia where the condyles of the femur articulate. Their outer borders are thick and convex and are attached to the tibia by the coronary ligament. The inner edges of the menisci are paper thin and lie freely (unattached) on the floor of the condylar surfaces of the tibia. The inner surfaces of the menisci are concave to accommodate the condyles of the femur. The anterior and posterior aspects of each meniscus are often referred to as the *anterior and posterior horns of the meniscus*.

Fundamental Movements of the Knee and Lower Leg

Because the knee joint is a uniaxial joint capable of movement in the sagittal plane about an ever-changing frontal horizontal axis, the only two movements the joint is capable of are flexion and extension. However, because of the sizes and shapes of the femoral condyles and the soft-tissue configurations, when the knee flexes and extends, the lower leg (tibia and fibula) rotates. When the knee extends, the leg externally rotates. When the knee flexes, the leg internally rotates. As the knee joint "locks" into extension and "unlocks" when moving into flexion, the rotation of the leg is difficult to see when weight is not being borne by the leg. The locking into full extension is often referred to as the "screw home" movement or mechanism. This action allows for full extension of the joint. The last 15° to 30° (various authorities differ) of knee extension will see the tibia externally rotate approximately 10° (in the open chain). If the foot is planted and the lower leg is unable to rotate, the femur will internally rotate approximately 10° (in the closed chain). The vastus medialis and vastus lateralis muscles function to allow the knee joint to screw home on knee extension, and the popliteus muscle "unlocks" the knee joint upon flexion (return from extension). You might question how these two muscles, which seem to be working in opposition to each other (one lateral and one medial), can together produce this motion. Once you observe the points where these muscles insert and their angle of pull, you should be able to understand the action of these muscles regarding the external rotation.

Hands On

The screw home mechanism (figure 12.16) explains the movement between the tibia and the femur during the end range of knee extension. On one hand, in an open kinetic chain position of being seated, extending the knee from a flexed position, and reaching the end range of knee extension, the tibia externally rotates on the femur to lock the knee into full extension. To unlock the knee, the popliteus contracts to internally rotate the tibia and allow the knee to flex. On the other hand, in a closed kinetic chain position of rising up from a squat position, the femur rolls on the tibia. When the knee reaches terminal extension, the femur internally rotates on the tibia to lock the knee into extension. To unlock the knee in a closed kinetic chain position, the popliteus contracts to externally rotate the femur and allow the knee to flex. Figure 12.16 shows the femur rolling into extension on a fixed tibia.

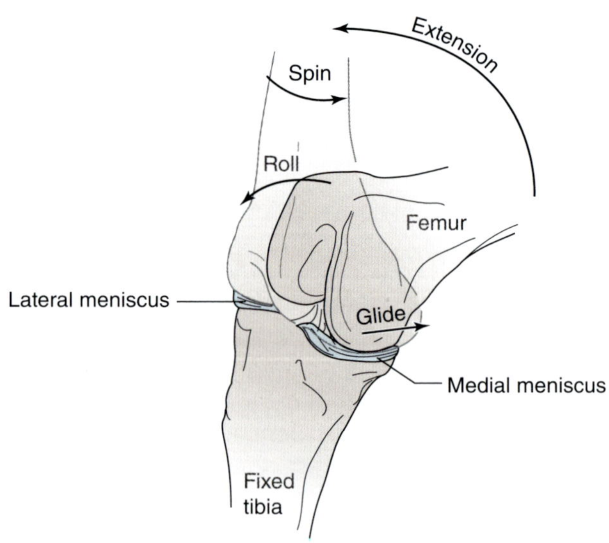

Figure 12.16 The screw home mechanism.

Hands On

Sit in a chair with your knees flexed to 90°. With your shoes off, place your feet on a smooth surface (e.g., tile or cement, not carpeting). Make a fist with one hand and place it between your knees. Internally rotate your lower legs so that your toes (first metatarsophalangeal joints; see chapter 13) on both feet push against each other. As you internally rotate your lower legs to push your feet against each other, with your free hand, feel the semimembranosus (medial thigh, just proximal to the knee) muscles posterior to your knee joints. These muscles (and those of the pes anserinus; see later in this chapter) are contracting to cause the internal rotation of the lower leg. Now, from the same starting position, have your partner place their feet along the sides, or the lateral aspect, of both of your feet. Find the biceps femoris tendon on the posterior lateral aspect of your knee with your free hand. Now, try to externally rotate your lower legs to force your feet to push your partner's feet away. What did you feel? As you review the muscles crossing the knee joint, their bony attachments will help identify what, if any, movement of the leg they perform.

Muscles of the Knee and Lower Leg

The muscles crossing the knee joint can easily be divided into those crossing the joint anteriorly and those crossing posteriorly. The anterior group is reviewed first.

Anterior Muscles

Because of their position anterior to the knee joint (figure 12.17), the logical conclusion is that these muscles extend the joint. This is primarily true, but other functions are also involved, as noted in the following discussions.

- *Sartorius:* This muscle, the longest in the body, is examined in the hip chapter because it is a flexor of the hip. The sartorius originates on the anterior superior iliac spine (ASIS), crosses the hip joint, passes posterior to the medial condyle of the femur, and inserts just inferior to the proximal end of the medial surface of the tibia (see figure 12.17). At the knee joint, the sartorius flexes the knee and internally rotates the lower leg.

The quadriceps femoris, frequently referred to as "the quads," is a group of four muscles (see figures 12.17 and 12.18). One of these muscles, the rectus femoris, crosses both the knee and the hip joint and is reviewed in the hip chapter as it pertains to hip joint action. Here, we look at its function as a knee extensor. The other three muscles—the vastus intermedius, vastus medialis, and vastus lateralis—cross only the knee joint and have only one function: extension at the knee.

- *Rectus femoris:* This is the one quad muscle that crosses both the hip and knee joints. It is the most superficial of the anterior thigh muscles. The muscle's straight head originates on the anterior inferior iliac spine, and the reflected head originates on the acetabulum, and it inserts on the base of the patella (see figure 12.17). The

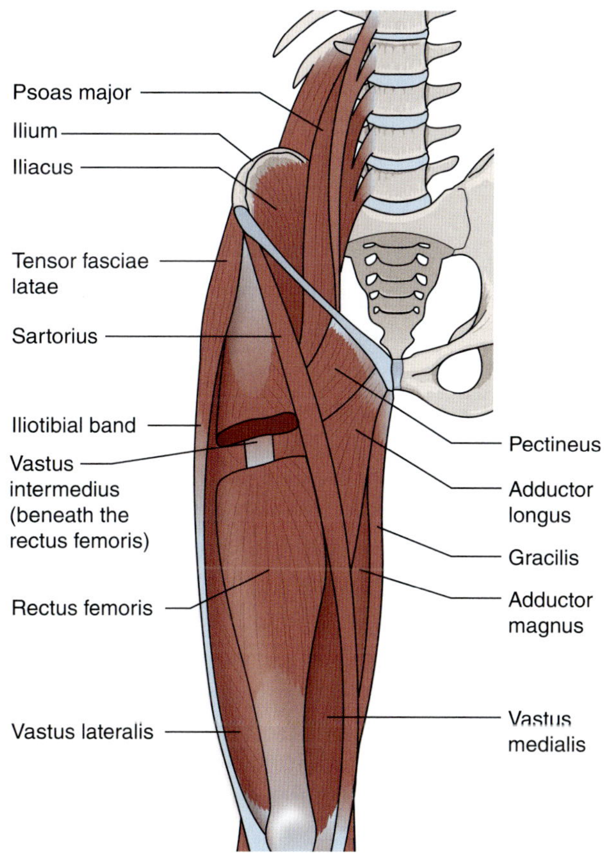

Figure 12.17 Anterior muscles of the knee.

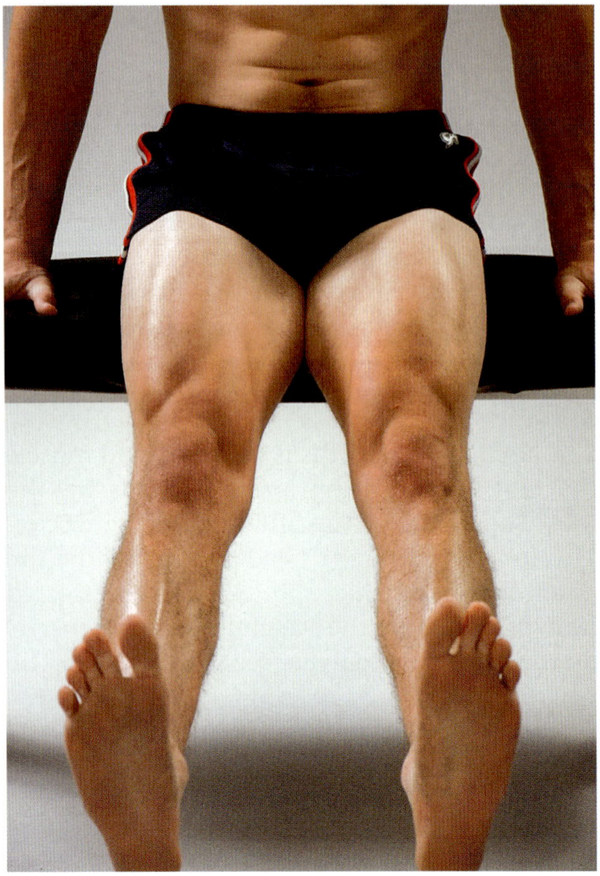

Figure 12.18 Viewing the quadriceps femoris muscle group.

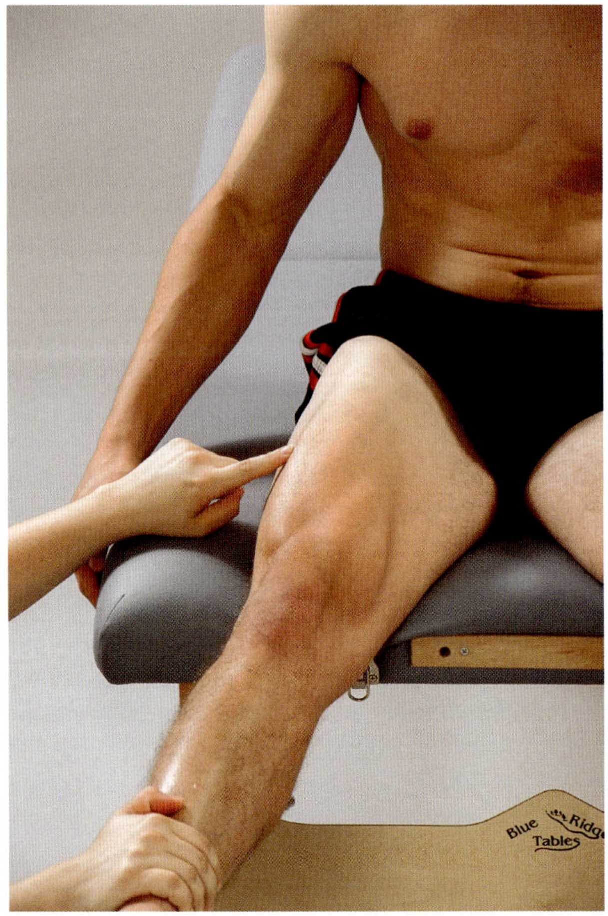

Figure 12.19 Locating the rectus femoris.

rectus femoris is an extensor of the knee joint. This muscle is frequently subject to direct trauma in activities such as American football, which justifies the use of protective thigh pads.

- *Vastus lateralis:* The largest of the three vastus muscles, the vastus lateralis, originates on the proximal half of the linea aspera, the intertrochanteric line, and the greater trochanter of the femur and inserts on the lateral border of the patella. The vastus lateralis extends the knee joint.
- *Vastus medialis:* Originating on the medial lip of the linea aspera, the vastus medialis inserts on the medial border of the patella. The vastus medialis extends the knee joint.

Hands On

With resistance to the lower leg (from either a partner, an iron boot, or a flexion–extension machine), extend your knee (see figure 12.19). Observe and palpate the medial and lateral aspects of your thigh. Identify the musculature you are observing.

- *Vastus intermedius:* Beneath the rectus femoris lies the vastus intermedius. This muscle originates on the proximal two-thirds of the anterior surface of the femur and inserts on the inferior surface of the patella. The vastus intermedius extends the knee joint.

Beneath the quadriceps femoris muscles lies another anterior muscle, the genu articularis.

- *Genu articularis:* This muscle, deep beneath the vastus intermedius, originates on the anterior surface of the femur just proximal to the condyles and inserts not on another bone but instead on the synovial membrane of the knee joint. As the knee moves into extension, this muscle contracts, pulling the articular capsule of the knee proximally to prevent the

FOCUS ON

Q Angle and Patellofemoral Pain

Chapter 8 points out the differences in the shape of the female and male pelvis. If you draw a line through the patella and the tibial tuberosity, and another line between the patella and the ASIS, you will note an angle between these two lines (figure 12.20). This angle is called the **Q angle** and is defined as the angle between the line of the quadriceps muscle pull and the line of insertion of the patellar tendon. In females, this angle normally is between 15° and 20°, whereas in males it ranges from 10° to 15°. A Q angle greater than 20° can result in any of a multitude of problems that cause pain in and around the patellofemoral joint. This all-inclusive evaluation of pain is often identified as *patellofemoral pain* (PFP). As mentioned, there are multiple reasons for anterior knee joint pain, with an excessive Q angle as a possible contributing factor. We will not go into an extended medical diagnosis in an entry-level anatomy textbook, but we will say that those experiencing PFP often complain about pain while climbing stairs. Consider the relationship of the patella and the femur when the knee is moved during stair climbing.

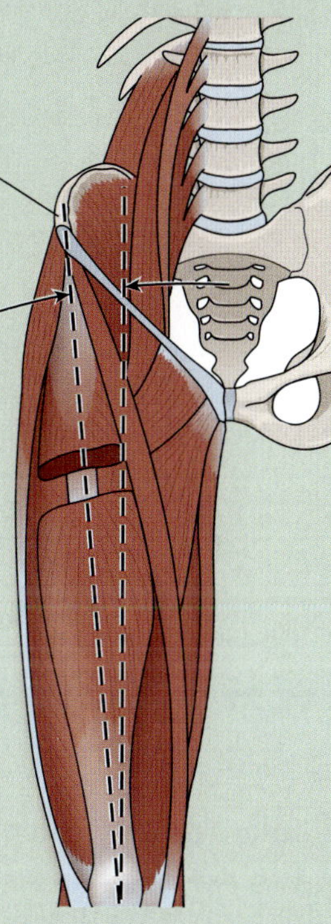

Figure 12.20 The Q angle.

synovial membrane from becoming impinged between the femur, the patella, and the tibia (figure 12.21).

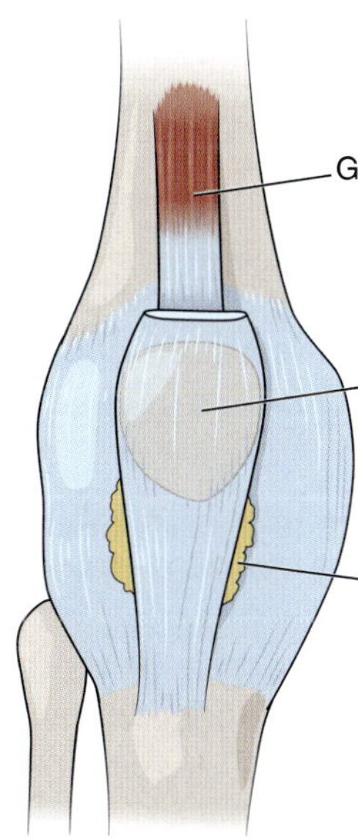

Figure 12.21 The genu articularis.

Posterior Muscles

The posterior muscles of the knee joint (figure 12.22) include muscles, such as the hamstrings and the gracilis, whose actions at the hip joint are reviewed in the discussion of muscles in chapter 11. Here, we look at their actions at the knee joint and at the other muscles that cross the knee posteriorly.

- *Biceps femoris:* This hamstring muscle has two heads: One, the long head, originates on the ischial tuberosity, and the other, the short head, originates on the lateral aspect of the linea aspera (see figure 12.22). The muscle inserts on the head

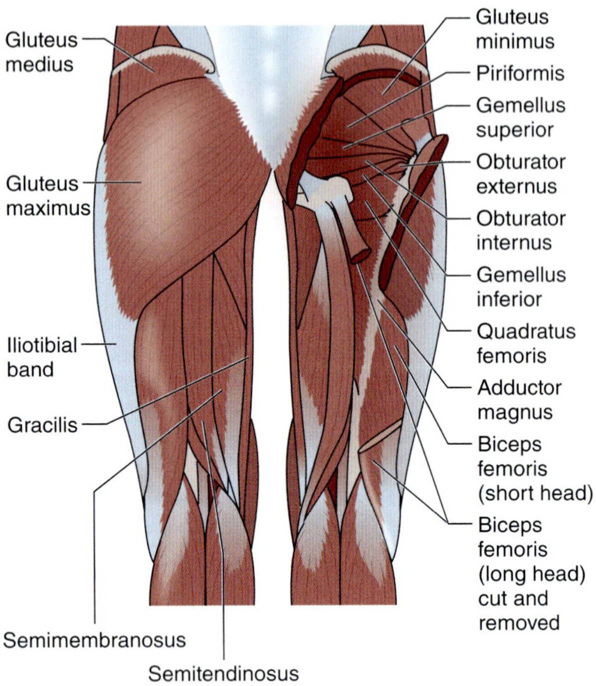

Figure 12.22 Posterior muscles of the knee.

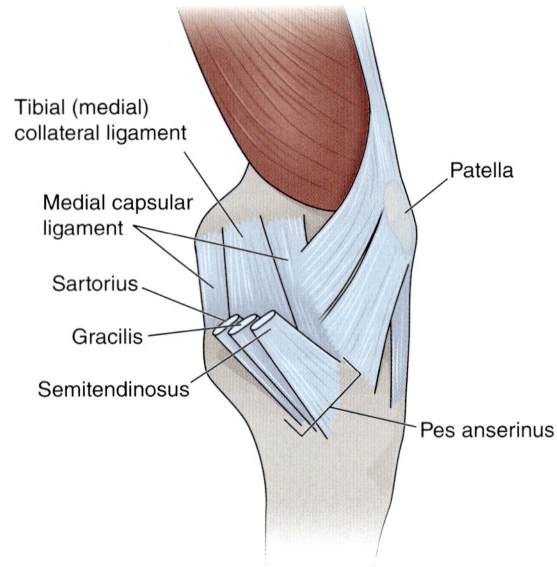

Figure 12.23 The insertion point of the three grouped tendons (gracilis, semitendinosus, and sartorius) is known as the *pes anserinus*.

of the fibula. The biceps femoris flexes the knee, and, as the knee reaches active full flexion, it externally rotates the lower leg.

• *Semitendinosus:* The second hamstring muscle, the semitendinosus, originates on the ischial tuberosity and inserts on the proximal aspect of the medial tibia (see figure 12.22). The semitendinosus flexes the knee and internally rotates the lower leg.

• *Semimembranosus:* The third hamstring muscle, the semimembranosus, originates on the ischial tuberosity and inserts on the posterior medial aspect of the medial condyle of the tibia (see figure 12.22). The semimembranosus flexes the knee and assists with internal rotation of the lower leg.

• *Gracilis:* The gracilis is the only adductor muscle of the hip joint that also crosses the knee joint. It originates on the inferior surface of the pubic symphysis, runs posterior to the medial condyle of the femur, and inserts just posterior to the medial aspect of the proximal end of the tibia (see figure 12.22). The gracilis flexes the knee joint and internally rotates the lower leg.

The gracilis, semitendinosus, and sartorius insert in the same general area, just below the proximal end of the tibia on its medial aspect. The insertion of the three closely grouped tendons is commonly identified as the *pes anserinus* (figure 12.23). All three components of the pes anserinus flex the knee joint and internally rotate the lower leg.

• *Popliteus:* Diagonally crossing the popliteal space of the knee joint, the popliteus runs between the lateral aspect of the lateral condyle of the femur and the popliteal line on the proximal third of the posterior surface of the tibia (figure 12.24). Observe the arrangement of the fibers of this muscle. Besides flexion of the knee, what obvious motion of the lower leg is likely to occur when this muscle contracts? (Answer: internal rotation)

• *Iliotibial band:* This structure (figure 12.25*a*), also discussed in the hip chapter, is a combination of the gluteus maximus and tensor fasciae latae tendons of insertion (see figure 12.22). It crosses the knee in the area of the lateral condyle of the femur and inserts onto a bony prominence just inferior and anterior to the lateral condyle of the

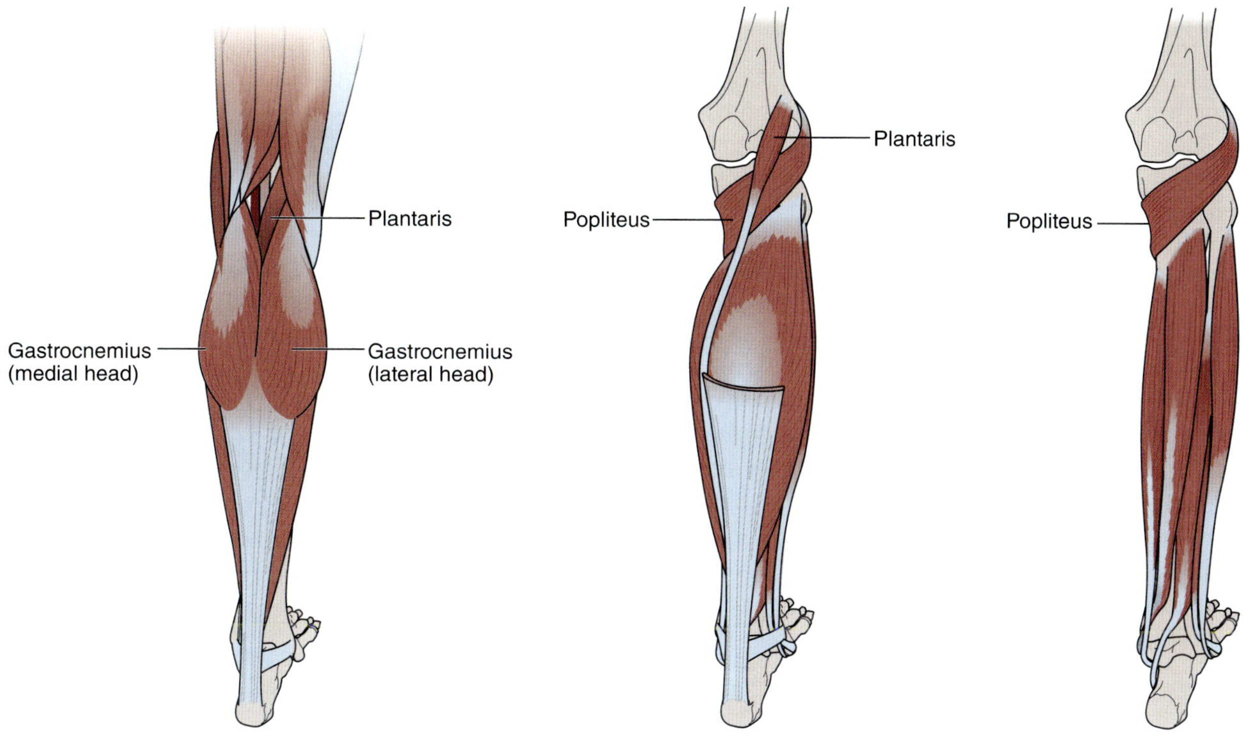

Figure 12.24 Posterior muscles of the knee and lower leg.

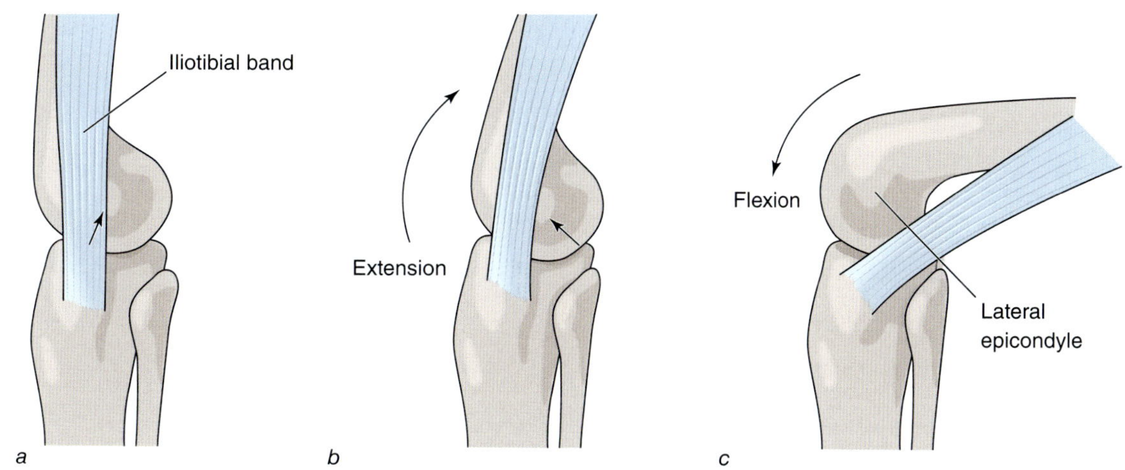

Figure 12.25 *(a)* The iliotibial band as *(b)* the knee joint approaches full extension and as *(c)* the knee joint moves into flexion.

tibia known as *Gerdy's tubercle* (see figure 12.7). This structure both flexes and extends the knee joint, depending on the angle of the knee joint at any particular moment. When the knee joint is between full extension and 10° to 15° of flexion, the iliotibial band is anterior to the lateral femoral condyle and assists with extension of the knee joint (figure 12.25*b*). As the knee joint continues to flex beyond 10° to 15°, the iliotibial band shifts to a position posterior to the lateral femoral condyle and becomes a flexor of the knee joint (figure 12.25*c*).

Hands On

Palpate the lateral side of your knee joint. Move your hand proximally, just above the lateral condyle of the femur. Flex and extend your knee joint, and feel the iliotibial band move anterior and posterior to the femoral condyle.

Two muscles of the lower leg originate above the knee joint and play a role in knee joint function, although their primary functions involve the ankle joint.

- *Gastrocnemius:* This muscle has two heads: one originating on the posterior aspect of the lateral condyle of the femur and the other originating on the posterior aspect of the medial condyle of the femur (see figure 12.24). Both heads combine into a single tendon of insertion that attaches to the calcaneus (heel bone). The gastrocnemius flexes the knee joint. In the rare instance where the gastrocnemius cannot perform its primary function (plantar flexion of the ankle joint) because the foot is held in a fixed position and cannot move, contraction of the gastrocnemius can cause extension of the knee joint.
- *Plantaris:* This short-bellied muscle with a long tendon of insertion originates on the lateral linea aspera and the oblique popliteal ligament and inserts on the calcaneus (see figure 12.24). The plantaris muscle assists with knee flexion but is of little importance in humans compared with the other muscles that perform the same function. Often, a rupture of this tendon is thought to be a strained Achilles tendon because of its position. A common sign of such a rupture is the victim's complaining that someone must have shot them based on the sudden feeling and, occasionally, the sound.

Visit the web resource to complete the activities for this chapter.

CHAPTER 13

The Lower Leg, Ankle, and Foot

This chapter may seem familiar because the lower leg, ankle, and foot are very similar anatomically to the forearm, wrist, and hand. The bones, ligaments, and muscles may or may not have similar names, but their structures are similar. In contrast to the upper extremity, the lower extremity is constructed to bear the weight of the body and absorb the force applied by that body weight with each foot strike.

Bones of the Lower Leg

The two bones of the lower leg are the tibia (medially) and the fibula (laterally). Much debate has been conducted over the weight-bearing aspects of both bones, but in this text, we assume that the tibia is the major weight-bearing bone and the fibula has little or no weight-bearing function.

The prominent bony markings of the proximal end of these bones are presented in chapter 12 on the knee. At the distal ends of the shafts of both the tibia and the fibula are large bumps known as *malleoli* (figures 13.1 and 13.2). The prominence at the distal end of the fibula is known as the *lateral malleolus*, and the prominence at the distal end of the tibia is known as the *medial malleolus*. Palpate these bumps on either side of the distal end of your lower leg, and note which one is more distal than the other. This is important to remember later, when the fundamental movements of the ankle joint are discussed.

On the lateral surface of the tibia, just proximal to the distal end, is a notch known as the *fibular notch of the tibia*, where the tibia and fibula articulate to form the distal tibiofibular joint.

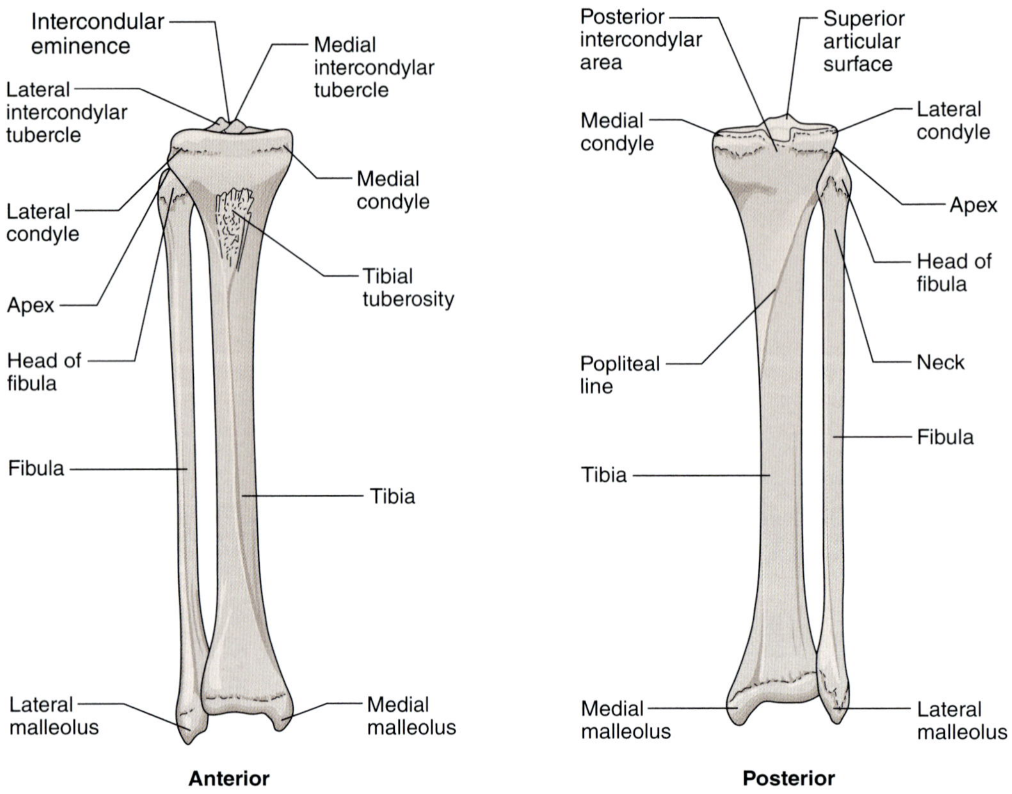

Figure 13.1 The tibia and fibula.

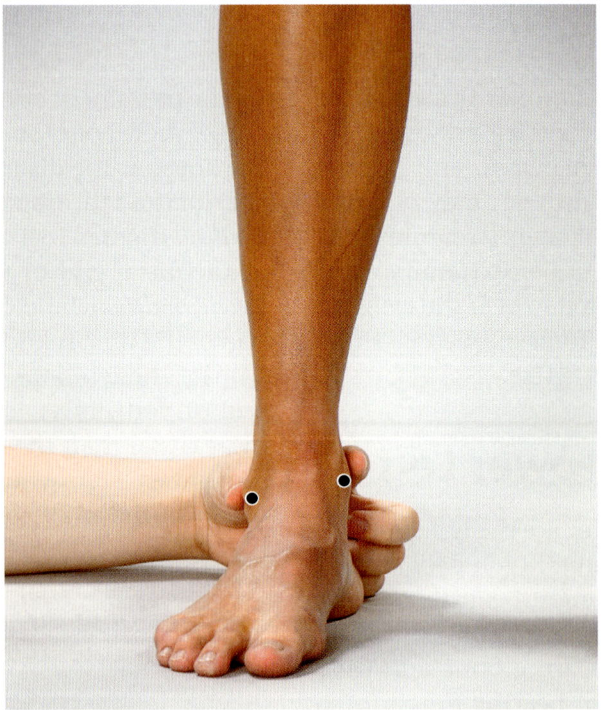

Figure 13.2 Locating the malleoli.

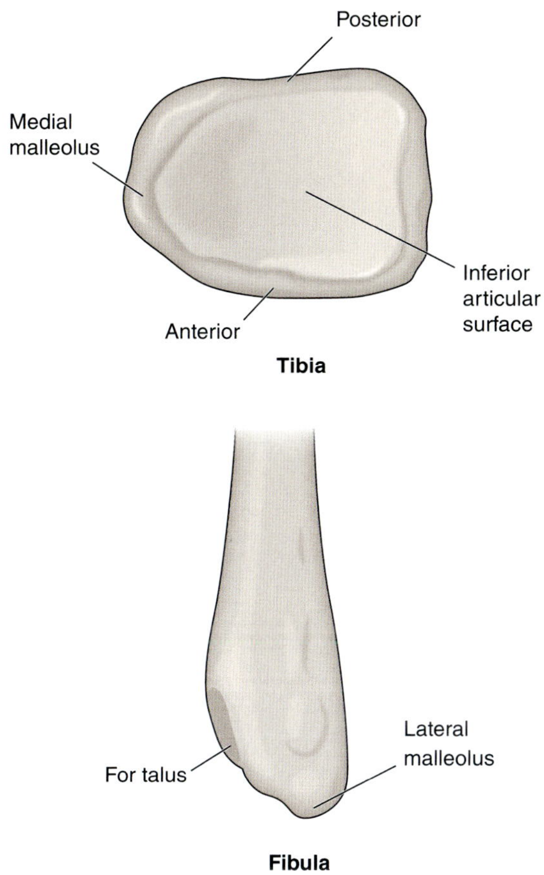

Figure 13.3 Inferior view of the distal ends of the tibia, and posterior view of the fibula.

The distal surfaces of both the tibia and the fibula have facets (or smooth surfaces) that articulate with the talus, one of the bones of the foot (figure 13.3).

Bones of the Foot

The 26 bones of the foot (figure 13.4) are usually separated into 3 distinct segments: the forefoot (19), the midfoot (5), and the hindfoot (2).

The forefoot consists of 14 phalanges: three per toe (proximal, middle, and distal phalanges), except for the great toe, which has only a proximal and a distal phalanx. The phalanges and the metatarsal bone of the first, or great, toe are larger than those of the other four toes (second, middle, fourth, and little) for a specific purpose. When the foot bears the weight of the body, as in walking, the great toe must bear most of the weight. This is the same arrangement we found in the hand: three phalanges per finger and only two in the thumb. The other five bones of the forefoot are the five long bones of the foot, known as the *metatarsal bones*. Again, these bones are similar to the five metacarpal bones of the hand. The metatarsal bones consist of a head (distal end), a shaft, and a base (proximal end). The most prominent of these structures is the base of the fifth metatarsal bone.

Hands On

Run your finger along the lateral edge of your foot, and palpate a rather large bump about half to two-thirds of the length of your foot from your fifth (little) toe. This is the base (also known as the *tuberosity*) of the fifth metatarsal bone (see figure 13.4).

The remaining seven bones of the foot are collectively known as the *tarsal bones*. Five of these tarsal bones (the cuboid; navicular; and medial, intermediate, and lateral cuneiforms) make up the midfoot, and the remaining two tarsal bones (talus and calcaneus) make up the hindfoot.

Just proximal to the medial metatarsal bones are three tarsal bones known as the *cuneiform bones* (figures 13.4 and 13.5). These bones are often referred to as the *first*, *second*, and *third cuneiforms*, but they are easier to remember by their anatomical position: the medial, intermediate, and lateral cuneiform bones. Proximal to the cuneiform bones is the fourth midfoot bone, the navicular. As in the wrist, the navicular bone is also known as the *scaphoid*. The fifth tarsal bone of the midfoot is also the most lateral of the five midfoot tarsals: the cuboid bone. Careful examination of this bone reveals that its lateral border is concave. This groove provides a space for a tendon of the foot and ankle (the fibularis longus, discussed later in this chapter).

The final two tarsal bones, the talus and the calcaneus, make up the bones of the hindfoot (see figures 13.4, 13.6, and 13.7). These two bones articulate with the tibia and fibula, similarly to how the scaphoid and lunate bones articulate with the ulna and radius of the forearm. The talus literally sits on top of (proximal to) the calcaneus, which is often referred to as the *heel bone* and is the largest of the tarsal bones. On the posterior view of the calcaneus (figure 13.8), note a bony prominence extending medially from its superior

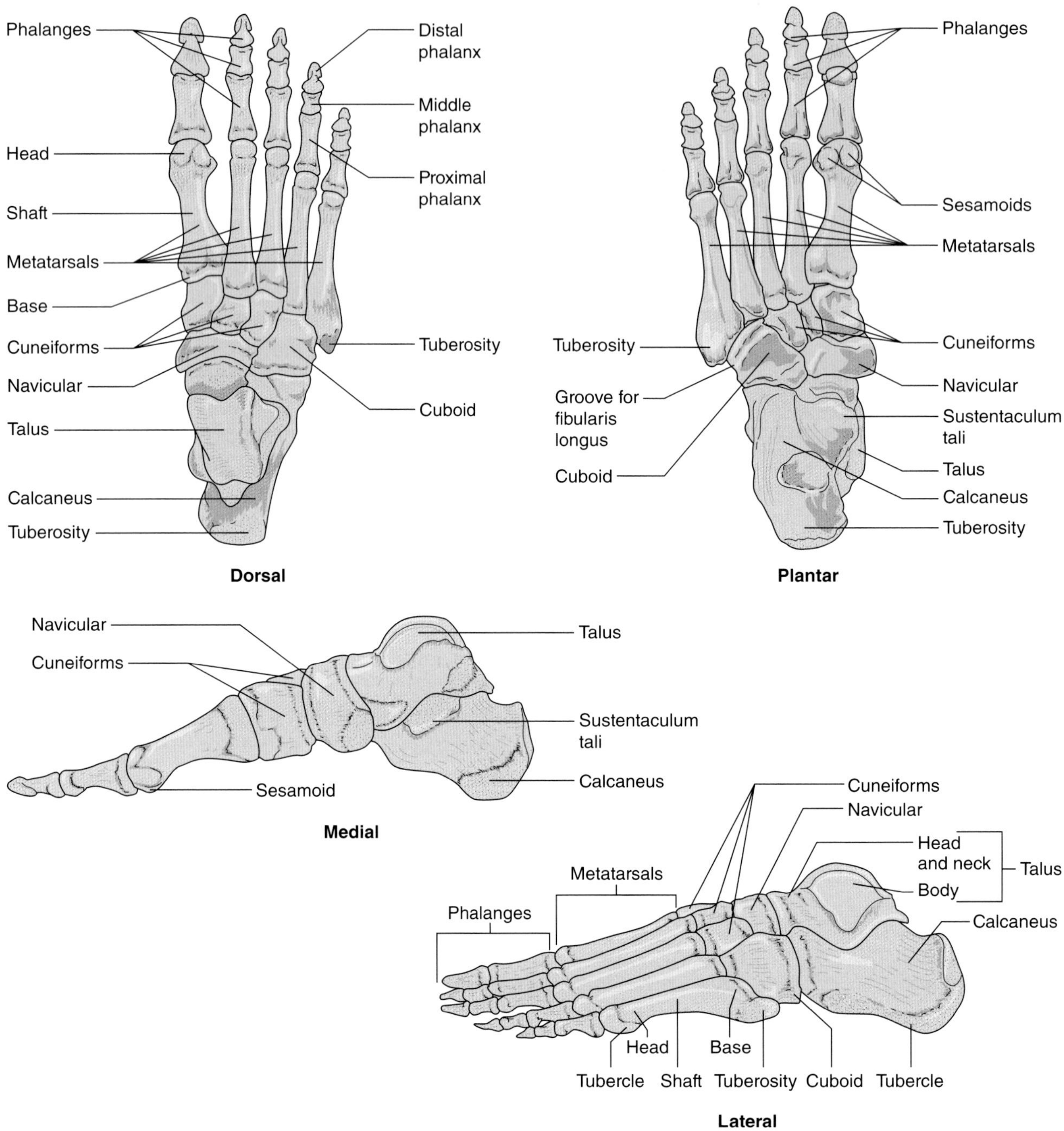

Figure 13.4 Bones of the foot.

surface. This prominence, the sustentaculum tali, serves as a platform for a portion of the talus to sit on. These two bones, articulating together and with the lower-leg bones, form the two joints (talocrural and talocalcaneal) that we refer to as the *ankle joint*. Movement in these two joints (two movements in each joint) results in the four movements attributable to the ankle joint. These joints and movements are further discussed later in this chapter.

On the plantar surface of the head of the first metatarsal bone are 2 sesamoid bones (see figure 13.4), which are not typically counted with the other 26 bones of the foot. Like the patella and sesamoid bones in general, by definition, these sesamoid bones are embedded in the

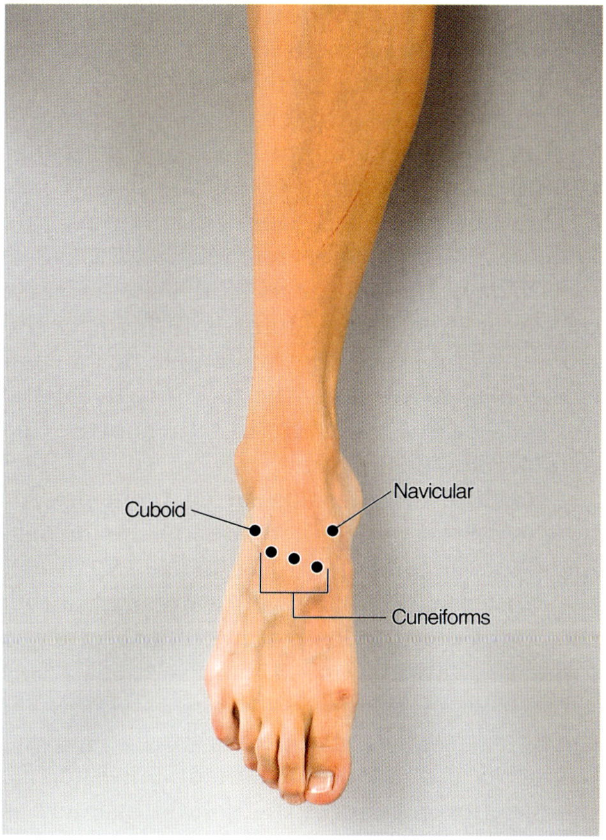

Figure 13.5 Dorsal view of the navicular, cuneiform, and cuboid bones.

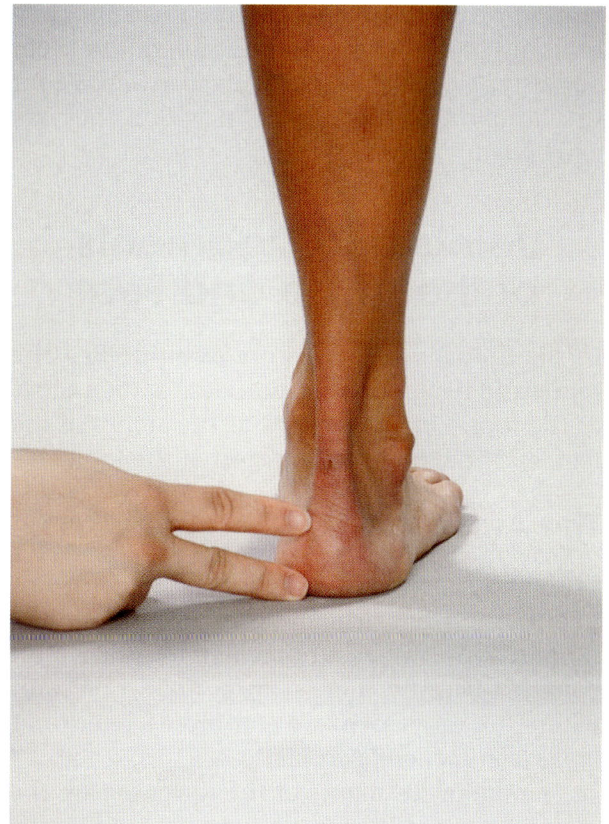

Figure 13.7 Locating the calcaneus.

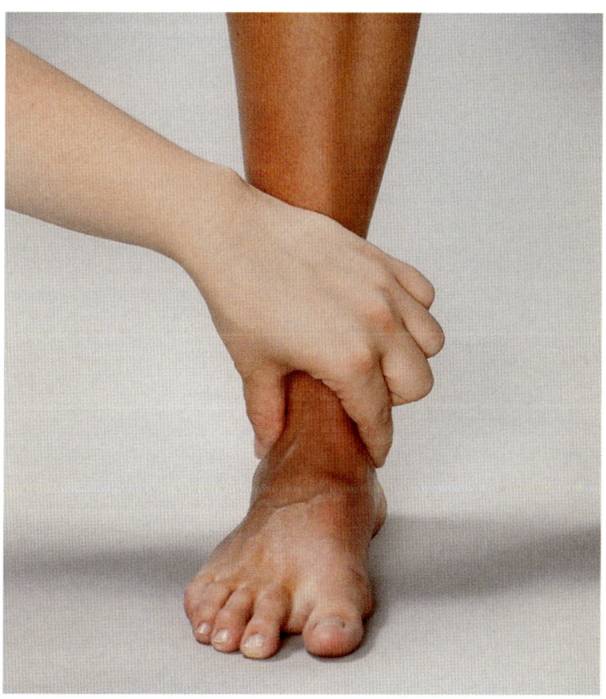

Figure 13.6 Locating the talus.

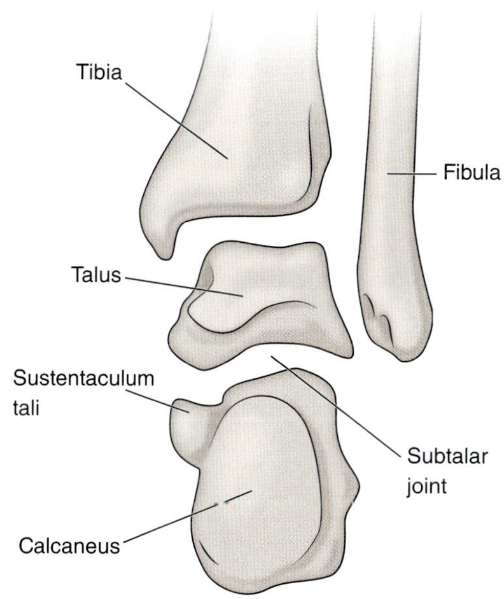

Figure 13.8 Posterior view of the ankle, illustrating the sustentaculum tali of the calcaneus and the subtalar joint.

tendon of a muscle and are free floating. They not only protect the structures superior to them but also provide a biomechanical advantage for the function of the muscle in which they are embedded.

Joints and Ligaments of the Ankle and Foot

In the following sections, we examine first the joints and ligaments of the ankle and then those of the foot. Finally, we look at the arches of the feet, essential to any discussion of the foot.

The Ankle

The ankle is not really a single well-defined joint like many other articulations throughout the human body. Some authors call it the ankle joint complex because there is more than one joint where the movement we commonly refer to as *ankle joint motion* takes place.

The major ligaments of the lower leg are the interosseous ligament (interosseous membrane, similar to the interosseous ligament of the forearm; figure 13.9) found between the medial border of the shaft of the fibula and the lateral border of the shaft of the tibia (from proximal to distal ends) and the anterior and posterior tibiofibular ligaments at the distal end of the leg (figure 13.10). The interosseous ligament serves as a source of attachment for numerous anterior and posterior muscles of the lower leg.

The talocrural aspect of the talocrural–talocalcaneal (ankle) joint is a hinge joint that permits movement in the sagittal plane (dorsiflexion and plantar flexion). In anatomy, the talocrural joint is often referred to as a *loosely formed mortise-and-tenon joint*. The mortise (recess or hole) of the joint is formed by the lateral malleolus of the fibula and the medial malleolus of the tibia. The tenon (peg) of the joint is the talus, which fits into the mortise (see figures 13.8 and 13.11).

In addition to the capsular ligament present in all synovial joints, there are four major ligaments of the ankle joint (see figure 13.10). For the most part, the names of ligaments indicate the bones that the ligaments bring together to form articulations.

Medially, the major ankle ligament is known as the *deltoid ligament*, made up of three superficial ligaments and one deep ligament. Superficially, the anterior portion of the deltoid ligament is the tibionavicular ligament, the middle portion is the calcaneotibial ligament, and the posterior portion is the posterior talotibial ligament. The deep ligament of the deltoid ligament is the anterior talotibial.

Laterally, the ankle has three major ligamentous structures. The shortest of these three

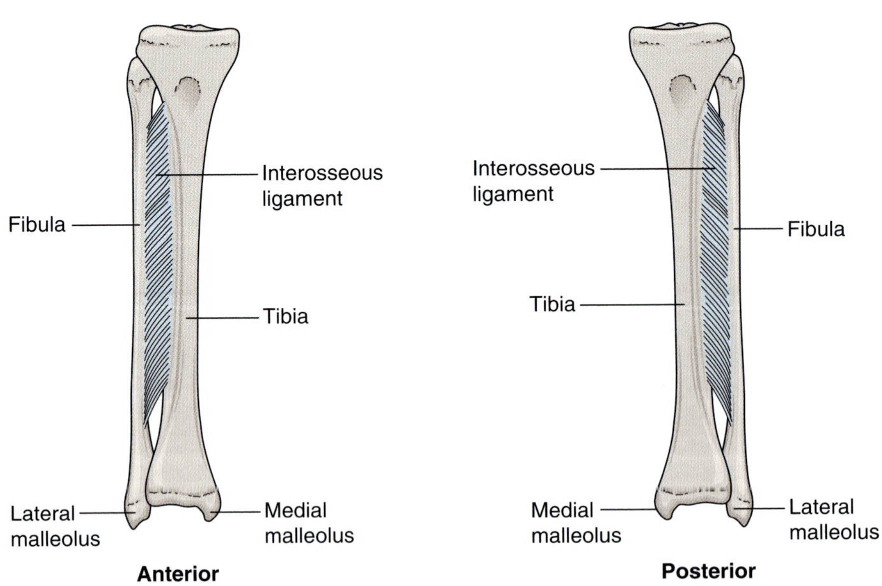

Figure 13.9 The interosseous ligament of the tibiofibular joint.

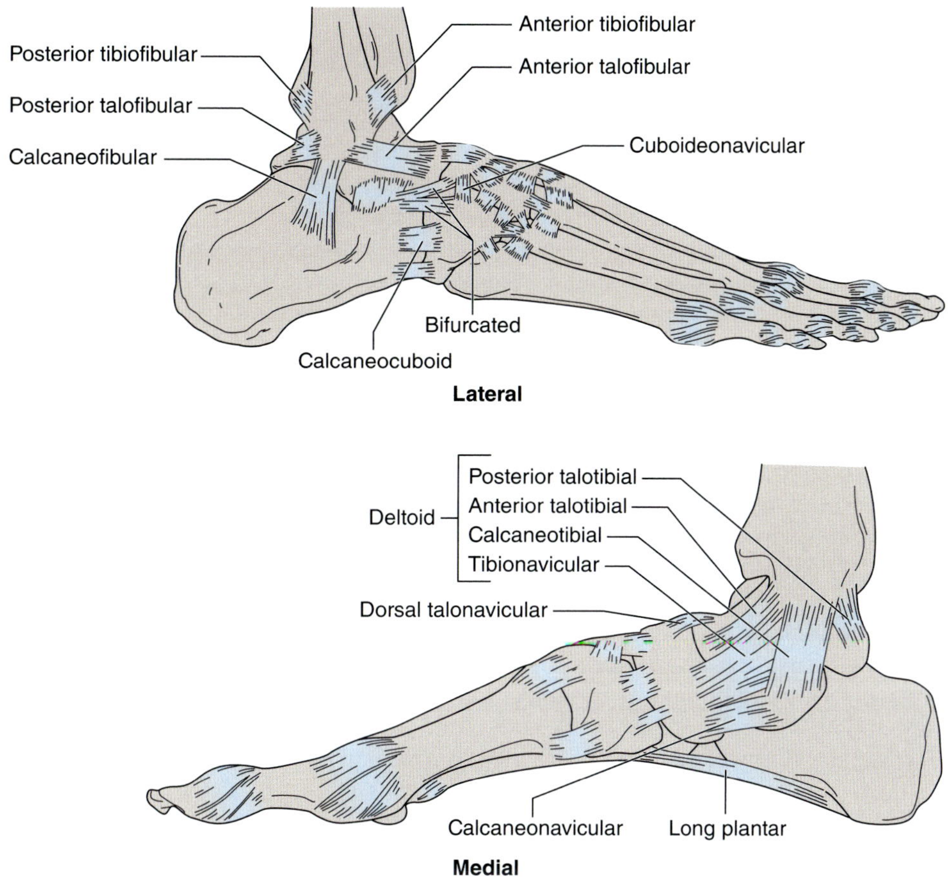

Figure 13.10 Major ligaments of the foot and ankle.

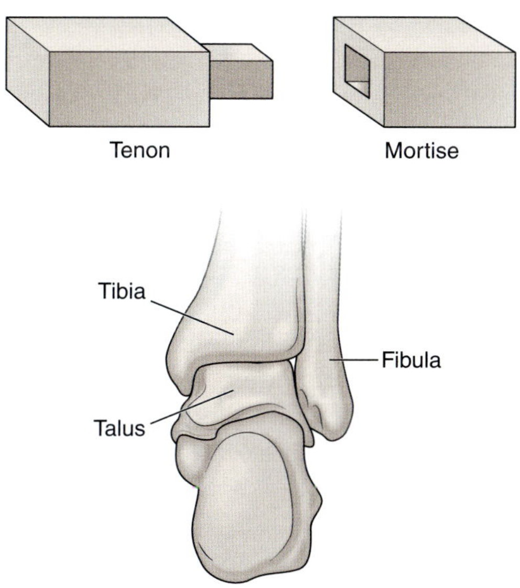

Figure 13.11 The mortise (tibia and fibula) and tenon (talus) of the talocrural joint.

ligaments is the anterior talofibular (ATF) ligament. It runs from the anterior aspect of the distal end of the lateral malleolus to the talus. The ATF ligament is the most commonly sprained ligament of the ankle joint.

Hands On

Placing your finger just anterior and distal to the end of your fibula, you will find a soft depression, the sinus tarsi (*sinus* = "cavity," *tarsi* = "tarsal bone"). The ATF ligament crosses through this area.

The strongest of the three lateral ankle ligaments is the posterior talofibular ligament, which runs from the posterior aspect of the lateral malleolus to the talus. The longest of the three lateral ankle joint ligaments is the calcaneofibular ligament. It runs from the lateral malleolus of the fibula to the lateral aspect of the calcaneus.

FOCUS ON

Ankle Joint Problems

Any of the ligaments of the ankle can be sprained, depending on the exact mechanics of the stress. However, the ATF ligament is by far the most commonly sprained ligament of the ankle joint. Rolling over on the ankle (often a combination of inversion and plantarflexion) places stress on all lateral ligaments of the ankle, but the ATF in particular is under the greatest amount of stress in this typical ankle sprain. The term *inversion sprain* or *lateral ankle sprain* is frequently used to describe this injury. Lateral ankle sprains are the most common musculoskeletal injury.

Although much less common than an inversion sprain, a high ankle sprain may occur from excessive eversion or dorsiflexion motion. A more anatomically correct term would be a *syndesmosis sprain*. The term **syndesmosis** is another way to define a joint: an articulation between bones tied together by ligaments. In the case of the ankle joint, when the ankle is excessively everted or dorsiflexed, the talus may be forced against the fibula, causing the joint between the fibula and tibia (the tibiofibular syndesmosis) to spread apart. This spreading can possibly sprain the anterior tibiofibular ligament, the posterior tibiofibular ligament, the interosseous membrane, or any combination of these. The spraining of any of these three structures can be defined as a *high ankle sprain*. X-rays often reveal not only the separation of the bones but also fractures because of the stress of bone on bone.

Preventing ankle joint problems is important for many people involved in physical activity. Prescribing specific exercises for strengthening or rehabilitating the ankle, designing footwear, using preventive measures (e.g., taping, wrapping, or bracing), and understanding the effects of specific playing surfaces rely on knowledge of the anatomy of the ankle joint. Prevention strategies rely on knowing the types of structures that are stressed during injury, the types of forces involved, how to counteract those potentially damaging forces, and the musculature that can be strengthened to better resist the damaging forces. Another consideration when attempting to prevent ankle injuries with taping, wrapping, or bracing techniques is stress on other structures. If a force is applied to the lower leg and ankle and there is no movement in the ankle joint, where might that force be dissipated? The knee? The disciplines of biomechanics, athletic training, sports medicine, physical therapy, and exercise science can provide the necessary background for addressing this important aspect of sport.

Hands On

After observing the anatomical arrangements of the bones and ligaments of the ankle, attempt to turn your ankle inward toward your other ankle (inversion), and then attempt to turn your ankle outward away from your other ankle (eversion). Which movement created the greatest motion? What structures limited the movement? Bones? Ligaments? Observing these actions and the structural arrangement of the bones and ligaments of the ankle should reveal why the ATF is the most likely ligament of the ankle joint to sprain. Inversion and eversion are discussed later in this chapter.

Ligaments of the Foot

The ligaments of the foot (see figures 13.10 and 13.12) can be divided into five groups: the intertarsal ligaments, tarsometatarsal ligaments, intermetatarsal ligaments, metatarsophalangeal (MTP) ligaments, and interphalangeal (IP) ligaments.

The intertarsal ligaments tie together the articulations between the tarsal bones of the hindfoot and the bones of the midfoot. The hindfoot joint between the talus and the calcaneus (talocalcaneal or subtalar joint, see figure 13.8) produces medial and lateral gliding movements in the hindfoot that are identified as inversion (movement of the foot toward the midline of the body) and eversion (movement of the foot away from the midline).

Intertarsal ligaments, like the intercarpals of the wrist, join the seven tarsal bones of the foot together. Most of these ligaments are identifiable by their names, which represent the bones they tie together. The talocalcaneonavicular joint has a capsular ligament and a dorsal talonavicular

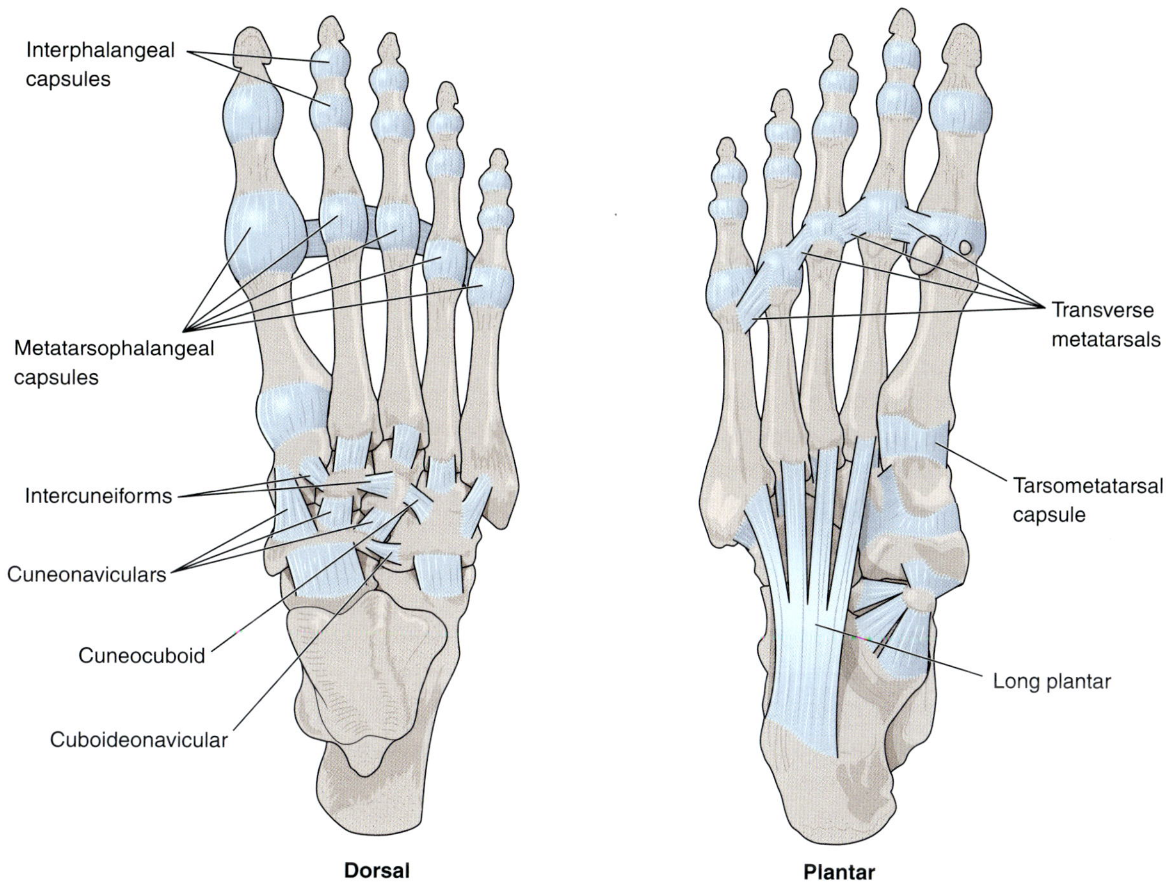

Figure 13.12 Ligaments of the foot.

POP QUIZ

Based on the position of the left lower leg and foot, with the ankle appearing to be in plantar flexion and inversion in this photo, which of the following ligaments is most likely to be sprained?

a. calcaneofibular

b. deltoid

c. anterior talofibular

d. posterior talofibular

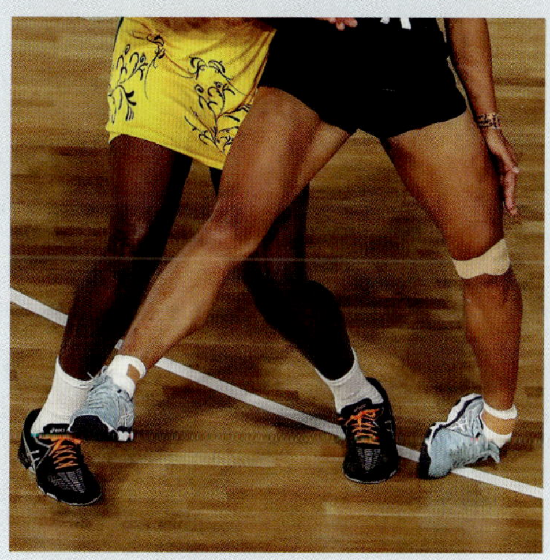

Scott Barbour/Getty Images

ligament. The calcaneocuboid joint has a capsular ligament and the calcaneocuboid ligament. The bifurcated ligament runs from the calcaneus and divides into fibers that run to the cuboid and the navicular bones. The long plantar ligament runs from the calcaneus to the cuboid, with fibers extending to the bases of the third, fourth, and fifth metatarsal bones (see figures 13.10 and 13.12). The long plantar ligament and the calcaneonavicular ligament are involved in one of the arches of the foot (see figure 13.10). Additional ligaments of the intertarsal group include three dorsal and three plantar cuneonavicular ligaments; the dorsal, plantar, and interosseous cuboideonavicular ligaments; the dorsal, plantar, and interosseous intercuneiform ligaments; and the dorsal, plantar, and interosseous cuneocuboid ligaments.

The tarsometatarsal ligaments are the dorsal and plantar capsular ligaments and the interosseous (collateral) ligaments (under the capsules on the medial and lateral aspects of the joints) joining the five metatarsal bones to the tarsal bones of the midfoot to form the tarsometatarsal joints (see figures 13.10, 13.12, and 13.13). The intermetatarsal ligaments join the bases of the five metatarsal bones together at the tarsometatarsal joints.

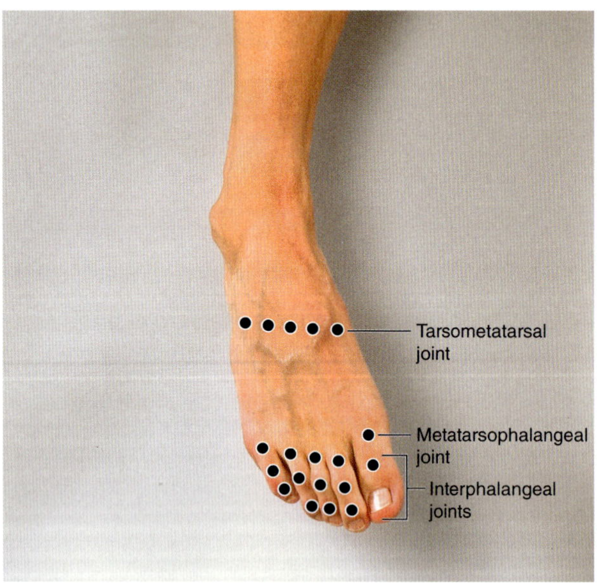

Figure 13.13 Locating the tarsometatarsal, metatarsophalangeal, and interphalangeal joints.

The MTP joints, which join the metatarsal bones with the proximal phalanges, also have dorsal and plantar capsular ligaments and interosseous (collateral) ligaments under the capsules on the medial and lateral aspects of the joints (see figures 13.12 and 13.13). Additionally, there is a transverse metatarsal ligament connecting all five heads of the metatarsal bones (see figure 13.12). Note that the first and fifth MTP joints are often referred to as the *large ball* and the *small ball of the foot*.

Hands On

Observe the plantar surface of your foot, and determine for yourself which ball is which.

The IP joints include the four proximal interphalangeal (PIP) joints and four distal interphalangeal (DIP) joints of the four toes and the single IP joint of the great toe. The IP joints are joined by dorsal and plantar capsular ligaments and interosseous (collateral) ligaments.

Arches

Anatomically, an arch is defined as the structures forming a curved or bow-shaped object. The foot contains either two or three arches, depending on one's viewpoint.

The longitudinal arch runs from the calcaneus to the heads of the metatarsal bones on the plantar surface of the foot (figure 13.14). The arch is formed by a combination of the shapes of the bones and the ligamentous structures supporting the bones. The space created beneath the arch allows the muscles, tendons, blood vessels, and nerves of the plantar surface of the foot to pass without being crushed against the ground.

The metatarsal bones form an arch (or arches) as a result of their ligamentous attachments to both the phalanges and the tarsal bones. An anterior view of the foot reveals that the MTP joints form the metatarsal arch. This arch appears at the other (proximal) end of the metatarsal bones as they articulate with the tarsal bones. Some authors also refer to this arch as the *metatarsal arch*, whereas others refer to it as the *transverse arch* of the foot (see figure 13.14).

Five other structures (similar to the two found in the wrist) in the ankle and foot are the extensor

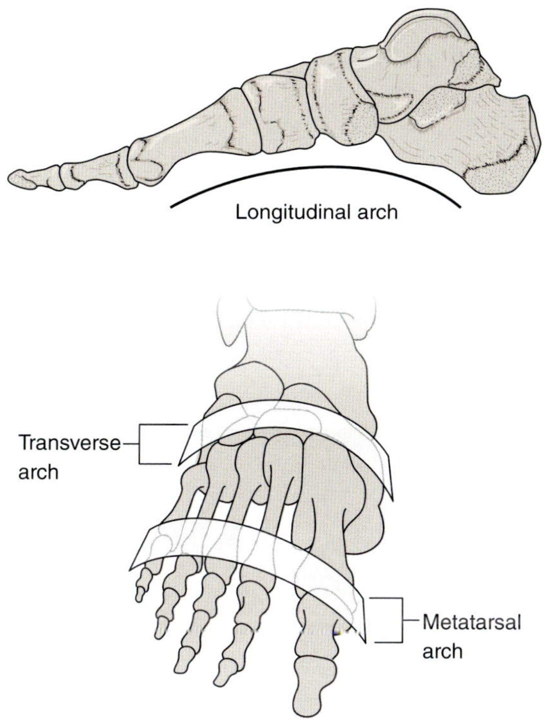

Figure 13.14 The longitudinal, transverse, and metatarsal arches.

retinaculum (superior and inferior), the fibular retinaculum (superior and inferior), and the flexor retinaculum (see figures 13.19 and 13.24). All these structures function primarily to keep the tendons of muscles in their appropriate positions.

Fundamental Movements of the Lower Leg, Ankle, and Foot

Upward movement of the foot toward the anterior leg is known as *dorsiflexion* of the ankle joint. Downward movement of the foot is known as *plantar flexion* (figure 13.15). Lateral (outward) movement of the foot at the talocalcaneal (subtalar) joint produces a movement known as *eversion*. Medial (inward) movement of the foot at the talocalcaneal joint produces a movement known as *inversion*. The motion between the hindfoot (talus) and midfoot (navicular) bones also contributes to these movements (figure 13.16).

Although the talocrural joint is considered the true ankle joint, most authors cite plantar flexion, dorsiflexion, inversion, and eversion as the four

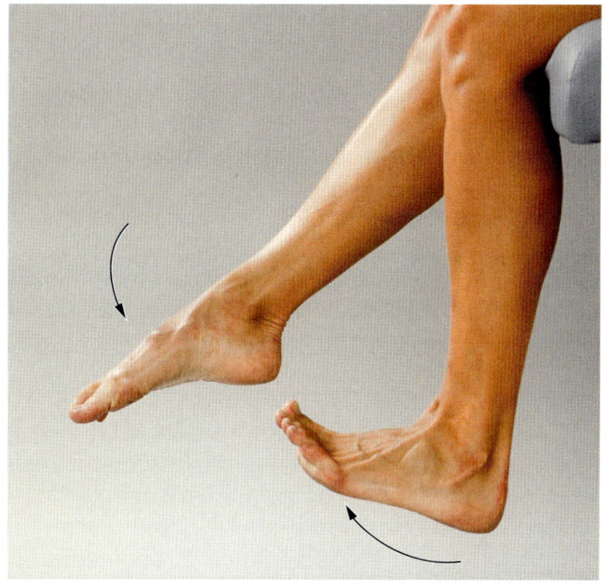

Figure 13.15 Plantar flexion of the right ankle and dorsiflexion of the left ankle.

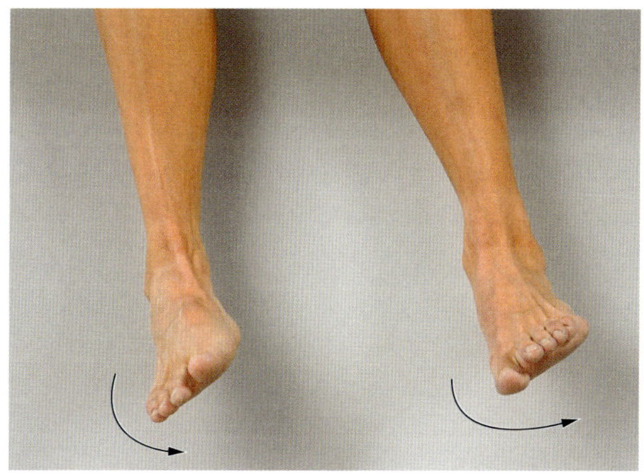

Figure 13.16 Inversion of the right ankle and eversion of the left ankle.

fundamental movements of the ankle joint. Because plantar flexion and dorsiflexion occur in the sagittal plane and eversion and inversion in the frontal plane, the ankle joint is a biaxial joint. As a biaxial joint, the ankle is capable of circumduction (a combination of the fundamental movements of a biaxial or triaxial joint).

Movement between the tarsal bones of the foot is similar to that of the carpal bones of the wrist. The bones glide over each other to produce slight movement. These gliding movements of the tarsals, combined with the movements of the ankle joint, result in either supination of the foot (inversion of the ankle and adduction of the foot) or pronation of the foot (eversion of the ankle and abduction of the foot). Again, the motion between the hindfoot and midfoot bones also contributes to these movements.

Movements at the tarsometatarsal, MTP, PIP, IP, and DIP joints are limited to flexion and extension, except that the MTP joints are also able to abduct and adduct (figure 13.17).

FOCUS ON

Turf Toe

Spraining the ligaments of the first MTP joint (great toe), often from hyperextension, is frequently referred to as *turf toe*, depending on the mechanism of injury. Knowledge of anatomy, physics, and biomechanics, combined with the experience of engineers familiar with the effects of the interaction of footwear and the playing surface, can assist in preventing a sprain of the first MTP joint.

danp68/iStock/Getty Images Plus

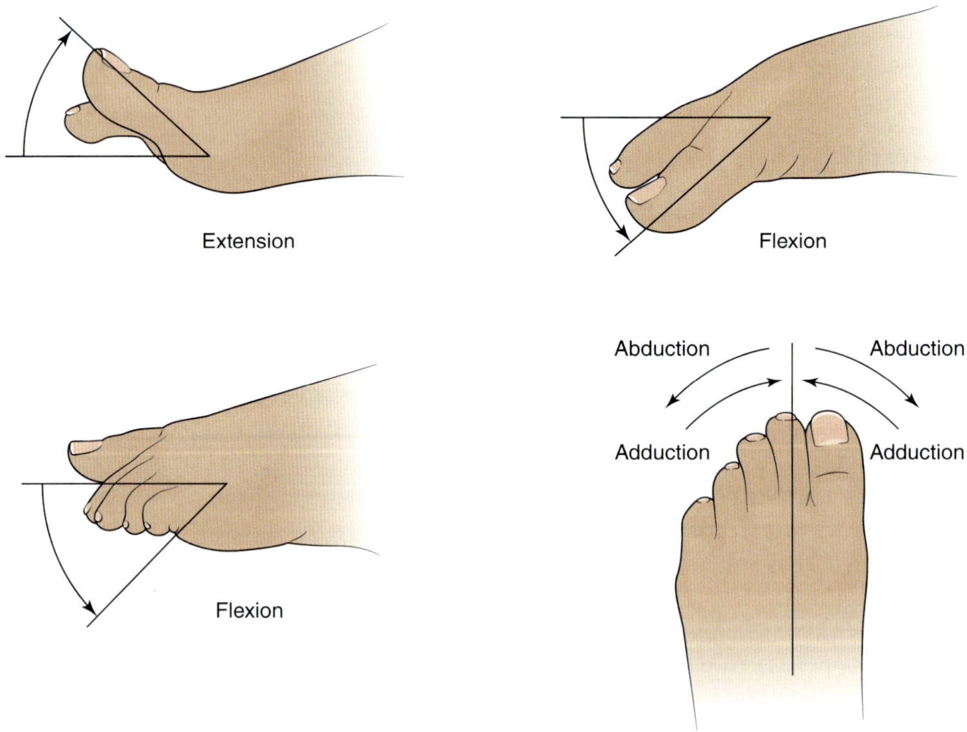

Figure 13.17 Extension and flexion of the toe joints. Abduction and adduction of the metatarsophalangeal joints.

Muscles of the Lower Leg, Ankle, and Foot

The muscles of the leg, ankle, and foot, like those of the hand, are typically divided into the extrinsic muscles (those originating outside the foot and inserting within the foot) and the intrinsic muscles (those originating and inserting within the foot).

Now that we have considered the movements possible in the ankle (talocrural and talocalcaneal) joint and the MTP, PIP, IP, and DIP joints, learning the actions and locations of these muscles should be easier.

Extrinsic Muscles

There are 12 extrinsic muscles of the foot and ankle that are contained in 4 well-defined compartments of the lower leg (figure 13.18). Four muscles are found in the anterior compartment, two in the lateral compartment, three in the superficial posterior compartment, and three in the deep posterior compartment.

Anterior Compartment

The four muscles of the lower-leg anterior compartment are the tibialis anterior, the extensor digitorum longus, the extensor hallucis longus, and the fibularis tertius (figure 13.19).

- *Tibialis anterior:* The tibialis anterior originates from the upper two-thirds of the lateral side of the tibia and inserts on the medial side of the medial cuneiform and the base of the first metatarsal bone (see figure 13.19). If you consider the origin and insertion of the tibialis anterior, what actions of the ankle joint are produced by contraction of this muscle? (Answer: dorsiflexion and inversion)

Hands On

Find your anterior tibial shaft and move your fingers just lateral to the tibia. The first soft tissue you palpate is your tibialis anterior muscle (figure 13.20). Additionally, as you dorsiflex your ankle, observe the tendon of the tibialis anterior on the anterior medial surface of the ankle joint.

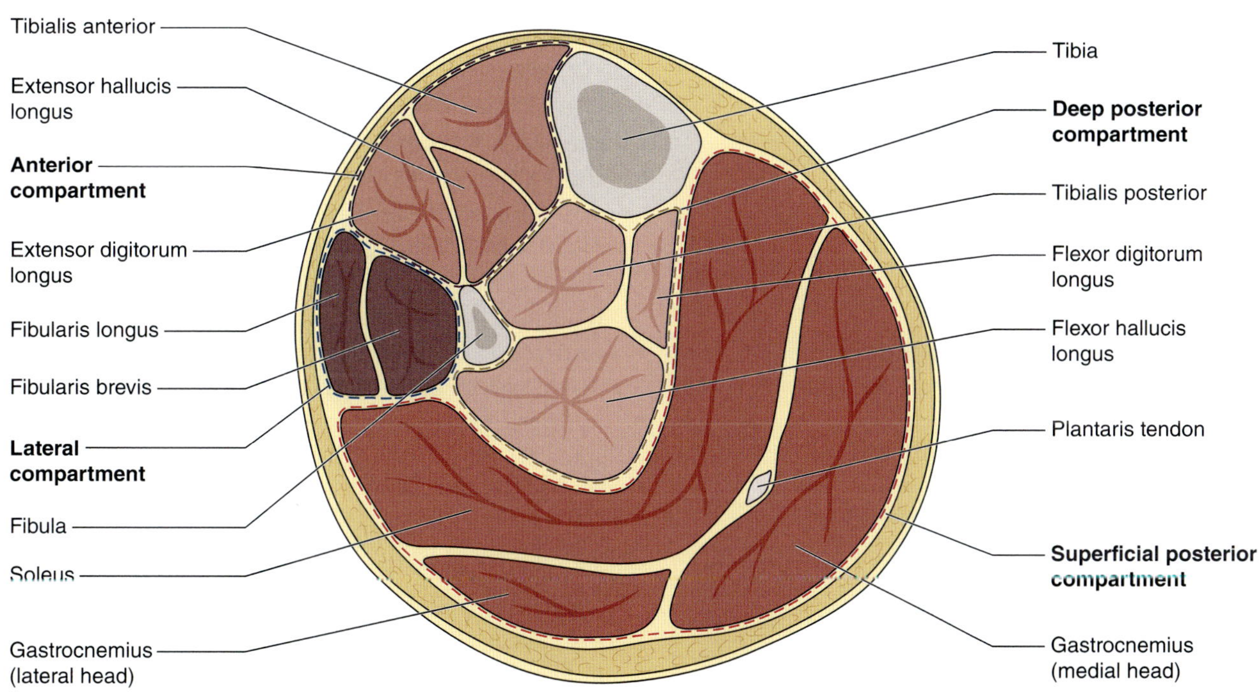

Figure 13.18 Cross section of the 4 compartments of the lower leg, which contain the 12 extrinsic muscles of the ankle and foot.

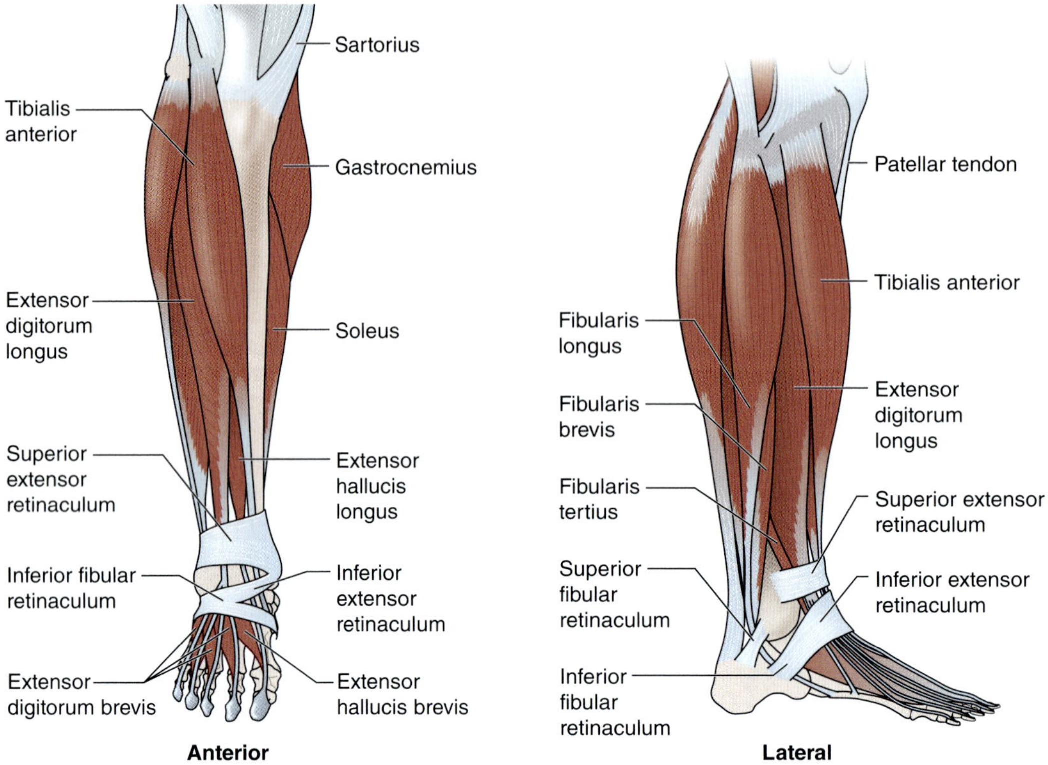

Figure 13.19 The retinacula and anterior and lateral muscles of the ankle and foot.

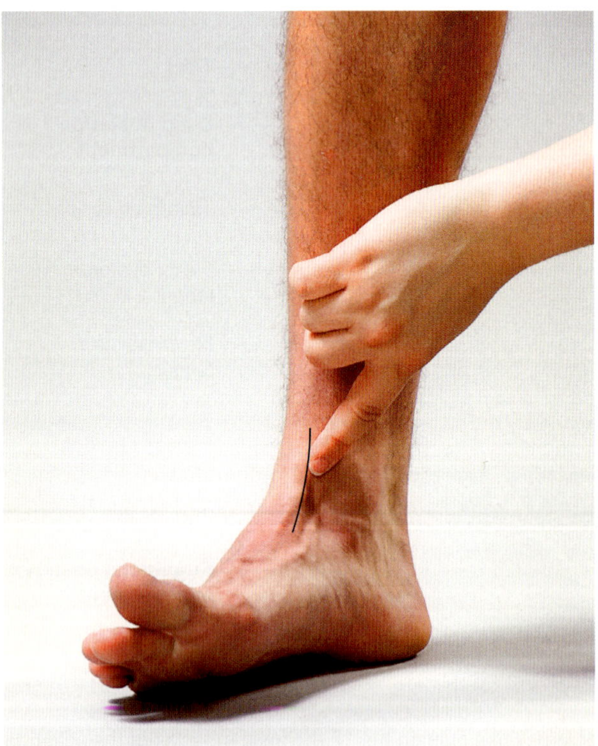

Figure 13.20 Locating the tibialis anterior.

- *Extensor digitorum longus:* The second extrinsic muscle of the anterior compartment is the extensor digitorum longus (see figure 13.19). The extensor digitorum longus originates from the lateral condyle of the tibia, the proximal three-quarters of the fibula, and the interosseous membrane and inserts on the middle and distal phalanges of the lateral four toes. From this muscle's name and its origin and insertion, what are the actions of this muscle at the ankle joint and the MTP, PIP, and DIP joints? (Answer: dorsiflexion of the ankle and extension of the MTP, PIP, and DIP joints of the four lateral toes)

Hands On

As you dorsiflex your ankle, observe the tendons of the extensor digitorum longus on the anterior surface of the ankle joint (lateral to the tibialis anterior; figure 13.21). Also, observe how the tendon splits into four tendons of insertion, which should be visible on the superior (dorsal) aspect of your foot.

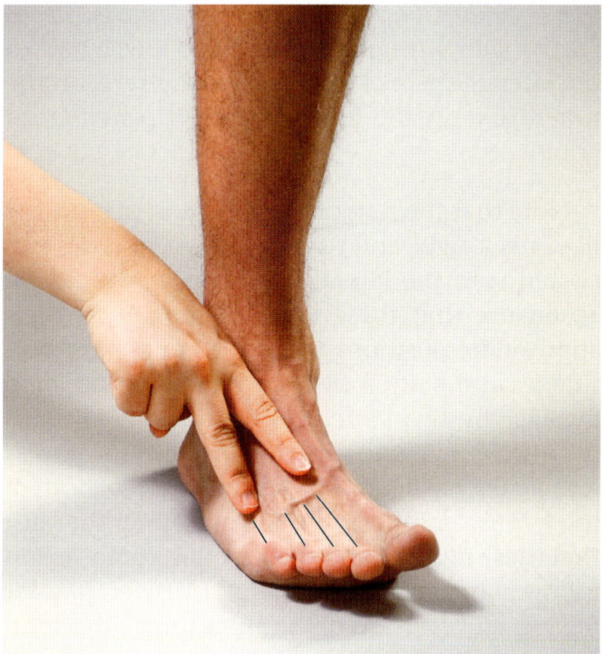

Figure 13.21 Locating the extensor digitorum longus.

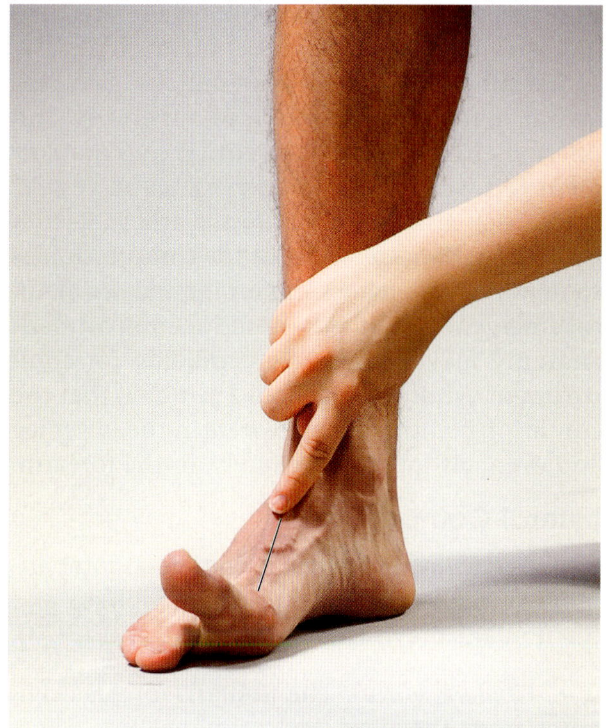

Figure 13.22 Locating the extensor hallucis longus.

• *Extensor hallucis longus:* The third muscle of the lower-leg anterior compartment is the extensor hallucis longus (see figure 13.19). Just as structures in the upper extremity use the term *pollicis* to mean thumb, in the lower extremity, the term *hallucis* refers to the great toe. The extensor hallucis longus originates on the middle half of the medial aspect of the fibula and the adjoining interosseous membrane and inserts on the distal phalanx of the great toe. As you observe the joints that this muscle crosses, what would you expect its actions to be at the ankle and the MTP and IP joints of the great toe? (Answer: dorsiflexion and inversion of the ankle and extension of the MTP and IP joint of the great toe)

Hands On

Extend your great toe, and observe the tendon of insertion of the extensor hallucis longus (figure 13.22). Tracing it from the great toe to the ankle joint, note that it quickly disappears as it moves beneath the muscle fibers of the two previously discussed anterior compartment muscles: the tibialis anterior and the extensor digitorum longus (see figure 13.19).

• *Fibularis tertius:* The fourth and final muscle of the anterior compartment of the lower leg is the fibularis tertius (one of three fibularis muscles in the lower leg; the term *tertius* means third). The fibularis muscles are sometimes referred to as *fibular muscles*. The fibularis tertius originates from the lower third of the fibula and adjoining interosseous membrane and inserts on the superior (dorsal) surface of the base of the fifth metatarsal bone (see figure 13.19). Because the fibularis tertius crosses the anterior aspect of the ankle and is the most lateral of the four anterior compartment muscles, what would you expect the actions of the fibularis tertius muscle to be? (Answer: dorsiflexion and eversion of the ankle joint)

The anterior compartment is prone to a traumatic condition known as *compartment syndrome*. As a result of direct trauma, overuse, or muscle fatigue, hemorrhaging into the compartment causes an acute emergency. Swelling that leads to hardness of the compartment, discoloration, and shiny appearance of the skin from being stretched indicates bleeding into the compartment. The compartment is enclosed laterally by skin and fascia, medially by bone, and superiorly and inferiorly by bone and ligament. As the compartment fills with blood, the nerves, blood vessels, and muscle tissue are compressed and subject to permanent damage. Relief of the compression as soon as possible makes this an immediate medical emergency. While the anterior compartment anatomy makes it a likely candidate for compartment syndrome, it should be noted that the condition could occur elsewhere in the lower leg and even, although more rarely, in the forearm.

Lateral Compartment

Two muscles are contained in the lateral compartment of the lower leg: the fibularis longus and the fibularis brevis (see figures 13.19 and 13.23).

- *Fibularis longus:* The fibularis longus originates from the head and proximal two-thirds of the lateral aspect of the fibula and the lateral condyle of the tibia and inserts on the inferior lateral surfaces of the medial cuneiform and first metatarsal bones. It was previously noted in this chapter that the cuboid bone has a concavity (groove) along its lateral border. This concavity is where the tendon of insertion of the fibularis longus passes from the lateral aspect of the foot to the inferior plantar surface.

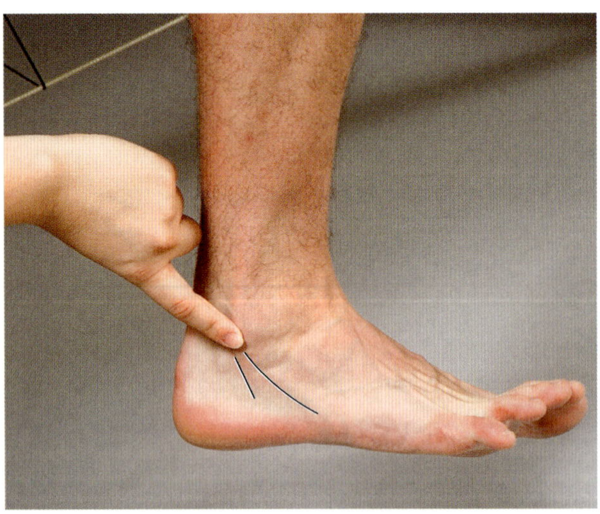

Figure 13.23 Locating the fibularis longus and fibularis brevis.

- *Fibularis brevis:* The fibularis brevis (*brevis* means short) originates from the lower half of the lateral aspect of the fibula and inserts on the tuberosity on the base of the fifth metatarsal bone.

Note that the tendons of insertion of both the fibularis longus and fibularis brevis muscles run posterior to the lateral malleolus of the fibula. To ensure that these tendons remain in their appropriate positions while the ankle joint dorsiflexes and plantar flexes, the tendons are held in place by two structures previously discussed in this chapter: the superior and inferior fibular retinacula (see figure 13.19). The action of both of these muscles is the same. If we consider the path that these muscles and their tendons of insertion follow, what are these actions? (Answer: plantar flexion and eversion of the ankle joint) Anatomically, it is important to note the involvement of these muscles in the prevention and rehabilitation of the classic ankle (ATF, or inversion) sprain.

Superficial Posterior Compartment

The three muscles of the superficial posterior compartment of the lower leg are the gastrocnemius, the soleus, and the plantaris. Collectively, these muscles are often referred to as the *muscles of the calf* (figure 13.24) and are also referred to as the *triceps surae*. Two of these muscles originate superior to the lower leg on the femur of the thigh (gastrocnemius and plantaris), whereas the third (soleus) originates on the lower leg (tibia only).

- *Gastrocnemius:* The gastrocnemius is a two-headed muscle (figures 13.24 and 13.25). Its lateral head originates on the popliteal surface of the femur just medial to the lateral condyle, whereas the medial head originates on the popliteal surface of the medial condyle of the femur. Both heads combine to form the belly of the muscle, and a common tendon of insertion, known as the *Achilles tendon*, attaches the muscle to the posterior surface of the calcaneus. Because the gastrocnemius crosses both the knee and ankle joints, extension of the knee with dorsiflexion of the ankle can contribute to an Achilles tendon strain.

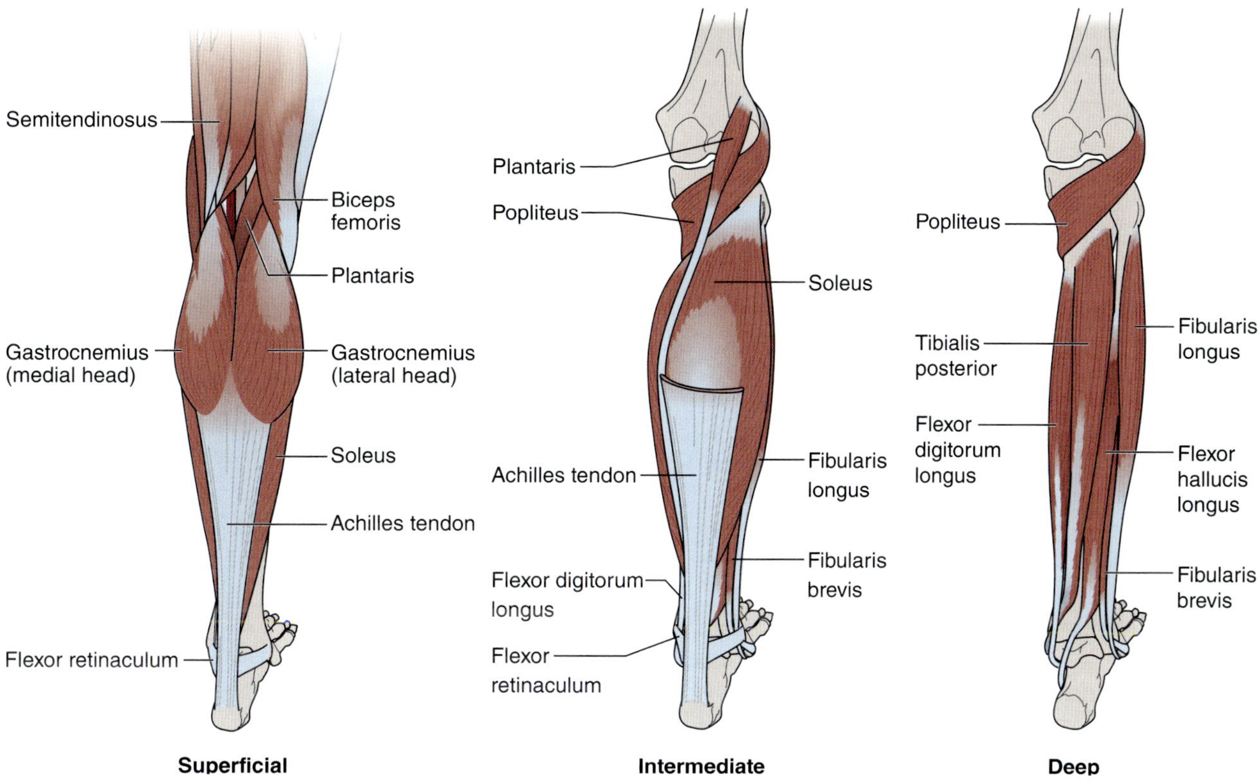

Figure 13.24 Posterior muscles of the ankle and foot.

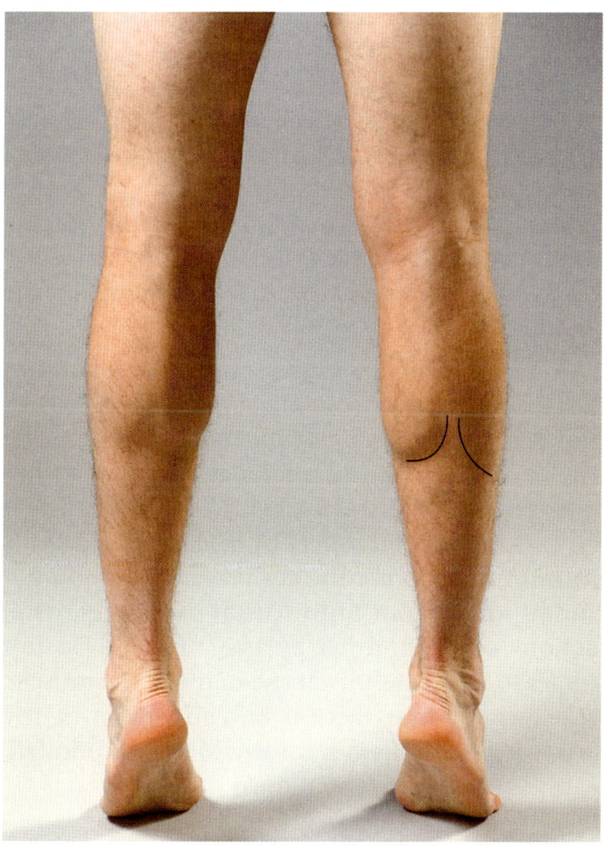

Figure 13.25 Locating the gastrocnemius muscle.

- *Soleus:* The soleus (see figure 13.24) originates from the head and proximal third of the posterior aspect of the fibula and the middle third of the posterior aspect of the tibia. It inserts on the calcaneus via the Achilles tendon, the same tendon of insertion as the gastrocnemius.

Hands On

You can easily palpate the Achilles tendon just proximal to your calcaneus.

- *Plantaris:* The third muscle that occurs in the superficial posterior compartment is the plantaris muscle (see figure 13.24). This muscle originates from the lateral linea aspera and the oblique popliteal ligament and inserts on the posterior aspect of the calcaneus (separately from the Achilles tendon). Although this short-bellied muscle lays claim to having the longest tendon of any muscle in the body, it is of minor significance in assisting other muscles of the knee and ankle joints. The plantaris crosses the knee and ankle in the same fashion as the gastrocnemius and is

FOCUS ON

Achilles Tendon

The tendon of insertion of both the gastrocnemius and soleus muscles is the Achilles tendon, which is often a source of concern in athletics. The gastrocnemius and soleus muscles are subject to strain as a result of overuse, and these strains are common in middle-aged tennis players, sports officials, and runners. Overuse injuries often result from increased activity (running or jumping), defined as an increase in frequency, intensity, duration, or any combination of these factors. Increased activity can lead to inflammation of the tendon, which is poorly vascularized to begin with, making recovery more complicated. Achilles tendinitis (inflammation) can lead to a rupture of the tendon as a result of a progressive degeneration of the tissue. Additionally, a rupture of the tendon can result from excessive force in the form of (1) direct force to, (2) forceful contraction of, or (3) forceful loading of the gastrocnemius and soleus muscles.

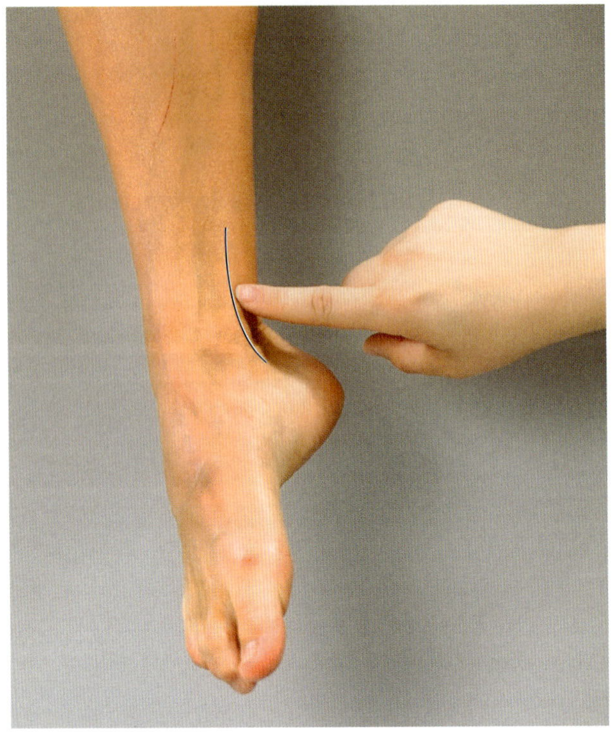

Figure 13.26 Locating the tibialis posterior.

strained by the same mechanisms. Its function is that of an assistant to the knee flexors and ankle plantar flexors. It is not a major joint mover and is analogous in function to the palmaris longus in the forearm and wrist.

Because all three muscles of the superficial posterior compartment insert on the posterior aspect of the calcaneus, their action at the ankle joint is limited to one movement. What action do all three of these muscles perform at the ankle joint? (Answer: plantar flexion of the ankle)

Deep Posterior Compartment

Three muscles, which bear names very similar to those of the anterior compartment, are found in the deep posterior compartment of the lower leg. These three muscles are the tibialis posterior, the flexor digitorum longus, and the flexor hallucis longus (see figure 13.24).

- *Tibialis posterior:* The tibialis posterior originates from the middle third of the posterior aspect of the tibia, the proximal two-thirds of the medial aspect of the fibula, and the interosseous membrane. It inserts on the inferior (plantar) surfaces of the navicular and medial cuneiform bones and the bases of the second, third, fourth, and fifth metatarsal bones (see figures 13.24 and 13.26). Observe that the tendon of insertion passes posterior to the medial malleolus of the tibia: What are the likely actions of this muscle? (Answer: plantar flexion and inversion of the ankle joint)
- *Flexor digitorum longus:* The flexor digitorum longus originates from the lower two-thirds of the posterior aspect of the tibia and inserts on the bases of the distal phalanges of the four lateral toes (see figures 13.24 and 13.27). Because it crosses the medial aspect of the ankle and splits into four tendons running through the plantar surface of the foot to the distal phalanges, what actions are performed by this muscle? Do not forget to look carefully at the name of the muscle. (Answer: plantar flexion and inversion of the ankle joint and flexion of the MTP, PIP, and DIP joints of the four lateral toes)

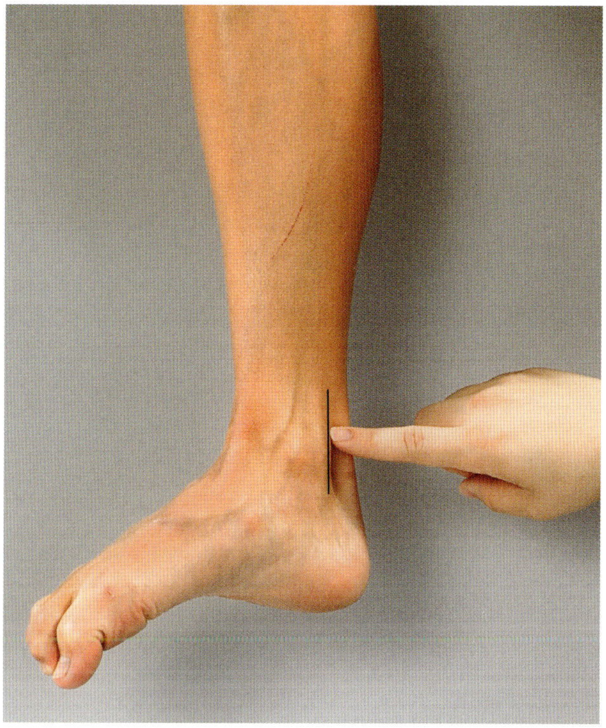

Figure 13.27 Locating the flexor digitorum longus.

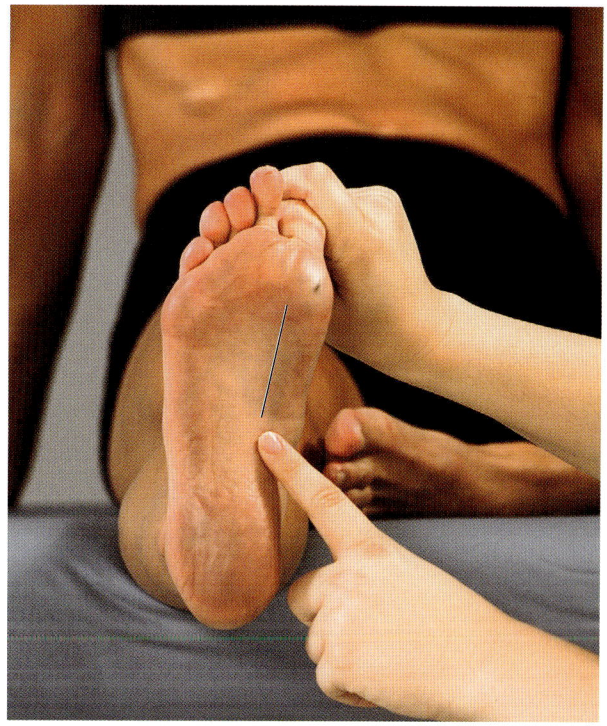

Figure 13.28 Locating the flexor hallucis longus.

• *Flexor hallucis longus:* The flexor hallucis longus originates from the lower two-thirds of the posterior aspect of the fibula and interosseous membrane and inserts on the base of the distal phalanx of the great toe (see figures 13.24 and 13.28). Because it crosses the medial aspect of the ankle and runs through the plantar surface of the foot to the distal phalanx, what actions are performed by this muscle? Do not forget to look carefully at the name of the muscle. (Answer: plantar flexion and inversion of the ankle joint and flexion of the MTP and IP joints of the great toe)

Intrinsic Muscles of the Foot

The majority of intrinsic muscles of the foot are found on the plantar surface, where the bony and ligamentous arrangements create space to accommodate these structures. These muscles are found in four distinct layers and are very similar in arrangement to the intrinsic muscles of the hand. Note that the names of many of these muscles indicate their actions.

Superficial to all muscles on the plantar surface of the foot is a structure known as the *plantar fascia* (figure 13.29). This fibrous band originates from the calcaneus; inserts on the five MTP joints; and helps to protect the muscles, blood vessels, and nerves running through the plantar surface of the foot.

First Layer

Beneath the plantar fascia, the first plantar layer contains three muscles (see figure 13.29): the flexor digitorum brevis, the abductor hallucis, and the abductor digiti minimi.

• *Flexor digitorum brevis:* The flexor digitorum brevis originates from the tuberosity on the plantar surface of the calcaneus and inserts on the middle phalanges of the four lateral toes. This muscle flexes the MTP and PIP joints of the four lateral toes.

• *Abductor hallucis:* The abductor hallucis originates from the medial aspect of the plantar surface of the calcaneus and inserts on the medial surface of the proximal phalanx of the great toe.

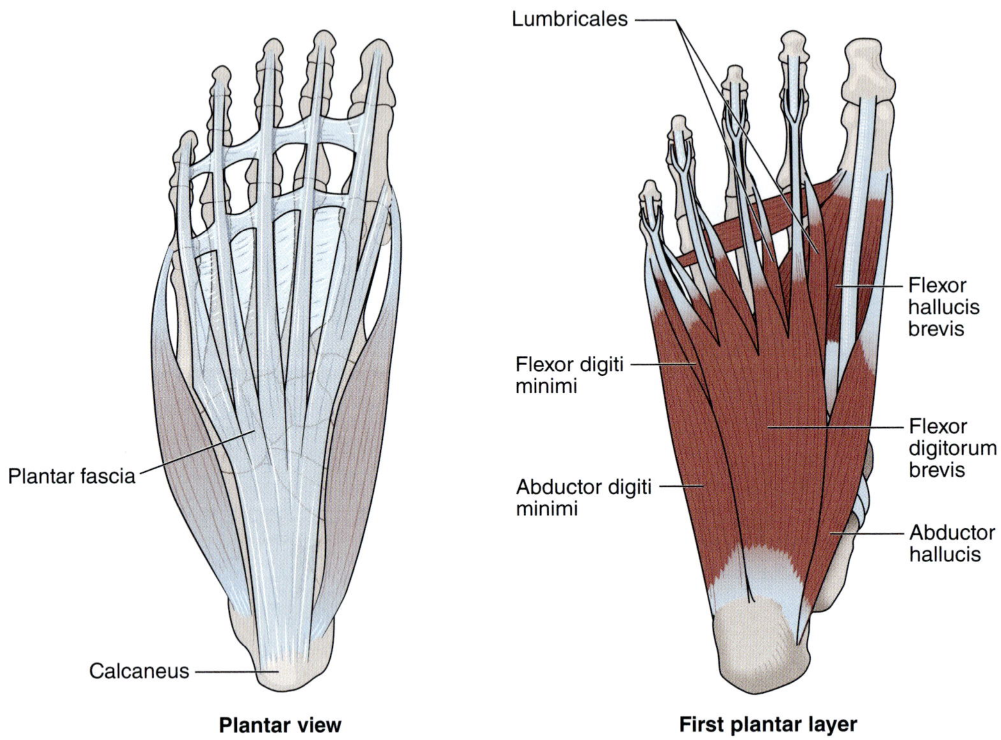

Figure 13.29 The plantar fascia and muscles in the first plantar layer of the foot.

FOCUS ON

Shin Splints

The term *shin splints* has been called a *wastebasket term* by many because it is used to describe a variety of conditions involving the lower leg, including inflammation of the tendons of both the anterior and posterior tibialis muscles, stress fractures of the tibia and fibula, inflammation of the periosteum of the tibia and fibula (periostitis), and inflammation of the interosseous membrane between the fibula and tibia. Another accepted description is *osteoperiostitis* (inflammation of the periosteum of the tibia) resulting from overuse (repetitive loading) of the posterior tibialis and soleus muscles. Causes include improper footwear, problems of the longitudinal arch, playing surfaces, increased training demand (intensity, frequency, duration, or some combination of these), and repetitive training activities (resulting in multiple symptoms referred to as an *overuse syndrome*). Whatever the definition or cause, attention must be given to lower-leg pain. Running it out is not the answer. Ignoring a leg injury can result in continuing inflammation that can develop into ischemia (anemia, a lack of blood, caused by obstruction of circulation), which can lead to serious conditions, such as necrosis (death of tissue) and stress fractures.

The muscle abducts (moves away from the second toe) the MTP joint of the great toe and assists with flexion of the MTP joint of the great toe.

- *Abductor digiti minimi:* The abductor digiti minimi originates from the tuberosity on the plantar surface of the calcaneus and inserts on the lateral aspect of the base of the proximal phalanx of the fifth toe. The muscle abducts the MTP joint of the fifth (little) toe and assists with flexion of the fifth MTP joint.

FOCUS ON

Plantar Fasciitis

A common problem for runners is inflammation of the area of origin of the plantar fascia from repeated stress. This condition is known as *plantar fasciitis*. The point of attachment of the plantar fascia is just anterior to the tubercle of the inferior aspect of the calcaneus. Stress placed on the foot through the repetitive action of running can cause inflammation of this area. Prevention and treatment of this condition are often concerns for orthopedists and podiatrists.

Second Layer

The second plantar layer contains two muscles, the quadratus plantae and the lumbricals (figure 13.30), with some unique characteristics.

- *Quadratus plantae:* The quadratus plantae muscle has two heads—one from the medial surface of the calcaneus and the other from the lateral surface of the calcaneus—which combine to attach to the tendons of the flexor digitorum longus. This muscle assists with flexion of the MTP, PIP, and DIP joints of the lateral four toes.
- *Lumbricals:* The other muscle of the second plantar layer is really a group of four muscles collectively known as the *lumbricals*. These four muscles have no bony attachments: They originate from the tendon of the flexor digitorum longus and insert on the tendon of the extensor digitorum longus. The lumbricals flex the MTP joints and extend the PIP and DIP joints of the lateral four toes.

Third Layer

The third plantar layer contains three muscles whose names indicate their functions (see figure 13.30): the flexor hallucis brevis, the adductor hallucis, and the flexor digiti minimi brevis.

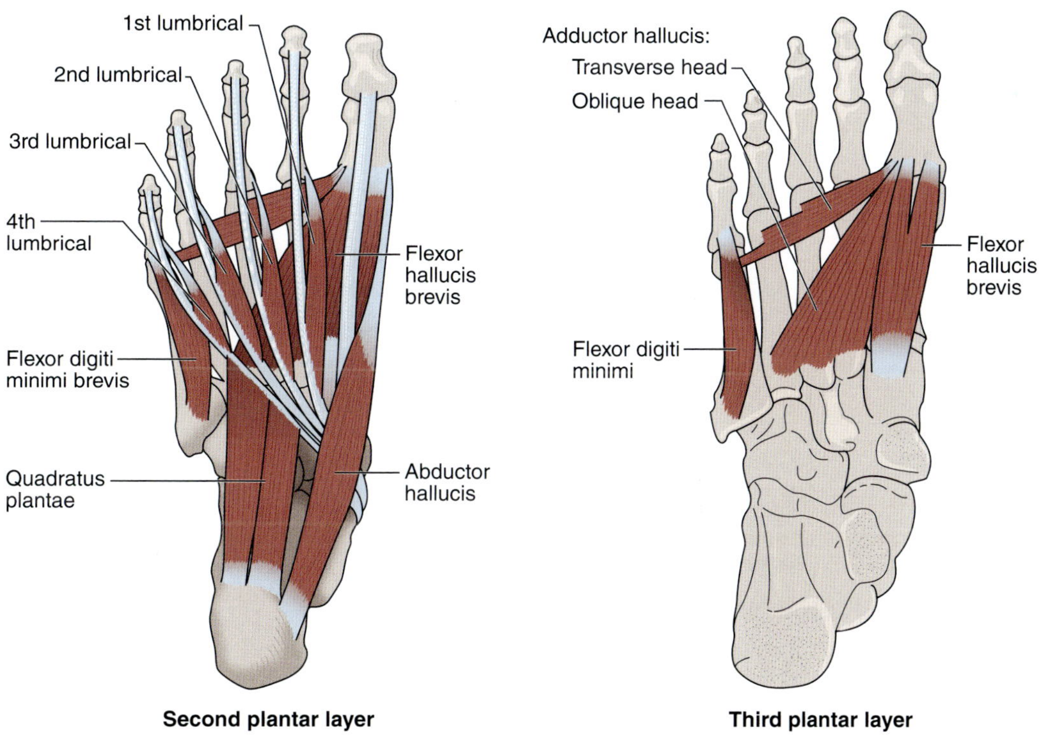

Figure 13.30 Muscles in the second and third plantar layers of the foot.

- *Flexor hallucis brevis:* The flexor hallucis brevis originates from the three cuneiform bones and inserts on the base of the proximal phalanx of the great toe. It flexes the MTP joint of the great toe, as its name indicates.
- *Adductor hallucis:* The adductor hallucis has two heads: One is an oblique head that originates from the bases of the second, third, and fourth metatarsals and inserts on the proximal phalanx of the great toe. The second head is a transverse head that originates from the plantar surface of the MTP capsular ligaments of the third, fourth, and fifth toes and inserts on the proximal phalanx of the great toe. As its name indicates, this muscle adducts the great toe.
- *Flexor digiti minimi brevis:* The flexor digiti minimi brevis originates from the base of the fifth metatarsal bone and inserts on the lateral aspect of the base of the proximal phalanx of the fifth (little) toe. Its function, as its name indicates, is to flex the MTP joint of the little toe.

Fourth Layer

The fourth layer on the plantar surface of the foot contains two groups of muscles: the dorsal interossei and plantar interossei found between the bones (interosseous) of the foot (figure 13.31).

- *Dorsal interossei:* There are four dorsal interossei muscles. The dorsal interossei originate from the adjacent sides of all metatarsal bones and insert on the bases of the second, third, and fourth proximal phalanges. Note that the midline of the foot is considered the second toe (as opposed to the middle finger in the hand). The action of the dorsal interossei results in abduction of the third and fourth MTP joints. Because there are dorsal interossei muscles on either side of the second toe, contraction of the dorsal interossei results in no movement of the second toe.
- *Plantar interossei:* There are three plantar interossei muscles. The plantar interossei originate from the medial surfaces of the third, fourth, and fifth metatarsal bones and insert on the medial aspect of the bases of the proximal phalanges of the third, fourth, and fifth toes. The action of the plantar interossei results in the adduction of the MTP joints of the third, fourth, and fifth toes. Note again that the second toe does not move on contraction of the plantar interossei muscles. Is this for the same reason that the second toe did not move when the dorsal interossei contracted? If not, why does the second toe not move when the plantar interosseous muscles contract?

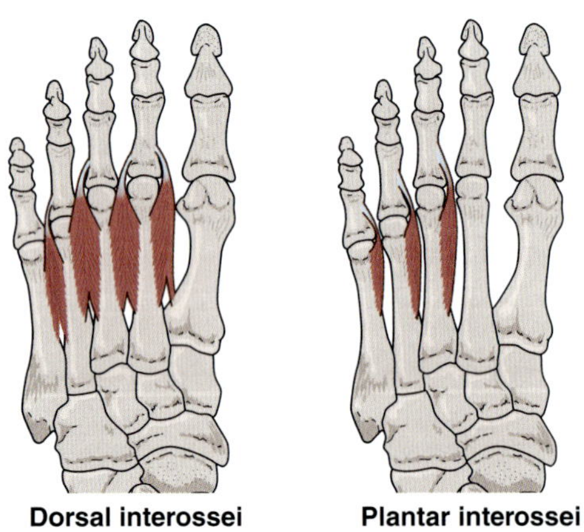

Figure 13.31 The dorsal and plantar interossei.

Dorsal Intrinsic Muscles of the Foot

Although there is not as much space on the dorsal aspect of the foot as there is on the plantar aspect, there are two intrinsic muscles (extensor digitorum brevis and extensor hallucis brevis) on the dorsal aspect (see figure 13.19).

- *Extensor digitorum brevis:* The extensor digitorum brevis (figure 13.32) originates from the calcaneus and inserts on the tendons of the extensor digitorum longus, which in turn inserts on the second, third, and fourth toes.
- *Extensor hallucis brevis:* A fourth tendon of the extensor digitorum brevis branches to the distal phalanx of the great toe and is often called the *extensor hallucis brevis muscle.*

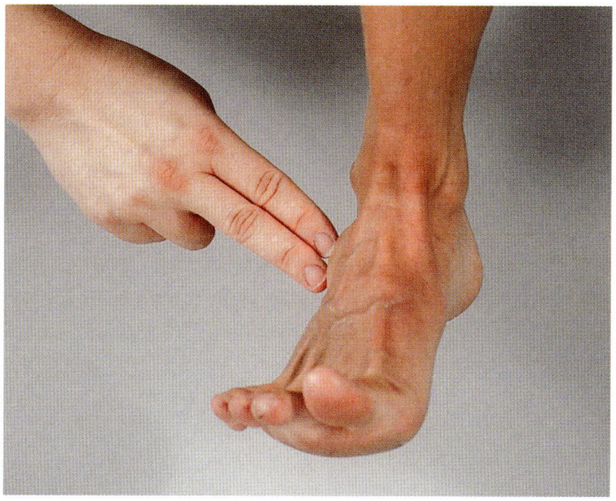

Figure 13.32 Locating the extensor digitorum brevis.

As their names indicate, these two muscles assist with extension of the MTP and IP joints of the great toe and the MTP, PIP, and DIP joints of the second, third, and fourth toes. Table 13.1 summarizes the primary muscles that create movement in the joints of the lower extremity. It presents a review and additional learning source for what performs the fundamental movements of the lower extremity.

Table 13.1 Primary Muscles That Create Movement in the Lower Extremity

Primary joint movements	Agonists
Hip flexion	Rectus femoris, iliacus, psoas major, pectineus, sartorius, tensor fasciae latae
Hip extension	Gluteus maximus, semitendinosus, semimembranosus, biceps femoris (long head)
Hip abduction	Gluteus medius, gluteus minimus, sartorius
Hip adduction	Pectineus, adductor magnus, adductor longus, adductor brevis, gracilis
Hip internal rotation	Gluteus minimus, semitendinosus, semimembranosus
Hip external rotation	Piriformis, superior gemellus, internal obturator, inferior gemellus, external obturator, quadratus femoris
Knee extension	Rectus femoris, vastus lateralis, vastus medialis, vastus intermedius
Knee flexion	Biceps femoris, semitendinosus, semimembranosus, gracilis, popliteus, gastrocnemius, plantaris
Knee internal rotation	Semitendinosus, semimembranosus, gracilis, sartorius, popliteus
Knee external rotation	Biceps femoris
Ankle dorsiflexion	Tibialis anterior, extensor digitorum longus, extensor hallucis longus, fibularis tertius
Ankle plantarflexion	Gastrocnemius, soleus, plantaris, tibialis posterior, flexor digitorum longus, flexor hallucis longus, fibularis longus, fibularis brevis
Ankle eversion	Fibularis longus, fibularis brevis, fibularis tertius, extensor digitorum longus
Ankle inversion	Tibialis anterior, extensor hallucis longus, tibialis posterior, flexor digitorum longus, flexor hallucis longus
Great toe flexion	Flexor hallucis longus, flexor hallucis brevis
Great toe extension	Extensor hallucis longus, extensor hallucis brevis
Toe flexion (2-5)	Flexor digitorum longus, flexor digitorum brevis, lumbricals, quadratus plantae
Toe extension (2-5)	Extensor digitorum longus, extensor digitorum brevis

Visit the web resource to complete the activities for this chapter.

CHAPTER 14

Nerves and Blood Vessels of the Lower Extremity

As in both the upper extremity and the spinal column and thorax, the musculature of the lower extremity is innervated by spinal nerves that are formed into plexuses. The spinal nerves that innervate the muscles of the lower extremity arise from the lumbosacral plexus, which is typically divided into the lumbar plexus (T12, L1, L2, L3, and L4; see figure 14.1), the sacral plexus (L4, L5, S1, S2, and S3; see figure 14.1), and the pudendal (coccygeal) plexus (S2, S3, S4, S5, C1, and C2). The pudendal plexus, which is often considered part of the sacral plexus, innervates many of the structures of the abdominal cavity and the reproductive systems, but none of the muscles of the lower extremity. Therefore, the pudendal plexus is not considered in this discussion of the lower extremity.

Nerves of the Lumbosacral Plexus

The nerves of the lumbosacral plexus are divided into three parts according to their location relative to the spinal column (superior, middle, and inferior).

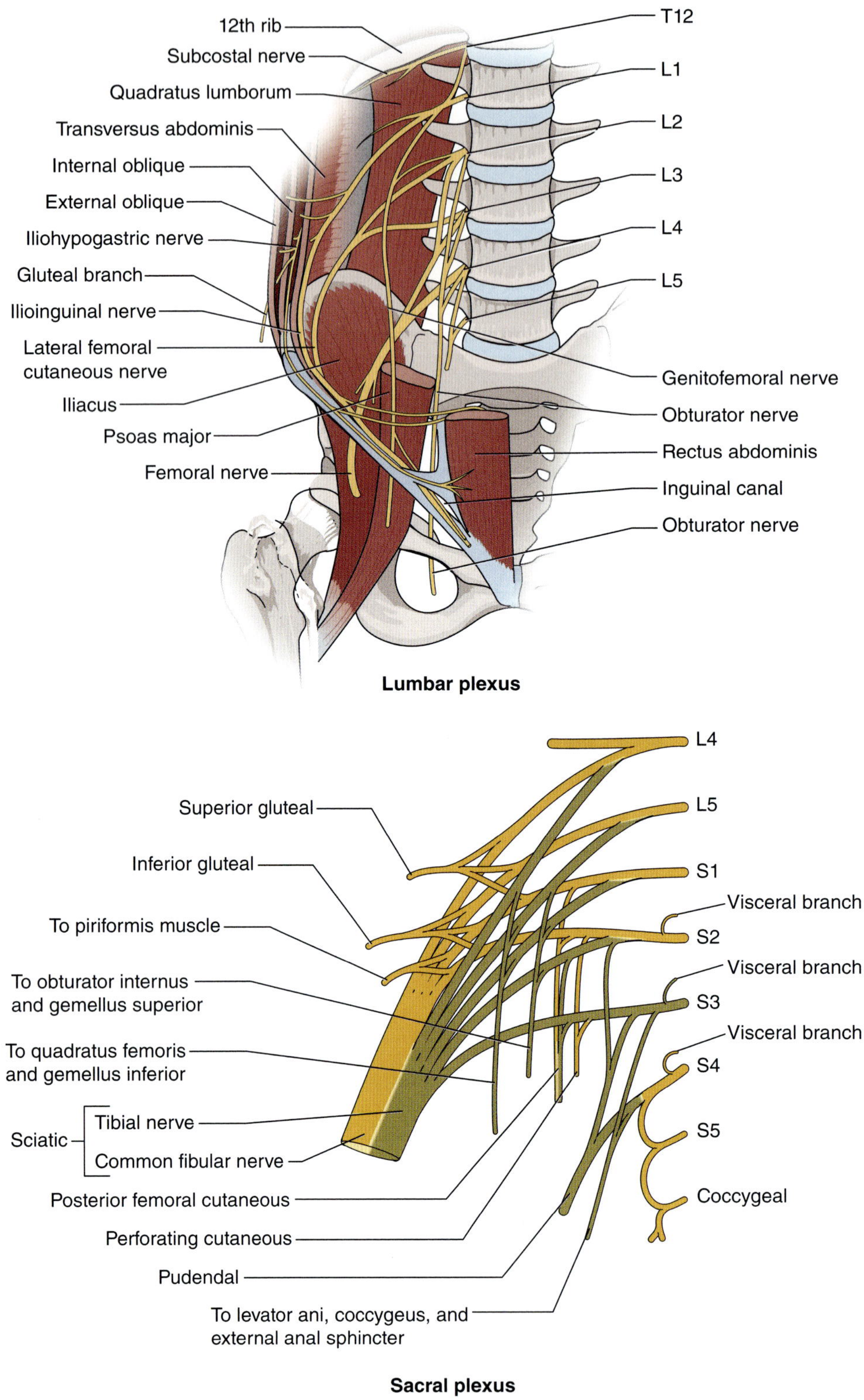

Figure 14.1 The lumbar and sacral plexuses.

Superior Portion

The most superior portion of the lumbosacral plexus, known as the *lumbar plexus*, consists of the femoral, anterior femoral cutaneous, saphenous, genitofemoral, iliohypogastric, ilioinguinal, and obturator nerves and the lateral cutaneous nerve of the thigh, the lateral femoral cutaneous (figure 14.2).

The femoral nerve (L2, L3, and L4) innervates the pectineus, the four muscles of the quadriceps group, and the sartorius. The cutaneous branches of the femoral nerve, the anterior femoral cutaneous nerve, and the saphenous nerve do not innervate muscles. The genitofemoral nerve (L1 and L2) supplies the male and female genitalia and has a muscular branch innervating the cremaster muscle in the scrotum and, therefore, is not considered a nerve of the lower extremity. Likewise, both the iliohypogastric (L1 and, in some people, T12) and ilioinguinal (T12 and, in some people, L1) nerves, although part of the lumbar plexus, have muscular branches that innervate only the abdominal muscles and are considered nerves of the trunk and not of the lower extremity. The lateral femoral cutaneous nerve (L1, L2, L3, and L4) has muscular branches to the psoas major and minor muscles, as well as the quadratus lumborum. Again, these muscles are typically considered muscles of the

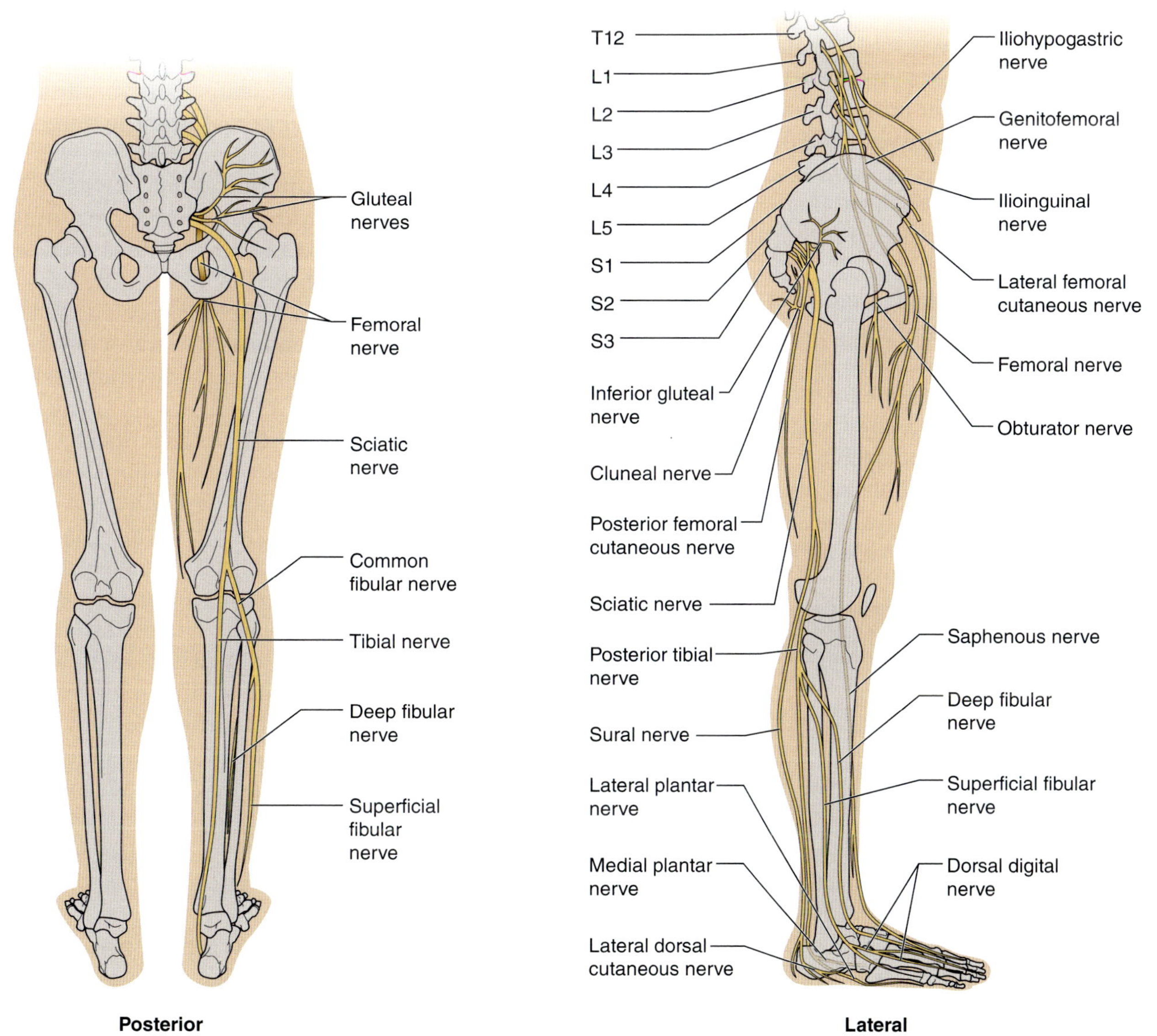

Figure 14.2 Nerves of the lower extremity.

trunk and not of the lower extremity. The obturator nerve (L2, L3, and L4) has anterior and posterior muscular branches that innervate the adductor magnus, adductor longus, adductor brevis, gracilis, pectineus, and external obturator muscles.

Knowledge of the various structures that are innervated by the nerves of the lumbar and sacral plexuses can help in evaluating problems, particularly those at the lumbosacral level of the spinal column. The discussion of the spinal column in chapter 8 pointed out that the cervical and lumbar areas are where the greatest amount of movement takes place. The lumbar area of the spinal column often is an area of concern (i.e., low-back pain). One major problem mentioned in chapter 8 was the space between the vertebral lamina, where nerve roots exit the spinal column and innervate areas of the trunk and lower extremities. Another problem could be impingement of a nerve by the vertebral disc moving into the space where the nerve root exits the spinal column. In evaluating a problem, pain, muscle atrophy, and no history of physical trauma of a lower extremity should point you in the direction of possible nerve damage, either obviously at the site or somewhere along the nerve path, including the level at which the nerve root exits the spinal column. Low-back pain can be caused by several problems, and knowledge of the nervous system is necessary for evaluation. While not the only levels where problems might occur, L3, L4, L5, and S1 seem to be the levels of most concern when evaluating low-back pain. Knowledge of the sensory and motor areas serviced by these plexuses can help in your evaluations.

FOCUS ON

Sciatic Neuritis

Because of its position and length, the sciatic nerve is often subjected to trauma in several places as a result of a direct force, stretching, or impingement. Impingement at the lumbar or sacral disc space and muscular strains in the low back and lower extremities often cause an inflammation of the sciatic nerve called *sciatic neuritis*. Rest, anti-inflammatory agents, and eventually exercise are often prescribed for this condition.

Hands On

Ask a partner to lie supine on a table or the floor. Raise one of your partner's legs as high as you can raise it with the knee flexed. Repeat the leg raise with the knee now in full extension. Repeat the leg raise again with the knee extended. Once you reach the point where you can no longer raise your partner's leg, hold that position. Then, put your hand on your partner's foot and manually dorsiflex the ankle joint. What have you done to your partner regarding the sciatic nerve and its branches?

Middle Portion

The middle portion of the lumbosacral plexus, known as the *sacral plexus*, consists of the superior gluteal, inferior gluteal, sciatic, common fibular, tibial, and sural nerves and the nerve to the piriformis.

The superior and inferior gluteal nerves (L4, L5, S1, and S2) innervate the three gluteal muscles (the gluteus maximus, the gluteus medius, and the gluteus minimus) as well as the tensor fasciae latae muscle. The sciatic nerve (L4, L5, S1, S2, and S3) is the longest and largest nerve in the body and is a combination of two nerves: the tibial (L4, L5, S1, S2, and S3; figure 14.3) and the common fibular (L4, L5, S1, and S2; figure 14.4). A cutaneous branch of the common fibular nerve, the lateral sural nerve, combines with the medial sural branch of the tibial nerve to form the sural nerve. The sural nerve and these branches do not innervate any lower-extremity muscles.

Just distal to the head of the fibula, the common fibular nerve divides to form three terminal branches: the deep fibular nerve, the superficial fibular nerve, and the recurrent tibial nerve (figure 14.5). The deep fibular nerve has muscular branches innervating the four muscles of the anterior compartment in the lower leg, the extensor digitorum brevis, and the interosseous muscles of the foot. The superficial fibular nerve innervates the fibularis longus and fibularis brevis muscles of the lateral compartment of the lower leg. The recurrent tibial nerve innervates the proximal portion of the tibialis anterior muscle. The tibial nerve has muscular branches

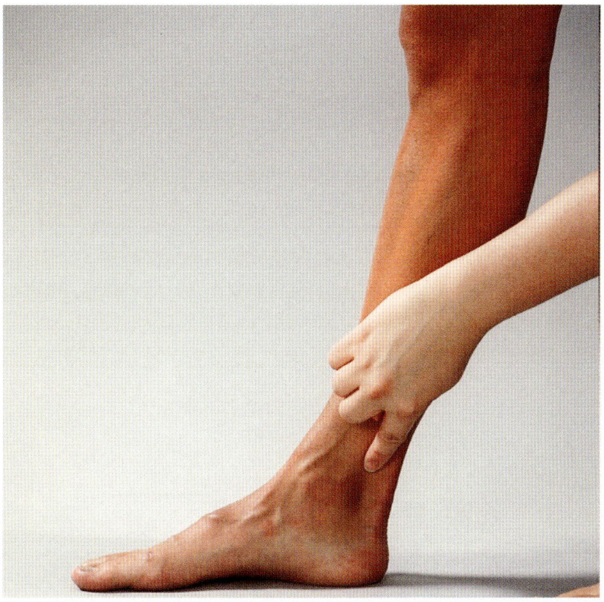

Figure 14.3 Locating the tibial nerve.

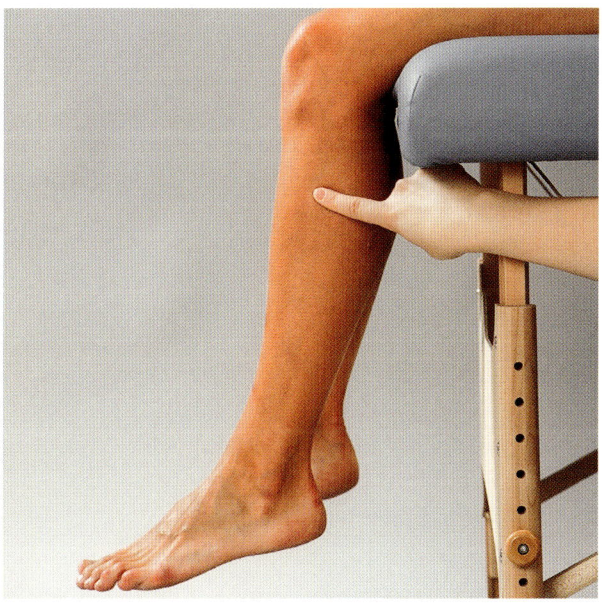

Figure 14.4 Locating the common fibular nerve.

innervating the three hamstring muscles (but not the short head of the biceps femoris). It also innervates four of the six deep external rotators of the hip joint (the superior and inferior gemelli muscles, the internal obturator muscle, and the quadratus femoris) and the muscles in both the superficial and deep posterior compartments of the lower leg.

FOCUS ON

Lower-Leg Trauma

Trauma (e.g., contusion or overuse) to the anterior compartment of the lower leg may cause so much swelling within the compartment that pressure on the deep fibular nerve reduces its ability to transmit an impulse to the muscles of the compartment. Loss of strength and atrophy (wasting away) of the muscle tissue can lead to complications, such as a "dropped foot," resulting from an inability to properly dorsiflex the ankle. In addition to deep fibular nerve trauma, the function of blood vessels in the compartment may be compromised, which can result in possible necrosis (death) of the tissues within the compartment.

The tibial nerve has two main terminal branches: the lateral plantar nerve and the medial plantar nerve. The lateral plantar nerve has branches innervating the abductor digiti minimi, the quadratus plantae, the adductor hallucis, the lumbricals, the flexor digiti minimi brevis, and the interosseous muscles of the foot. The medial plantar nerve has branches innervating the abductor hallucis, the flexor digitorum brevis, and the lumbricals. The last nerve of the sacral plexus, the nerve to the piriformis (S2), innervates the piriformis muscle, one of the six deep external rotators of the hip joint.

Inferior Portion

The most inferior portion of the lumbosacral plexus, known as the *pudendal plexus* (see figure 14.1), consists of the posterior femoral cutaneous nerve (S2 and S3), the anococcygeal nerve (S4, S5, and C1), the perforating cutaneous nerve (S2 and S3), and the pudendal nerve (S2, S3, and S4). None of these nerves innervates muscles of the lower extremity, and only the pudendal nerve and anococcygeal nerve innervate muscles at all (those of the abdominal cavity and genitalia).

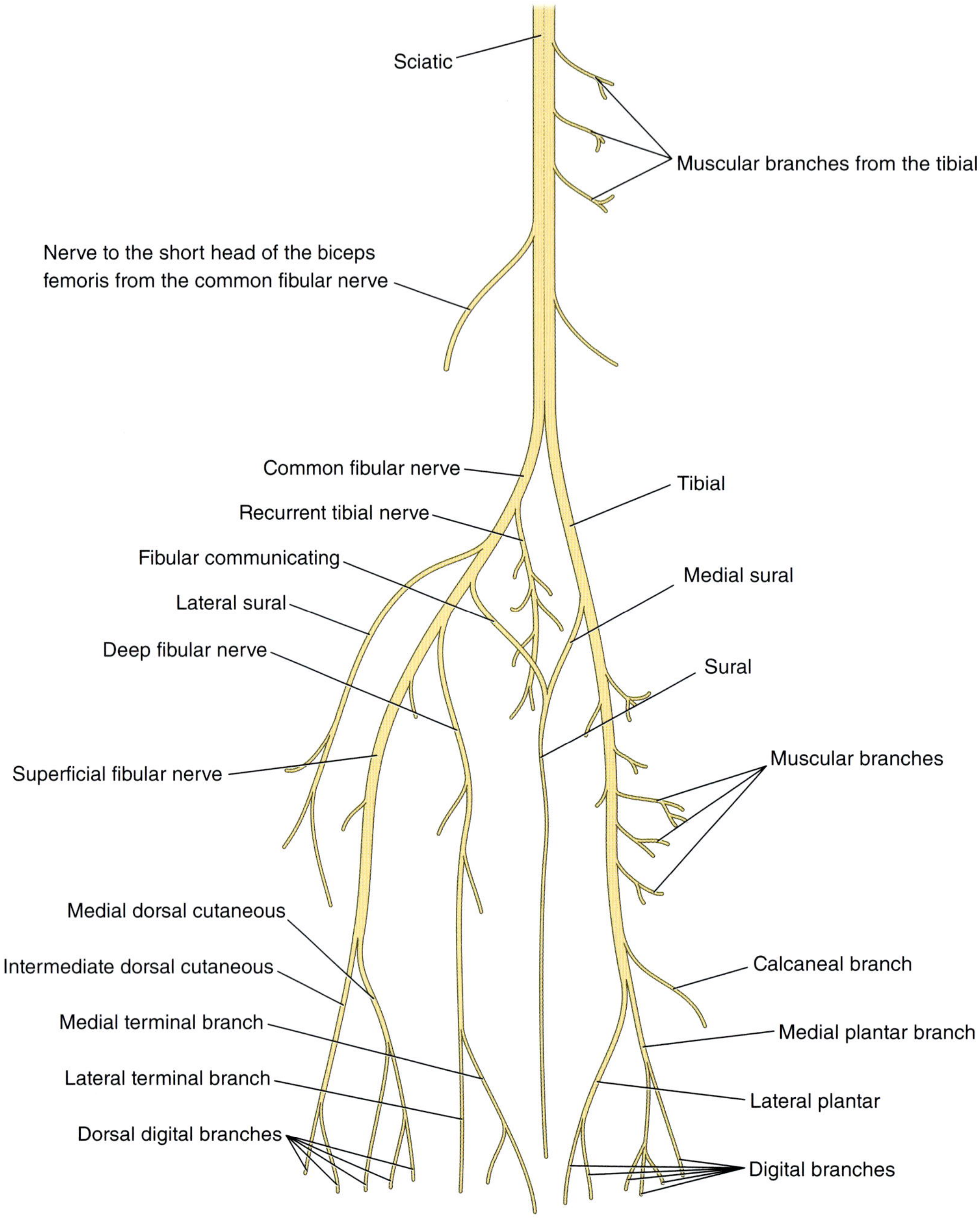

Figure 14.5 Anterior view of the sciatic nerve and its branches (on the right side).

POP QUIZ

Observe the position of the foot and ankle. Injury to which nerve would cause this position?

a. tibial
b. common fibular
c. obturator
d. femoral

Toa55/iStock/Getty Images Plus

Major Arteries of the Lower Extremity

As discussed in chapter 6, the blood vessels of the body are divided into arteries, arterioles, capillaries, venules, and veins (figures 14.6 and 14.14).

The major arteries of the lower extremity include the femoral, popliteal, anterior tibial, and posterior tibial, all with numerous branches. These arteries supply blood to the lower extremity. The femoral artery has many branches, including the genicular arteries, which supply the adductor magnus, the gracilis, the three vastus muscles, the adductor group, the sartorius, and the vastus medialis. Another branch of the femoral artery, the superficial circumflex iliac artery, supplies the iliacus, tensor fasciae latae, and sartorius muscles. The largest branch of the

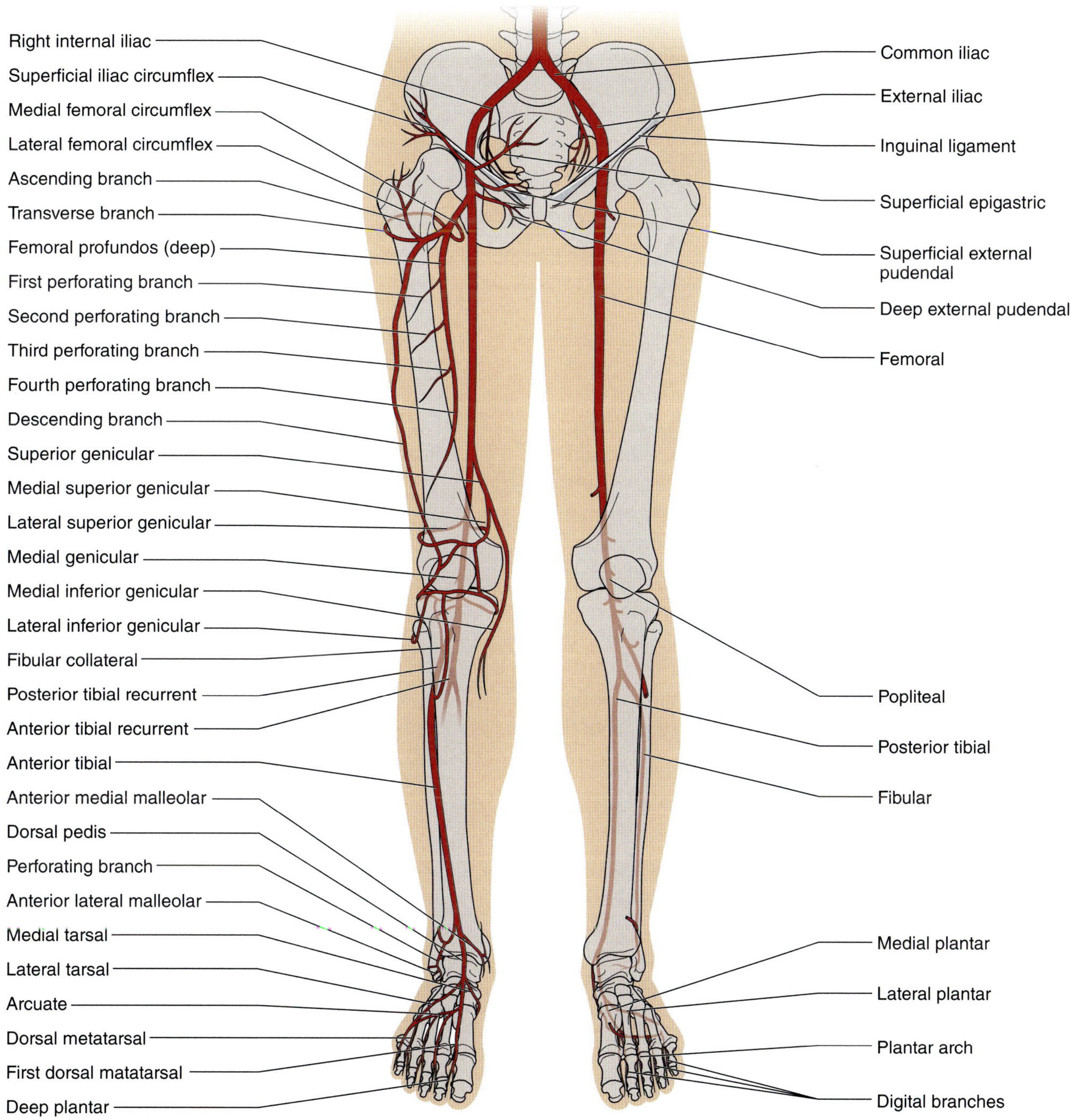

Figure 14.6 Major arteries of the lower extremity.

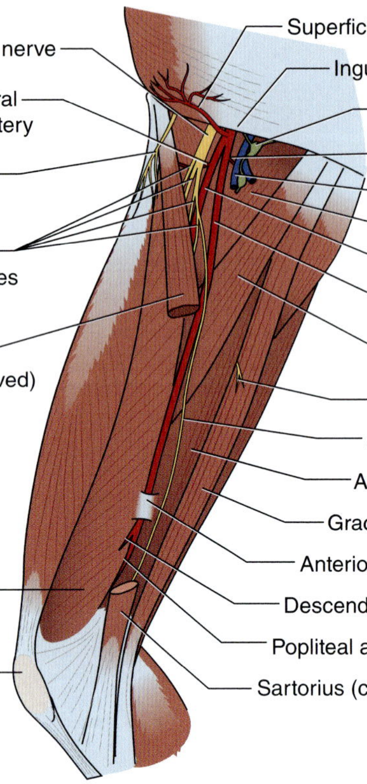

Figure 14.7 The femoral triangle.

femoral artery is the profunda artery. It has three branches: the lateral femoral circumflex, medial circumflex, and four perforating arteries, which supply the three gluteal muscles, tensor fasciae latae, sartorius, vastus lateralis and intermedius muscles, rectus femoris, adductors, external obturator, pectineus, and hamstrings (figure 14.7).

Hands On

Apply pressure with your fingertips on your femoral triangle, and attempt to feel your pulse from the femoral artery (figure 14.8).

At the level of the knee joint, the femoral artery becomes known as the *popliteal artery*. The branches of the popliteal artery (the lateral superior genicular and lateral inferior genicular arteries, medial superior genicular and medial inferior genicular arteries, and middle genicular artery) and its muscular branches supply the vastus lateralis, adductor magnus, biceps femoris, semimembranosus, semitendinosus, popliteus, gastrocnemius, plantaris, and soleus muscles. As the popliteal artery passes the popliteal muscle, the artery divides into the anterior and posterior

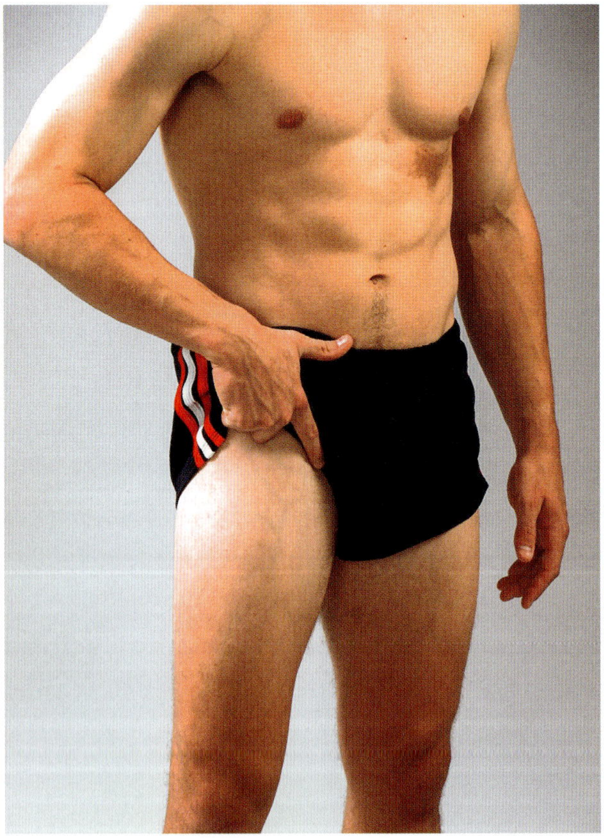

Figure 14.8 Locating the femoral artery.

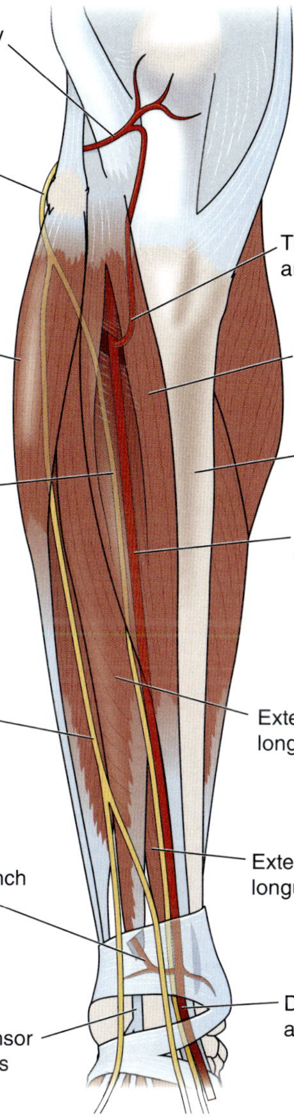

Figure 14.9 Anterior view of the arteries of the lower leg.

tibial arteries. The anterior tibial artery has branches—anterior and posterior tibial recurrent (see figures 14.6 and 14.9) and fibular—that supply the popliteus, extensor digitorum longus and brevis, tibialis anterior, fibularis longus, and soleus muscles. Below the ankle and into the foot, the anterior tibial artery becomes the dorsal pedis artery and ends on the dorsal (top) side of the foot (see figures 14.6, 14.9, and 14.10). The branches of the dorsal pedis artery (arcuate, deep plantar, first dorsal metatarsal, lateral tarsal, and medial tarsal) supply the dorsal interosseous and extensor digitorum brevis muscles (figure 14.10).

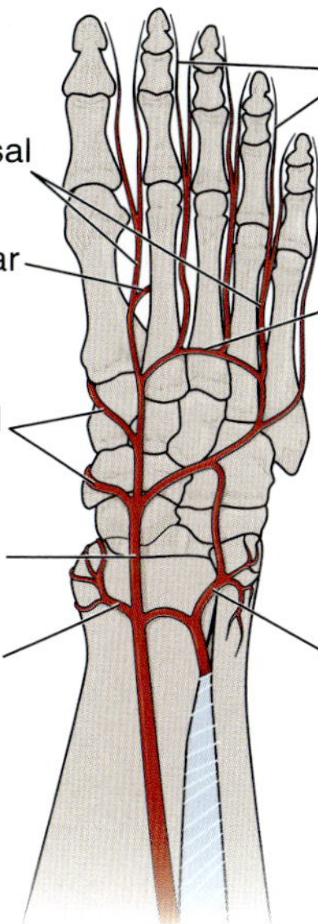

Figure 14.10 Branches of the dorsal pedis artery.

FOCUS ON

Femoral Triangle

One of the pressure points stressed in first-aid classes for control of bleeding in the lower extremity is the femoral artery. The femoral artery is easily located within the anatomical structure known as the *femoral triangle* (see figure 14.7). The triangle is formed by the inguinal ligament (superior border), the sartorius muscle (lateral border), and the adductor longus muscle (medial border). Also passing through this area are the femoral vein and the femoral nerve.

Hands On

In trying to establish whether the blood supply of the foot is impeded, health care professionals attempt to palpate a dorsal pedis pulse. This pulse is normally found on the dorsal side of the foot between the proximal ends of the first and second metatarsal bones (between the extensor hallucis longus and extensor digitorum longus tendons). Use figure 14.11 to assist you in checking your dorsal pedis pulse. Not everyone can easily feel this pulse because it is absent in 10% to 15% of people.

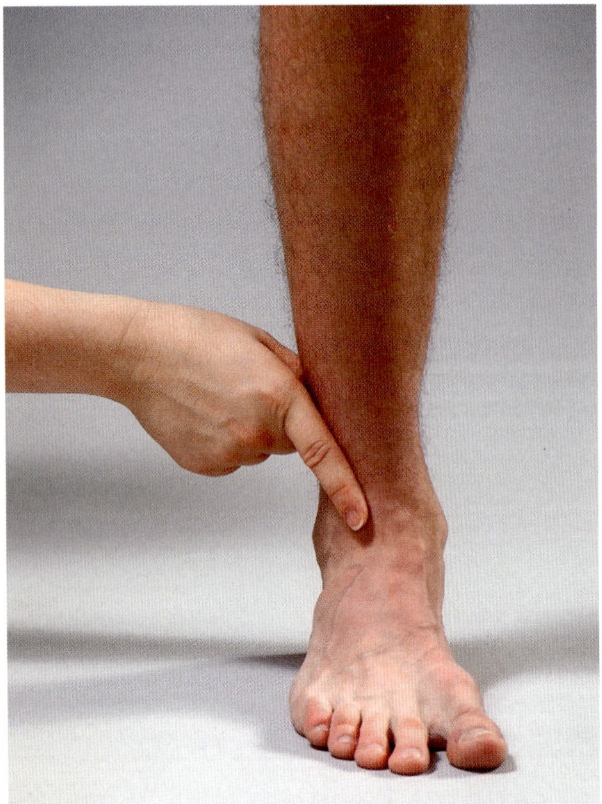

Figure 14.11 Locating the dorsal pedis artery.

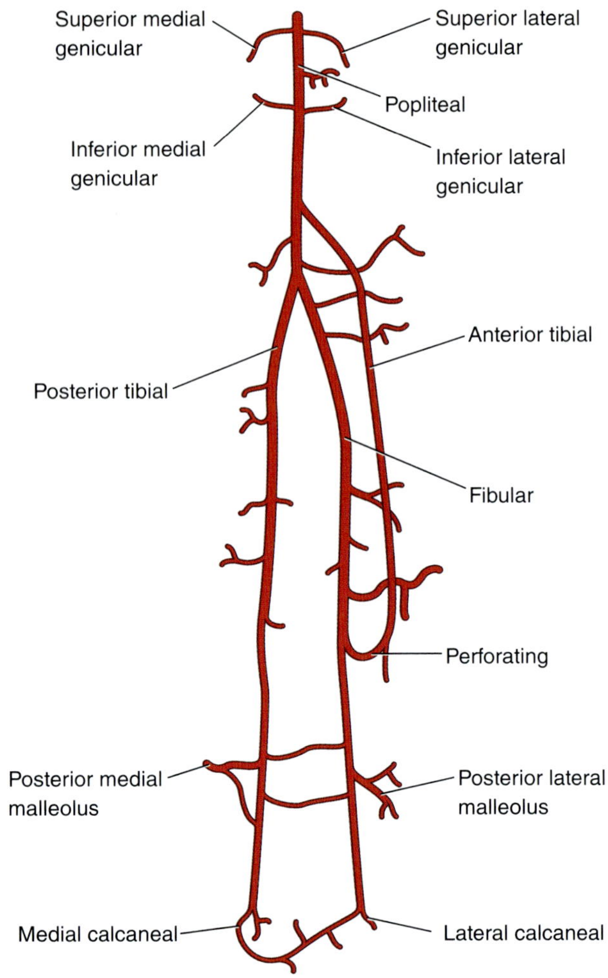

Figure 14.12 Branches of the posterior tibial artery.

The posterior tibial artery and its branches (medial calcaneal, fibular, and perforating branches; figure 14.12) supply the posterior and lateral lower-leg muscles, most of the muscles of the foot, and the fibularis tertius in the lower portion of the anterior lower-leg compartment. The posterior tibial artery has two terminating branches (the lateral and medial plantar arteries) that supply most of the muscles of the foot.

Hands On

When you are trying to determine blood flow to the foot, a more reliable pulse to take than the dorsal pedis pulse is the posterior tibial pulse. The artery is found between the flexor digitorum longus and flexor hallucis longus tendons just posterior to the medial malleolus of the tibia. Refer to figure 14.13 to palpate your posterior tibial pulse.

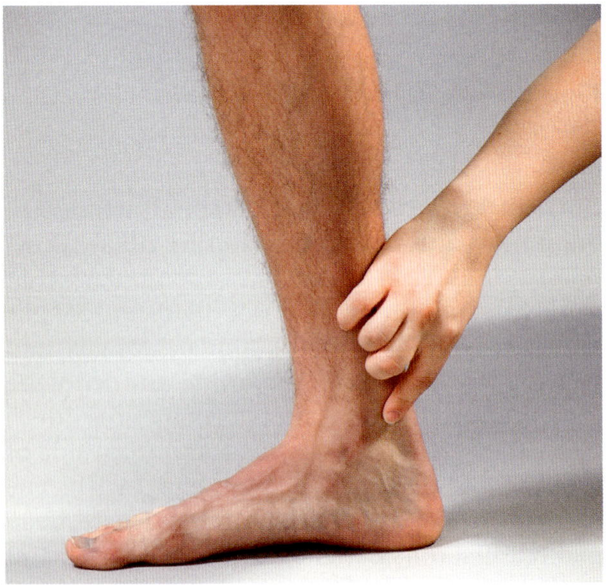

Figure 14.13 Locating the posterior tibial artery.

Major Veins of the Lower Extremity

The veins that return blood to the heart are commonly divided into deep veins and superficial veins (figure 14.14). With a few exceptions, the deep veins have the same names as the arteries they parallel, such as the femoral, popliteal, and tibial. The superficial veins have specific names and are located near the skin. Unlike arteries, veins do not appear exactly where one might expect, and they often are absent altogether.

The major deep veins of the lower extremity are the femoral and popliteal. The femoral vein drains the entire thigh and, in the area of the inguinal ligament, becomes the external iliac vein. The popliteal vein, which unites with the anterior and posterior tibial arteries, drains the structures from the foot to the lower edge of the popliteus muscle. It has tributaries known as the *anterior and posterior tibial veins* and the *lesser saphenous vein*, and it has branches that correspond to the anterior and posterior tibial arteries and their branches.

The major superficial veins of the lower extremity include the great saphenous and the lesser saphenous veins (figures 14.15 and 14.16). The great saphenous vein runs from the medial aspect of the dorsal venous arch in the foot to the femoral triangle at the groin, making it the longest vein in the body. Major tributaries of

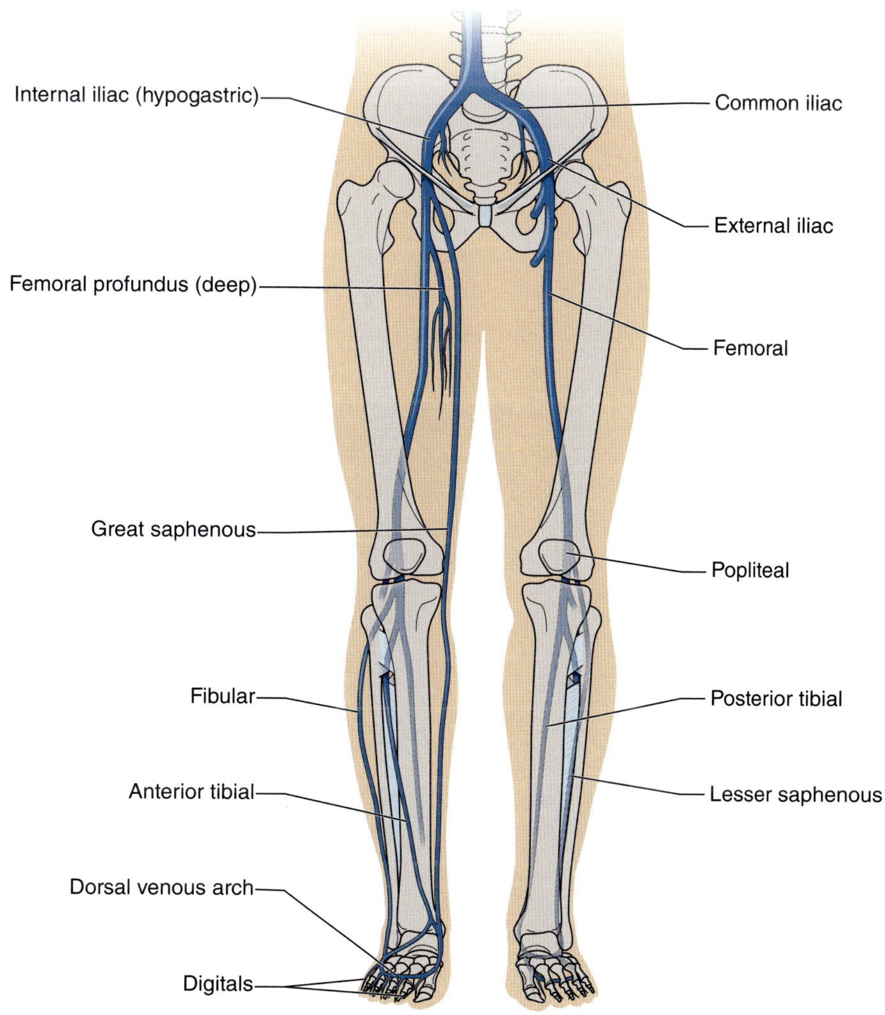

Figure 14.14 Major veins of the lower extremity.

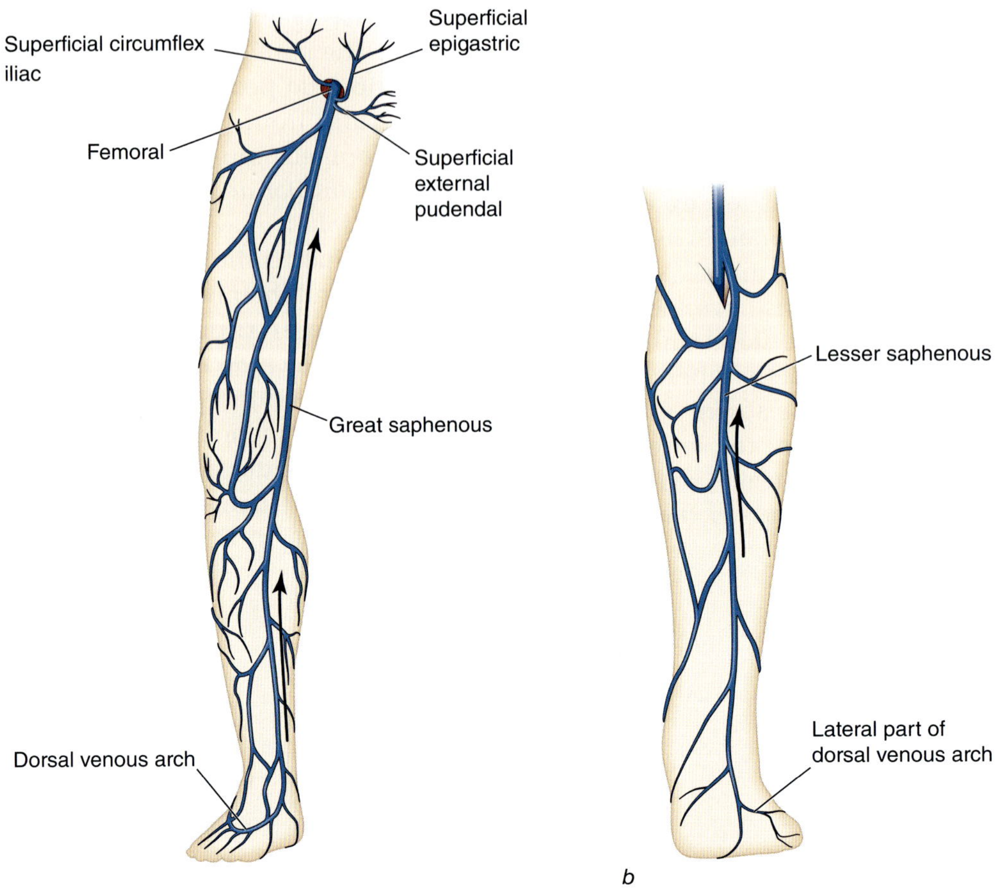

Figure 14.15 The *(a)* great and *(b)* lesser saphenous veins.

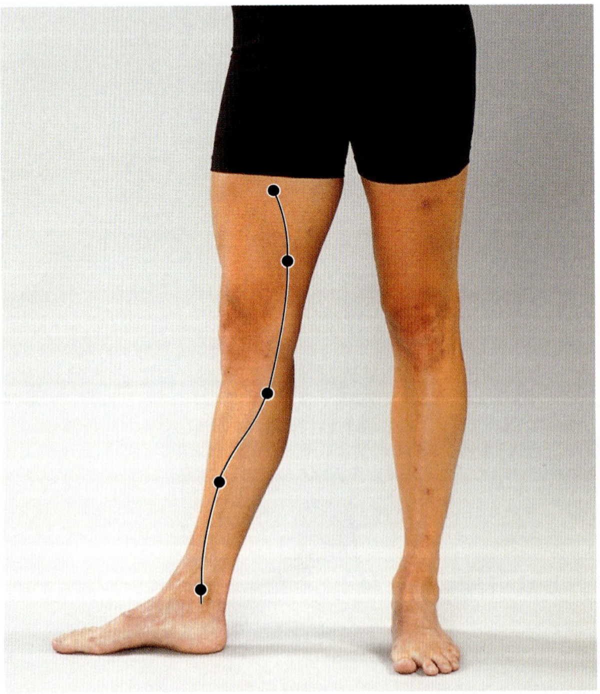

Figure 14.16 The path of the great saphenous vein.

the great saphenous vein in the lower extremity include the lateral and medial superficial femoral veins and the superficial circumflex iliac vein. The lesser saphenous vein is found at the lateral aspect of the dorsal venous arch and empties into the popliteal vein in the area of the popliteal space posterior to the knee joint. The lesser saphenous vein drains the lateral and posterior aspects of the foot and lower leg. Interestingly, sections of the saphenous veins are often surgically removed and used to replace damaged or diseased sections of other blood vessels. The many peripheral branches of the saphenous vein assume the function of the removed portion.

FOCUS ON

Varicose Veins

Varicose veins are simply defined as *enlarged veins*. The condition can appear in any vein in the body, but it is mentioned in this chapter because it most often is found in the lower extremity (thigh and lower leg). There are many causes for the condition to develop, such as heredity, pregnancy, obesity, and prolonged standing. Any of these causes can lead to pressure on the veins of the lower extremity, particularly in the groin area, which can restrict blood flow return from the legs. This restriction can cause a pooling of blood in the lower extremities that can weaken valves in the veins and actually stretch the walls of the veins. Varicose veins usually involve incompetent valves. Blood tends to pool in the lower extremities, weakening the valves and eventually stretching the venous walls.

Visit the web resource to complete the activities for this chapter.

Articulations of the Lower Extremity

Joint	Type	Bones	Ligaments	Movements
Hip	Ball and socket	Ilium, ischium pubic, and femur	• Capsule • Acetabular labrum (glenoid lip) • Iliofemoral (Y) • Pubofemoral • Transverse acetabular • Ligamentum capitis femoris (teres)	Flexion, extension, abduction, adduction, internal (or medial or inward) rotation, external (or lateral or outward) rotation
KNEE				
Tibiofemoral	Modified hinge	Femur and tibia	• Capsule • Medial (tibial) collateral • Lateral (fibular) collateral • Anterior cruciate • Posterior cruciate • Oblique popliteal • Arcuate (popliteal) • Wrisberg • Coronary • Transverse • Patellar (considered a continuation of the quadriceps tendon)	Flexion, extension (lower leg internally rotates on knee flexion and externally rotates on knee extension in non-weight-bearing movements)
Patellofemoral	Plane	Patella and femur	Patellar	Gliding
LOWER LEG				
Proximal tibiofibular	Diarthrodial (plane)	Tibia and fibula	• Capsule • Anterior of head of fibula • Posterior of head of fibula	Slight gliding
Middle tibiofibular	Synarthrodial	Tibia and fibula	Interosseous	None
Distal tibiofibular	Syndesmodial	Tibia and fibula	• Interosseous • Anterior tibiofibular • Posterior tibiofibular • Inferior transverse (anterior to the posterior tibiofibular)	Slight give on ankle dorsiflexion

Joint	Type	Bones	Ligaments	Movements
ANKLE				
Talotibial (talocrural)	Hinge	Tibia and talus	• Capsule • Deltoid • Tibionavicular • Calcaneotibial • Anterior talotibial • Posterior talotibial	Dorsiflexion, plantar flexion
Talofibular	Hinge	Talus and fibula	• Capsule • Calcaneofibular • Anterior talofibular • Posterior talofibular	Dorsiflexion, plantar flexion
INTERTARSAL JOINTS				
Subtalar* (talocalcaneal)	Arthrodial	Talus and calcaneus	• Capsule • Anterior talocalcaneal • Posterior talocalcaneal • Lateral talocalcaneal • Medial talocalcaneal • Interosseous talocalcaneal	Gliding
Transverse tarsal*	Arthrodial	Talus, navicular, calcaneus, and cuboid	• Bifurcated (the calcaneonavicular part) • Plantar calcaneonavicular (or spring ligament)	Gliding
Talocalcaneonavicular	Arthrodial	Talus, navicular, and calcaneus	• Capsule • Dorsal talonavicular • Plantar calcaneonavicular	Gliding
Calcaneocuboid	Arthrodial	Calcaneus and cuboid	• Capsule • Plantar calcaneocuboid (short plantar) • Dorsal calcaneocuboid • Bifurcated • Long plantar	Gliding
Cuneonavicular	Arthrodial	Cuneiforms (3) and navicular	• Dorsal cuneonaviculars • Plantar cuneonaviculars	Gliding
Cuboideonavicular	Arthrodial	Cuboid and navicular	• Dorsal cuboideonavicular • Plantar cuboideonavicular • Interosseous cuboideonavicular	Gliding
Intercuneiform and cuneocuboid	Arthrodial	Cuneiforms (3) and cuboid	• Dorsal intercuneiforms • Plantar intercuneiforms • Interosseous intercuneiforms	Gliding

(continued)

Articulations of the Lower Extremity *(continued)*

Joint	Type	Bones	Ligaments	Movements
TARSOMETATARSOPHALANGEAL JOINTS				
Tarsometatarsals	Arthrodial (plane)	Cuneiforms (3), cuboid, and bases of metatarsals	• Capsules • Plantar tarsometatarsal and dorsal tarsometatarsal • Interosseous collaterals	Gliding
Intermetatarsals	Arthrodial (plane)	Bases of metatarsals	• Dorsal capsules • Plantar capsules • Interosseous collaterals	Gliding
Metatarsophalangeals (MTP) (5)	Condyloid	Heads of metatarsals and bases of proximal phalanges	• Dorsal capsules • Plantar capsules • Interosseous collaterals • Transverse metatarsal	Flexion, extension, abduction, adduction
Interphalangeals (four proximal interphalangeals [PIPs], four distal interphalangeals [DIPs], and one interphalangeal)	Hinge	Proximal, middle, and distal phalanges	• Dorsal capsules • Plantar capsules • Interosseous collaterals	Flexion, extension

*The combined movements of the subtalar and transverse tarsal joints produce inversion (supination, adduction, and plantar flexion) and eversion (pronation, abduction, and dorsiflexion) of the ankle joint.

Muscles, Nerves, and Blood Supply of the Lower Extremity

Muscle	Origin	Insertion	Action	Nerve	Blood supply
HIP, ANTERIOR					
Psoas major Iliopsoas	L1-L5 transverse processes, T12 and L1-L5 intervertebral discs and bodies	Lesser trochanter of femur	Flexion of spinal column when femur is fixed; flexion of hip	Femoral	External iliac, internal iliac, and lumbar
Psoas minor	T12 and L1 intervertebral discs and bodies	Iliopectineal line	Flexion of pelvis	Lumbar	Lumbar
Iliacus Iliopsoas	Iliac fossa	Just distal to lesser trochanter of femur	Flexion of hip, flexion of spinal column when femur is fixed	Femoral	External iliac and hypogastric
Sartorius	Anterior superior iliac spine	Inferior to medial condyle of tibia on pes anserinus	Flexion, abduction, and external rotation of hip; flexion of knee; internal rotation of lower leg	Femoral	Profunda and femoral
Rectus femoris	Anterior inferior iliac spine and area superior to the acetabulum	Patella via quadriceps tendon, tibial tuberosity via patellar ligament	Flexion of hip, extension of knee	Femoral	Profunda
Tensor fasciae latae	Outer lip of iliac crest and between anterior superior and anterior inferior iliac spines	Greater trochanter of femur and (as iliotibial band) inferior and anterior to lateral condyle of tibia	Flexion, abduction, and rotation of hip	Superior gluteal	Lateral circumflex
Pectineus	Iliopectineal line and superior aspect of pubic bone	Pectineal line on femur	Flexion, adduction, and external rotation of hip	Femoral	Femoral and femoral circumflex
HIP, LATERAL					
Gluteus medius	Between iliac crest and anterior and posterior gluteal lines	Posterolateral edge of greater trochanter	Abduction of hip; anterior portion: flexion and internal rotation of hip; posterior portion: extension and external rotation of hip	Superior gluteal	Superior gluteal
Gluteus minimus	Anterior and inferior gluteal lines of ilium	Anterior edge of greater trochanter	Abduction and internal rotation of hip; anterior portion: flexion of hip; posterior portion: extension of hip	Superior gluteal	Superior gluteal

(continued)

Muscles, Nerves, and Blood Supply of the Lower Extremity *(continued)*

Muscle	Origin	Insertion	Action	Nerve	Blood supply
HIP, POSTERIOR					
Biceps femoris (long head)	Ischial tuberosity	Head of fibula	Extension and external rotation of hip, flexion of knee	Tibial	Profunda branch of femoral
Biceps femoris (short head)	Linea aspera	Head of fibula	Flexion of knee, external rotation of lower leg	Fibular	Profunda branch of femoral
Semitendinosus	Ischial tuberosity	Inferior to medial tibial condyle on pes anserinus	Extension of hip, flexion of knee, internal rotation of lower leg	Tibial	Profunda branch of femoral and popliteal
Semimembranosus	Ischial tuberosity	Medial tibial condyle	Extension of hip, flexion of knee, internal rotation of lower leg	Tibial	Profunda branch of femoral and popliteal
Gluteus maximus	Sacrum and coccyx, posterior gluteal line, and iliac crest	Gluteal line of femur and (as iliotibial band) inferior and anterior to lateral condyle of tibia	Extension, adduction, and external rotation of hip; abduction of flexed hip	Inferior gluteal	Superior and inferior gluteals and medial circumflex
Piriformis	Superior 2/3 of sacrum and greater sciatic notch	Greater trochanter	External rotation, abduction, and extension of hip	First and second sacral	Superior and inferior gluteals
Superior gemellus	Ischial spine	Greater trochanter	External rotation of hip	First and second sacral	Inferior gluteal and obturator
Internal obturator	Inner rim of obturator foramen	Greater trochanter	External rotation of hip	First and second sacral	Inferior gluteal and obturator
Inferior gemellus	Ischial tuberosity	Greater trochanter	External rotation of hip	Lumbosacral to quadratus femoris	Inferior gluteal and obturator
External obturator	Outer aspect of pubic and ischial bones at the obturator foramen	Trochanteric fossa of femur	External rotation of hip	Obturator	Inferior gluteal and obturator
Quadratus femoris	Ischial tuberosity	Intertrochanteric crest of femur	External rotation, adduction, and extension of hip	Branch from lumbosacral plexus	Inferior gluteal and medial femoral circumflex

Muscle	Origin	Insertion	Action	Nerve	Blood supply
HIP, MEDIAL					
Adductor longus	Pubic bone	Middle 1/3 of linea aspera	Adduction, flexion, and internal rotation of hip	Obturator	Femoral
Adductor brevis	Pubic bone	Proximal 1/3 of linea aspera	Adduction and flexion of hip	Obturator	Femoral
Adductor magnus	Pubic bone and ischial tuberosity	Linea aspera and adductor tubercle	Anterior portion: adduction and flexion of hip, posterior portion: adduction and extension of hip	Obturator	Femoral
Gracilis	Anterior inferior aspect of pubic symphysis and inferior pubic bone	Distal to medial tibial condyle on pes anserinus	Adduction, flexion, and external rotation of hip; flexion of knee; internal rotation of lower leg	Obturator	Femoral
KNEE, ANTERIOR					
Sartorius			*See:* Hip, anterior		
QUADRICEPS FEMORIS					
Rectus femoris			*See:* Hip, anterior		
Vastus lateralis	Greater trochanter; intertrochanteric line, proximal to lateral lip of linea aspera	Patella via quadriceps tendon, tibial tuberosity via patellar ligament	Extension of knee	Femoral	Lateral circumflex
Vastus intermedius	Proximal 2/3 of anterior of femur	Patella via quadriceps tendon, tibial tuberosity via patellar ligament	Extension of knee	Femoral	Lateral circumflex and profunda
Vastus medialis	Intertrochanteric line and linea aspera	Patella via quadriceps tendon, tibial tuberosity via patellar ligament	Extension of knee	Femoral	Femoral
KNEE, POSTERIOR					
Biceps femoris (hamstring)			*See:* Hip, posterior		
Semimembranosus (hamstring)			*See:* Hip, posterior		
Semitendinosus (hamstring)			*See:* Hip, posterior		
Gracilis			*See:* Hip, medial		

(continued)

Muscles, Nerves, and Blood Supply of the Lower Extremity *(continued)*

Muscle	Origin	Insertion	Action	Nerve	Blood supply
KNEE, POSTERIOR					
Popliteus	Lateral femoral condyle	Proximal tibia at popliteal line	Flexion of knee; internal rotation of lower leg	Tibial	Popliteal
Gastrocnemius (lateral head)	Popliteal area medial to lateral femoral condyle	Calcaneus	Flexion of knee; plantar flexion of ankle	Tibial	Posterior tibial and fibular
Gastrocnemius (medial head)	Medial condyle of femur	Calcaneus	Flexion of knee; plantar flexion of ankle	Tibial	Posterior tibial and fibular
Plantaris	Lateral linea aspera and oblique popliteal ligament	Calcaneus	Flexion of knee; plantar flexion of ankle	Tibial	Popliteal
ANKLE AND FOOT (EXTRINSIC), ANTERIOR					
Tibialis anterior	Lateral condyle of tibia and proximal 1/2 of lateral aspect of tibia	First cuneiform and base of first metatarsal	Dorsiflexion and inversion of ankle	Deep fibular	Anterior tibial
Extensor digitorum longus	Lateral condyle of tibia, proximal 3/4 fibula, and interosseous membrane	Phalanges of toes (2-5)	Dorsiflexion and eversion of ankle; extension of toes (2-5)	Deep fibular	Anterior tibial
Extensor hallucis longus	Middle 1/2 of fibula and interosseous membrane	Base of distal phalanx of great toe	Dorsiflexion and inversion of ankle; extension of great toe	Deep fibular	Anterior tibial
Fibularis tertius	Distal 1/3 of fibula and interosseous membrane	Dorsal aspect of base of fifth metatarsal	Dorsiflexion and eversion of ankle	Deep fibular	Anterior tibial and fibular
ANKLE AND FOOT (EXTRINSIC), LATERAL					
Fibularis longus	Lateral condyle of tibia, head, and proximal 2/3 of lateral fibula	Lateral aspect of base of first metatarsal and medial cuneiform	Eversion and plantar flexion of ankle	Superficial fibular	Fibular
Fibularis brevis	Distal 1/2 of fibula	Fifth metatarsal tuberosity	Eversion and plantar flexion of ankle	Superficial fibular	Fibular

Muscle	Origin	Insertion	Action	Nerve	Blood supply
ANKLE AND FOOT (EXTRINSIC), POSTERIOR					
Gastrocnemius			*See:* Knee		
Soleus	Head and proximal 1/3 of fibula, middle 1/3 of tibia	Calcaneus	Plantar flexion of ankle	Tibial	Posterior tibial and fibular
Plantaris			*See:* Knee, posterior		
Tibialis posterior	Middle 1/3 of posterior tibia, head and proximal 2/3 of medial aspect of fibula, and interosseous membrane	Navicular, medial cuneiform, and metatarsals (2-4)	Plantar flexion and inversion of ankle; adduction of foot	Tibial	Posterior tibial
Flexor digitorum longus	Posterior tibia: popliteal line to within a few inches of distal end of tibia	Base of distal phalanges of toes (2-5)	Plantar flexion and inversion of ankle; adduction of foot; flexion of toes (2-5)	Tibial	Posterior tibial
Flexor hallucis longus	Lower 2/3 of posterior fibula and interosseous membrane	Distal phalanx of great toe	Plantar flexion and inversion of ankle; flexion of great toe	Tibial	Posterior tibial
ANKLE AND FOOT (INTRINSIC), FIRST PLANTAR LAYER					
Flexor digitorum brevis	Calcaneal tuberosity	Sides of middle phalanges of toes (2-5)	Flexion of toes (2-5)	Medial plantar	Posterior tibial
Abductor hallucis	Medial aspect of calcaneus	Base of proximal phalanx of great toe	Abduction and flexion of great toe	Medial plantar	Posterior tibial
Abductor digiti minimi	Calcaneal tuberosity	Lateral aspect of base of proximal phalanx of fifth toe	Abduction and flexion of fifth toe	Lateral plantar	Posterior tibial
ANKLE AND FOOT (INTRINSIC), SECOND PLANTAR LAYER					
Quadratus plantae (medial head)	Medial aspect of calcaneus	Flexor digitorum longus tendon	Assists flexion of toes (2-5)	Lateral plantar	Posterior tibial
Quadratus plantae (lateral head)	Lateral aspect of calcaneus	Flexor digitorum longus tendon	Assists flexion of toes (2-5)	Lateral plantar	Posterior tibial
Lumbricals	Flexor digitorum longus tendon	Extensor digitorum longus tendon	Flexion of MTP joints (toes 2-5); extension of PIP and DIP joints (toes 2-5)	Lateral and medial plantar	Plantar

(continued)

Muscles, Nerves, and Blood Supply of the Lower Extremity *(continued)*

Muscle	Origin	Insertion	Action	Nerve	Blood supply
ANKLE AND FOOT (INTRINSIC), THIRD PLANTAR LAYER					
Flexor hallucis brevis	First, second, and third cuneiforms	Base of proximal phalanx of great toe	Flexion of MTP joint of great toe	Medial plantar	Posterior tibial
Adductor hallucis (transverse head)	Plantar surface of MTP ligament of toes (3-5)	Proximal phalanx of great toe	Adduction of great toe	Lateral plantar	Posterior tibial
Adductor hallucis (oblique head)	Bases of metatarsals (2-4)	Proximal phalanx of great toe	Adduction of great toe	Lateral plantar	Posterior tibial
Flexor digiti minimi brevis	Base of fifth metatarsal	Lateral aspect of base of proximal phalanx of fifth toe	Flexion of fifth toe	Lateral plantar	Posterior tibial
ANKLE AND FOOT (INTRINSIC), FOURTH PLANTAR LAYER					
Dorsal interossei (4)	Adjacent aspects of all metatarsals	Base of proximal phalanges of toes (2-4)	Abduction of toes three and four	Lateral plantar	Anterior tibial
Plantar interossei (3)	Inferior medial aspect of metatarsals (3-5)	Medial aspect of base of proximal phalanges of toes (3-5) and extensor digitorum longus tendon	Adduction of toes (3-5)	Lateral plantar	Deep and lateral plantars
ANKLE AND FOOT (INTRINSIC), DORSAL					
Extensor digitorum brevis	Calcaneus	Long extensor tendons of toes (2-5)	Extension of toes (2-5)	Deep fibular	Anterior tibial and fibular
Extensor hallucis brevis	Calcaneus	Distal phalanx of great toe	Extension and adduction of great toe	Deep fibular	Anterior tibial

GLOSSARY

all-or-none law of muscle contraction—If a stimulus from a nerve is intense enough to reach the threshold of a muscle fiber, all muscle fibers innervated by that nerve contract completely. There is no such thing as a partial contraction of a muscle fiber.

anastomoses—Connection between two structures.

anatomical neck—Landmark between the head of the humerus and the greater and lesser tubercles.

ball-and-socket joint—Consists of the rounded head of one bone fitting into the cuplike cavity of another bone.

biaxial joint—Permits movement in two planes, about two axes.

concentric contraction—Type of muscle contraction where the muscle shortens in overall length while generating force.

contractility—The ability of a muscle to shorten and develop tension when stimulated.

condyloid joint—Consists of one oval-shaped surface fitting into an oval-shaped socket or depression. These joints are also known as *ellipsoidal joints*.

diarthrodial joints—Distinguished by a separation of the bones and the presence of a joint cavity.

eccentric contraction—Type of muscle contraction where the muscle elongates in overall length while generating force.

elasticity—The ability of a muscle to return to its original length when stretching ceases.

excitability (irritability)—The ability of a muscle to be sensitive or responsive to chemical, electrical, or mechanical stimuli.

extensibility—The ability of a muscle to be stretched beyond its normal resting length.

fossa—A smooth, hollow surface on a bone that usually functions as a source of attachment for other structures.

hematoma—Well-formed blood tumor.

hemorrhage—A more diffuse discharge of arterial, venous, or capillary blood.

hernia—Abnormal protrusion of an organ.

hinge joint—One concave surface, with the other surface looking like a spool of thread.

irregular joint—Irregularly shaped surfaces that are typically either flat or slightly rounded.

isometric contraction—Type of muscle contraction where the muscle length does not change and there is no joint movement while generating force.

motor unit—Motor nerve and all the muscle fibers it supplies.

myology—The study of muscles.

neuromuscular junction—The space between the end brushes of the motor end plate and the muscle fibers. Also known as the *myoneural junction*.

nonaxial joint—Allows for only a gliding type of movement.

pivot joint—One bone rotates about the other bone.

Q angle—The angle between the line of the quadriceps muscle pull and the line of insertion of the patellar tendon.

saddle joint—Both bones have a surface that is convex in one direction and concave in the opposite direction, like a saddle.

surgical neck—Upper portion of the shaft of the humerus.

synarthrodial joints—Joints with no separation or joint cavity.

syndesmosis—An articulation between bones tied together by ligaments.

tetanus—Once the muscle is receiving impulses at such a rate that it cannot relax, it reaches a state of continuous contraction.

triaxial joint—Allows movement in three planes, about three axes, illustrated by the movements of the shoulder and hip joints, which are both ball-and-socket joints.

uniaxial joint—Permits movement in only one plane about one axis.

ANSWERS TO POP QUIZ QUESTIONS

Chapter 1

Page 17: Based on the views alone, the answer is a, because the runner in this photo most likely has more fast-twitch fibers in his lower extremities than the runner in b, who would have more slow-twitch fibers in her lower extremities.

Chapter 2

Page 35: Bicep curl exercise

Chapter 3

Page 45: The correct answer is b, but all three injuries are possible.
Page 57: 1. all four; 2. all four; 3. all four

Chapter 4

Page 71: Ulnar (medial) collateral ligament

Page 71: (left) ULC/medial collateral ligament; (center) supinator-extensor muscle group and common extensor tendon; (right) supinator-extensor muscle group and common extensor tendon

Chapter 5

Page 77: The correct answer is a.

Chapter 6

Page 94: The correct answer is d.

Chapter 7

Page 122: The correct answer is a.

Chapter 8

Page 136: Right internal oblique and left external oblique

Chapter 9

Page 157: The correct answers are a, b, and c.

Chapter 10

Page 163: The correct answers are (top) a; (middle) b; (bottom) c.

Chapter 11

Page 212: The muscles being stretched in the front leg are the hamstring group (biceps femoris, semitendinosus, semimembranosus). The muscles being stretched in the back leg are the iliopsoas, psoas major, iliacus, psoas minor, sartorius, rectus femoris, and pectineus.

Chapter 12

Page 221: Medial collateral ligament

Chapter 13

Page 239: The correct answer is c.

Chapter 14

Page 260: The correct answer is b.

SUGGESTED READINGS

Agur, A.M.R., and A.F. Daily. 2024. *Grant's atlas of anatomy*. 16th ed. Baltimore: Lippincott, Williams & Wilkins.

Biel, A. 2023. *Trail guide to the body*. 6th ed. Boulder: Books of Discovery.

Drake, R.A., W. Vogl, and A.W.M. Mitchell. 2023. *Gray's anatomy for students*. 5th ed. New York: Churchill Livingstone.

Gilroy, A., B. MacPherson, J. Wikenheiser, M. Schuenke, E. Schulte, and U. Schmacher. 2020. *Atlas of anatomy*. 4th ed. New York: Thieme.

Gray, H. 2000. *Anatomy of the human body*. Philadelphia: Lea & Febiger; Bartleby.com.

Logan, B.M., P. Reynolds, and R.T. Hutchings. 2009. *McMinn's color atlas of head and neck anatomy*. 4th ed. Philadelphia: Elsevier (Mosby).

Logan, B.M. and R.T. Hutchings. 2011. *McMinn's color atlas of foot and ankle anatomy*. 4th ed. Philadelphia: Elsevier (Saunders).

Marieb, E.N., L.W. Haynes, and M.A. Abbott. 2024. *Human anatomy and physiology*. 12th ed. San Francisco: Benjamin/Cummings.

Morton, D., K.B. Foreman, and K. Albertine. 2018. *Gross anatomy: The big picture*. 2nd ed. New York: McGraw-Hill Medical.

Netter, F.H. 2022. *Atlas of human anatomy*. 8th ed. Philadelphia: Elsevier (Saunders).

Whiting, W. 2019. *Dynamic human anatomy*. 2nd ed. Champaign, IL: Human Kinetics.

INDEX

Note: The italicized *f* and *t* following page numbers refer to figures and tables, respectively.

A

A-band (anisotropic band), 15, 16*f*
abdominal aorta, 22*f*, 176
abdominal muscles, 134
abdominal parietal branch, 176, 176*f*
abdominal visceral branch, 176, 176*f*
abducens nerve, 164*f*, 165, 166*t*
abduction, 31*t*, 32, 33*f*
  foot, 241, 243*f*
  hip, 33*f*, 205
  horizontal, 34
  shoulder, 33*f*, 47-49, 49*f*, 51, 52, 59
  thumb, 87, 88*f*
  wrist and hand, 77, 77*f*, 85*f*
abductor digiti minimi
  foot, 250-251, 250*f*
  hand, 79*f*, 83
abductor hallucis, 250, 250*f*
abductor pollicis brevis, 79*f*, 86
abductor pollicis longus, 81*f*, 85, 86*f*
AC (acromioclavicular) joint, 43, 44*f*, 45
accessory cephalic vein, 98, 99*f*
acetabular fossa, 139*f*
acetabular labrum, 203
acetabular notch, 139*f*
acetabulum, 139*f*, 141, 201, 202*f*, 204*f*, 242*f*
acetylcholine, 25
Achilles tendon, 247*f*
Achilles tendonitis, 247
ACL (anterior cruciate ligament), 220-222, 220*f*, 221*f*
acromial end of clavicle, 40, 40*f*
acromioclavicular (AC) joint, 43, 44*f*, 45
acromioclavicular ligament, 43, 44*f*
acromion process, 6*f*, 40*f*, 41, 41*f*
actin, 15, 16, 16*f*
Adam's apple (laryngeal prominence), 123
adduction, 31*t*, 32, 33*f*
  foot, 241, 243*f*
  hip and upper leg, 33*f*, 205
  horizontal, 34
  shoulder, 33*f*, 48-49, 49*f*, 51-52
  thumb, 87, 88*f*
  wrist and hand, 77, 77*f*, 85*f*
adductor (groin) muscles, 205-206, 211-212, 211*f*
adductor hallucis, 251*f*, 252
adductor pollicis, 79*f*, 86-87
adductor tubercle, 6*f*, 216*f*, 217
afferent (sensory) nerves, 20*f*, 21, 165, 166*t*
afterbrain (pons), 164
agonist muscles, 14
air exchange, 182*f*
air sacs, 156*f*, 157, 181, 182*f*
airway, 156-158, 157*f*
all-or-none theory of muscle contraction, 26
alveolar ducts, 156*f*, 157
alveolus, 156*f*, 182*f*
anastomosis, 179*f*
anatomical landmarks, 5, 6*f*-7. *See also specific bones or body parts*
anatomical location terms, 30
anatomical neck, 40, 42*f*
anatomical position, 29, 29*f*
anatomical snuffbox, 85, 85*f*
anatomy study, 3-4
anconeus, 66*f*, 68, 69*f*
angular vein, 171
ankle, 269*t*, 274*t*-276*t*. *See also* lower extremity
  bones, 231-233, 232*f*, 235*f*
  injuries, 238
  joints and ligaments, 234, 236-237, 237*f*
  movements, 241, 242*f*
  muscles, 242-246, 244*f*, 247*f*, 254*t*
annular ligament, 8*f*, 64, 64*f*
annulus fibrosus, 130, 130*f*
anococcygeal nerve, 259
antagonist muscles, 14
anterior band, 64
anterior cardiac veins, 181
anterior cerebral artery, 168, 169*f*
anterior chamber (eyeball), 123
anterior communicating artery, 169*f*, 170
anterior compartment (lower leg), 243-246, 244*f*, 259
anterior cruciate ligament (ACL), 220-222, 220*f*, 221*f*
anterior, defined, 30
anterior facial vein, 171, 171*f*
anterior femoral cutaneous nerve, 257, 257*f*
anterior inferior iliac spine, 139*f*, 140
anterior lateral malleolar artery, 261*f*, 263*f*
anterior longitudinal, 242*f*
anterior longitudinal ligament, 128, 129*f*, 142*f*
anterior medial malleolar artery, 261*f*, 263*f*
anterior pubic, 242*f*
anterior pubic ligament, 142*f*, 143
anterior sacroiliac, 242*f*
anterior sacroiliac ligament, 141, 142*f*
anterior superior iliac spine (ASIS), 139*f*, 140, 142*f*, 227, 228*f*, 242*f*
anterior talofibular (ATF) ligament, 236-237, 237*f*
anterior talotibial ligament, 236, 237*f*
anterior tibial artery, 22*f*, 261, 261*f*, 263*f*
anterior tibial recurrent artery, 261*f*, 263
anterior tibial vein, 23*f*, 265*f*

anterior tibiofibular ligament, 236, 237*f*
aorta, 155*f*, 176, 178, 179*f*, 180*f*
aortic arch, 22*f*, 175, 176*f*
aortic semilunar valve, 181
apex of fibula, 217, 218*f*
apex of patella, 218, 219*f*
aqueous humor, 123
arachnoid, 160, 160*f*
arch of foot, 240-241, 241*f*
arcuate (inferior pubic) ligament, 142*f*, 143
arcuate artery, 261*f*, 263, 263*f*
arcuate ligament (knee), 222, 222*f*
arm. *See* upper extremity
arteries
  diagram of, 22*f*
  function of, 21, 89
  head and brain, 166-170, 167*f*, 169*f*, 185*t*-192*t*
  heart, 178-181, 179*f*
  lower extremity, 261-264, 261*f*, 263*f*, 271*t*-276*t*
  respiratory, 181, 182*f*
  thorax and trunk, 175-176, 176*f*, 193*t*
  upper extremity, 95-97, 95*f*, 103*t*-108*t*
arterioles, 21
arthrology, 8
articular capsule, 10, 10*f*, 75*f*
articular cartilage, 5, 5*f*
articular disc, 43, 43*f*, 117
articular process, 125*f*, 127*f*
articular surface, 5*f*, 6
  patellar, 218, 219*f*
  tibial, 232*f*
aryepiglotticus muscle, 123
arytenoid oblique muscle, 123
arytenoid transverse muscle, 123
ascending aorta, 175, 176*f*
ASIS (anterior superior iliac spine). *See* anterior superior iliac spine (ASIS)
atlantoaxial ligaments, 128, 128*f*
atlantooccipital ligaments, 128, 128*f*
atlas vertebra, 124*f*, 125, 126*f*
atria (singular, atrium)
  heart, 153, 155*f*
  lungs, 156-158
atria myocardium, 177
atrioventricular bundle (AV node), 177, 178*f*
atrioventricular septum, 154
auditory canal (external meatus), 112*f*, 113*f*, 114, 115*f*
auditory center, 162, 163*f*
auditory nerve, 166*t*, 168
auricle (pinna), 114
auricle of left atrium, 154*f*
auricularis muscles, 121, 122*f*
autonomic nervous system, 19, 164, 165*f*
AV node (atrioventricular bundle), 177, 178*f*
axillary artery, 22*f*, 95, 95*f*
axillary nerve, 90*f*, 91*f*, 92*f*, 94
axillary vein, 23*f*, 97, 97*f*, 171*f*
axis (plural, *axes*), 30-31, 30*f*, 31*t*
axis vertebra, 124*f*, 125, 126*f*
axons, 19, 20*f*, 25*f*

B

ball-and-socket joint, 9, 9*f*
basal vein, 172
base of patella, 218, 219*f*
basilar artery, 167, 169*f*, 175
basilar plexus, 173
basilic cubital vein, 98
basilic vein, 23*f*, 97*f*, 98, 99*f*
biaxial joints, 10*f*, 111
biceps brachii, 53, 54*f*, 66*f*, 67, 68*f*, 70
biceps femoris ligament, 220*f*
biceps femoris muscle, 208, 208*f*, 209*f*, 227-228, 228*f*
bicuspid (mitral) valve, 155, 155*f*
bifid spinous process, 126, 126*f*
bifurcated ligament, 237*f*, 238
bipennate muscle fibers, 13, 13*f*
blood vessels, 21. *See also* arteries; veins
body of pubic bone, 139*f*, 141
bones
  ankle, 231-233, 232*f*, 269*t*
  classification of, 4-5, 4*f*
  elbow and forearm, 61-64, 62*f*, 100*t*-102*t*
  features of, 5, 6*f*-7*f*
  foot, 233-236, 234*f*, 269*t*-270*t*
  head, 111-114, 112*f*, 113*f*, 183*t*
  hip joint, 201-203, 202*f*, 268*t*
  knee, 215-219, 218*f*, 219*f*, 268*t*
  lower leg, 231-233, 232*f*, 268*t*
  osteoporosis, 5
  pelvis, 138-141, 139*f*, 184*t*, 202*f*
  shoulder, 39-42, 39*f*, 40*f*, 100*t*
  spinal column, 125-126, 183*t*
  thigh, 201-203, 203*f*, 204*f*
  thorax, 145-147, 146*f*, 184*t*
  wrist and hand, 73*f*, 74, 101*t*-102*t*
bow legs (genu valgum), 219
brachial artery, 22*f*, 95, 95*f*
brachial plexus, 20, 90*f*, 91-94, 91*f*, 92*f*, 175*f*
brachial vein, 97, 97*f*
brachialis, 66, 66*f*, 67*f*
brachiocephalic artery, 22*f*, 167*f*, 175, 176*f*
brachiocephalic vein, 23*f*, 97, 98*f*, 171*f*, 177, 178*f*
brachioradialis, 66, 66*f*, 67*f*
brain
  arteries, 166-170, 167*f*, 169*f*
  structure and functions, 159-164, 160*f*, 161*f*
  veins, 170-173
brain stem, 161*f*, 164
bronchial arteries, 181
bronchial veins, 181
bronchioles, 156*f*, 157
buccinator muscle, 121, 121*f*, 124*f*
bucket-handle tears of menisci, 223, 223*f*
bundle of His, 177, 178*f*
bursa, 21, 24*f*
bursitis, 24, 214
butt blocking injuries, 94

C

calcaneocuboid ligament, 237*f*, 240
calcaneofibular ligament, 237, 237*f*
calcaneonavicular ligament, 237*f*, 240
calcaneotibial ligament, 236, 237*f*
calcaneus, 233-234, 234*f*, 235*f*
calcium, 25
calf muscles, 247, 247*f*
caninus muscle, 121, 121*f*
capillaries, 21
capitate, 73*f*, 74
capitulum, 61, 62*f*
capsular carpometacarpal ligament, 76
capsular ligaments
  foot, 239*f*, 240
  forearm, 64-65
  hip joint, 203-204
  knee, 219, 220*f*
  shoulder joint, 46, 46*f*
  wrist and hand, 75, 75*f*
capsule (joint), 10, 10*f*
caput angulare muscle, 121
caput infraorbitale muscle, 121
caput zygomaticum muscle, 121
cardiac muscle, 12*f*, 13
cardiac notch, 157*f*, 158
cardiac veins, 180*f*, 181
cardinal (principal) planes, 30*f*, 31
cardioinhibitory center, 164
cardiopulmonary resuscitation (CPR), 147, 174, 175

carotid arteries, 22*f*, 166-170, 175-176, 176*f*, 177*f*
carpal bones, 74
carpal tunnel syndrome, 77
carpometacarpal joints, 75-76, 75*f*
cartilaginous joints, 11, 11*f*
cavernous sinus, 167*f*, 173
celiac trunk artery, 22*f*
cell body, nerve, 20, 20*f*, 25*f*
center of gravity, 31
central nervous system, 20
central tendon, 148*f*, 149*f*, 150
cephalic vein, 23*f*, 97, 97*f*, 98*f*, 99*f*
cerebellar veins, 173
cerebellum, 161*f*, 164
cerebral veins, 172
cerebrospinal fluid, 160
cerebrum, 161-164, 161*f*, 162*f*, 163*f*
cervical plexus, 20, 90*f*, 173, 174*f*, 175*f*
cervical spine injuries, 174
cervical vertebrae, 123, 124*f*, 125-126, 126*f*
cholesterol, 181
chondroglossus muscle, 123
chondrosternal joints, 147
chordae tendineae, 155, 155*f*
choroid, 123
ciliary muscle, 123
circle of Willis, 169, 169*f*
circular sinus, 173
circumduction, 34, 34*f*, 52, 77-78, 241
circumflex artery, 179*f*, 180*f*, 181
CKC (closed kinetic chain), 34-35
clavicle (collarbone), 39-41, 39*f*, 40*f*
clavicular notch, 43, 44, 145, 146*f*
closed kinetic chain (CKC), 34-35
cluneal nerve, 257*f*
coccygeal nerve, 256*f*
coccyx, 123, 124*f*, 126, 128*f*, 242*f*
cochlea, 114, 116*f*
cochlear nerve, 166*t*
collateral ligaments. *See also specific ligaments*
  elbow, 64-65, 64*f*
  foot, 240
  wrist and hand, 75*f*, 76
common carotid arteries, 166, 167*f*, 176*f*
common extensor tendon, 81
common facial vein, 171
common flexor tendon, 79, 79*f*
common groin strain, 214
common iliac arteries, 22*f*, 176*f*, 177, 261*f*
common iliac vein, 23*f*, 177, 178*f*, 265*f*
common fibular nerve, 257*f*, 259*f*, 260*f*
compartment syndrome, 246
concentric contraction, 15
concussion, 161, 170
condyles, 6, 7*f*
  femoral, 215, 218*f*
  fibular, 217, 218*f*
  mandibular, 116*f*
  occipital, 113*f*
  tibial, 217, 218*f*, 219*f*
condyloid joints, 9, 9*f*, 75
connective tissue, 3
conoid ligament, 44*f*, 45
conoid tubercle, 40*f*, 41
constrictor pharyngis muscles, 123
contraction
  all-or-none theory, 26
  heat from, 15
  motor units and, 26
  movements from, 14
  muscle viscosity and, 18
  tetanus, 27, 27*f*
  types of, 15
coracoacromial ligament, 44*f*, 46*f*, 47
coracobrachialis, 50*f*, 52, 53*f*
coracoclavicular joint, 44
coracoclavicular ligament, 44*f*, 45
coracohumeral ligament, 46, 46*f*
coracoid process, 6*f*, 40*f*, 41, 41*f*
cornea, 121, 122*f*
coronal suture, 112*f*
coronary arteries, 22*f*, 175, 176*f*, 179*f*, 181
coronary ligament, 220*f*, 223
coronary sinus, 180*f*, 181
coronoid fossa, 61, 62*f*
coronoid process, 7*f*, 62*f*, 64
corpocapitate articulation, 147
corpus callosum, 161, 162*f*
corrugator supercilii muscle, 121, 121*f*
cortex, 161, 163*f*
costal grooves, 157
costal surfaces, 157
costal tuberosity, 40, 40*f*
costochondral (capsular) ligament, 147, 148*f*
costochondral joints, 147, 148*f*, 149
costoclavicular ligament, 43, 43*f*
costosternal ligament, 147, 148*f*
costotransverse capsular articulation, 147
costotransverse capsular ligament, 147, 148*f*
costovertebral articulations, 147, 148*f*
costovertebral ligaments, 128
costovertebral radiate ligament, 147, 148*f*
CPR (cardiopulmonary resuscitation), 147, 174, 175
cranial muscles, 121
cranial nerves, 164-166, 164*f*, 166*t*
cranial sinuses, 172
craniofacial joints, 115
craniovertebral joints, 115
cranium. *See* head
crest of ilium, 139*f*, 140, 142*f*
crest of the ilium, 242*f*
cricoarytenoid lateralis muscle, 123
cricoarytenoid posterior muscle, 123
cricopharyngeus muscle, 124*f*
cricothyroid muscle, 123
cross-bridges, 15, 16*f*, 26, 26*f*
cross-control, 162
cuboid bone, 233, 234*f*, 235*f*
cuboideonavicular ligament, 237*f*, 240
cuneiform bones, 233, 234*f*, 235*f*
cuneocuboid ligaments, 239*f*, 240
cuneonaviculars ligaments, 239*f*, 240
cutaneous colli nerve, 90*f*

D

deep branch of ulnar artery, 97*f*
deep external rotators, 207
deep facial vein, 171
deep femoral artery, 22*f*, 261*f*
deep femoral vein, 265*f*
deep fibular nerve, 90*f*, 257*f*, 260*f*
deep plantar artery, 261*f*, 263, 263*f*
deep posterior compartment, 244*f*, 248-250
deep transverse metacarpal ligaments, 75*f*
deep veins, 97, 170, 265, 265*f*
deep volar arch, 95*f*, 97, 97*f*
deltoid ligament, 236, 237*f*
deltoid muscle, 50*f*, 53-54, 55*f*
deltoid tubercle, 40, 40*f*
deltoid tuberosity, 6*f*, 40*f*, 42-43, 43*f*
dendrites, 19, 20*f*, 25*f*
dens (odontoid process), 126, 126*f*, 128
depression of shoulder girdle, 47, 48*f*
depressor alaeque nasalis muscle, 121

depressor anguli oris muscle, 121*f*, 122*f*
depressor labii inferior muscle, 121*f*, 122*f*
descending aorta, 155*f*, 175, 180*f*
diaphragm, 148*f*, 149*f*, 150, 153*t*
diaphysis, 4, 5*f*
diarthrodial joints, 9-10, 9*f*, 10*f*
diastole, 12
diencephalon, 161*f*, 164
digastric muscles, 123, 124*f*
digital arteries, 97, 97*f*
digital nerves, 92*f*, 257*f*, 260*f*
digital veins, 97, 97*f*, 265*f*
dilator naris muscle, 121
dilator pupillae, 123
DIP (distal interphalangeal) joint, 76, 78, 84, 242
diploic veins, 172
discriminatory power, 162
dislocations
  hip, 141
  shoulder, 59, 60*f*
distal, defined, 30
distal facet, 7*f*, 42, 42*f*
distal interphalangeal (DIP) joint, 76, 78, 84, 242
distal phalanges, 74, 234*f*
distal radioulnar articular capsule, 75*f*
distal radioulnar joint, 61, 65
dorsal carpal arch, 97*f*
dorsal carpal branch, 96, 97*f*
dorsal carpometacarpal ligaments, 75*f*, 76
dorsal cutaneous nerve, 260*f*
dorsal, defined, 30
dorsal intercarpal ligaments, 75*f*
dorsal interossei
  foot, 252*f*, 253
  hand, 83-84, 84*f*
dorsal metacarpal ligaments, 75*f*
dorsal metatarsal artery, 261*f*, 263*f*
dorsal pedis artery, 261*f*, 263, 263*f*
dorsal radial carpal branch, 96
dorsal radiocarpal ligaments, 76
dorsal radioulnar ligament, 65, 65*f*
dorsal scapular nerve, 92, 92*f*
dorsal talonavicular ligament, 238, 240
dorsal thoracic nerve, 94
dorsal ulnar carpal branch, 96
dorsal venous arch, 265, 265*f*
dorsiflexion, 32, 241, 242*f*
downward rotation of scapula, 49
dropped foot, 259
dura mater, 160, 160*f*, 173
dura mater sinuses, 173

E

ear, 114, 115*f*, 116*f*, 173, 185*t*
eccentric contraction, 15
efferent (motor) nerves, 20*f*, 21, 25*f*, 165, 166*t*
elbow, 100*t*, 105*t*-106*t*
  anatomical landmarks, 7*f*, 62*f*
  bones, 61-64, 62*f*
  joints and ligaments, 8*f*, 64-65, 64*f*
  movements and muscles, 65-69, 65*f*, 66*f*
  pitcher's elbow, 69
  tennis elbow, 71
elevation of shoulder girdle, 48, 48*f*
end brushes, 20, 20*f*, 25, 25*f*
endocardium, 155*f*, 158
endomysium, 15, 16*f*
epicardium, 155, 155*f*
epicondyles, 6, 6*f*
  femoral, 215, 216*f*
  humeral, 61, 62, 62*f*
epicranius muscle, 121, 122*f*
epidural hematoma, 160, 160*f*, 170
epidural space, 160, 160*f*
epimysium, 15, 16*f*
epiphyses, 4, 5*f*
epithelial tissue, 3
equilibrium, 114, 164, 166
erector spinae muscles, 138, 138*f*, 153*t*
ethmoid bone, 111, 112*f*
ethmoid sinuses, 118, 118*f*
eustachian tube, 114, 115*f*
eversion, 32, 241, 242*f*
expiration, 178
extension, 31, 31*t*, 32*f*
  elbow, 65
  foot, 242, 243*f*
  hip, 32*f*, 205
  horizontal, 34
  knee, 32*f*, 224, 229*f*
  shoulder, 32*f*, 51
  spine, 130, 131*f*
  thumb, 87, 87*f*
  wrist and hand, 77*f*, 78, 85*f*
extensor carpi radialis brevis, 81, 81*f*
extensor carpi radialis longus, 81*f*, 82-83
extensor carpi ulnaris, 81, 81*f*, 82*f*
extensor digiti minimi (proprius), 81*f*, 82, 83*f*
extensor digitorum (communis), 81*f*, 82, 82*f*
extensor digitorum brevis, 244*f*, 253, 253*f*
extensor digitorum longus, 243, 244*f*, 245*f*
extensor hallucis brevis, 244*f*, 253
extensor hallucis longus, 244*f*, 245, 246*f*
extensor indicis, 81*f*, 83
extensor pollicis brevis, 81*f*, 85, 87*f*
extensor pollicis longus, 81*f*, 84-85, 85*f*
extensor retinaculum, 76, 76*f*, 244*f*
external carotid artery, 22*f*, 167, 167*f*, 175
external ear, 114
external iliac artery, 22*f*, 176, 176*f*, 261*f*
external iliac vein, 23*f*, 178*f*, 265, 265*f*
external jugular vein, 23*f*, 171, 171*f*, 177, 178*f*
external meatus (auditory canal), 112*f*, 113*f*, 114, 115*f*
external oblique muscle, 134, 135*f*, 136*f*, 153*t*
external obturator (obturator externus), 208*f*, 211
external occipital crest, 113*f*
external occipital protuberance, 112, 113*f*
external rotation, 32, 33*f*
extrinsic muscles. *See* muscles
eyeball muscles, 121-123, 122*f*, 188*t*

F

facets, 6, 7*f*
  articular, 126
  humeral, 7*f*, 42, 42*f*
  patellar, 216*f*, 217
  radial, 63*f*
facial artery, 22*f*, 167*f*
facial muscles, 121, 121*f*, 185*t*-187*t*
facial nerves, 164*f*, 166, 166*t*, 185*t*-187*t*
facial veins, 23*f*, 171, 171*f*
fascia, 16*f*, 21
fast-twitch fibers, 16, 16*f*
femoral artery, 22*f*, 176, 261, 261*f*, 262*f*
femoral condyle, 7*f*, 217
femoral fovea, 203*f*, 205
femoral nerve, 90*f*, 256*f*, 257, 257*f*
femoral notch, 7*f*, 217
femoral profundus artery, 22*f*, 261*f*

femoral profundus vein, 265*f*
femoral triangle, 262*f*, 263
femoral vein, 23*f*, 178*f*, 265, 265*f*
femur, 201-203, 203*f*, 204*f*, 215-217, 216*f*
fenestra ovalis, 114
fenestra rotunda, 114, 116*f*
fibrocartilage, 11, 11*f*
fibula, 217, 218*f*, 231, 232*f*, 233*f*
fibular collateral ligament, 220*f*, 222*f*, 261*f*
fibular notch, 231
fibularis tertius, 245
fingers. *See* wrist and hand
first-class lever, 17, 18*f*
first costal notches, 145
first dorsal metatarsal artery, 261*f*, 263, 263*f*
fissure of Rolando, 162
fissure of Sylvius, 162
fixator muscles, 15
flat bones, 4, 4*f*
flexion, 31, 31*t*, 32*f*
  elbow, 65
  foot, 241, 242*f*, 243*f*
  hip, 32*f*, 205
  horizontal, 34
  knee, 32*f*, 224, 229*f*
  shoulder, 32*f*, 51, 52, 60
  spine, 130, 131*f*
  thumb, 87, 87*f*
  wrist and hand, 77*f*, 78, 85*f*
flexor carpi radialis, 78, 79*f*, 80*f*
flexor carpi ulnaris, 79*f*, 80, 80*f*
flexor digiti minimi brevis
  foot, 251*f*, 252
  hand, 79*f*, 83
flexor digitorum brevis, 250, 250*f*
flexor digitorum longus, 244*f*, 247*f*, 249*f*, 250
flexor digitorum profundus, 79*f*, 80-81
flexor digitorum superficialis, 79*f*, 80-81, 80*f*
flexor hallucis brevis, 251*f*, 252
flexor hallucis longus, 247*f*, 249-250, 250*f*
flexor pollicis brevis, 79*f*, 86, 86*f*
flexor pollicis longus, 79*f*, 86, 86*f*
flexor retinaculum, 76, 76*f*, 241, 247*f*
floating ribs, 146, 146*f*
fontanels, 117
foot, 269*t*-270*t*, 274*t*-276*t*. *See also* lower extremity
  arches, 240-241, 241*f*
  bones, 233-235, 234*f*
  ligaments, 237*f*, 238-241, 239*f*
  movements, 241, 242*f*, 243*f*
  muscles, 242-254, 244*f*, 247*f*, 254*t*
  turf toe, 241
foramen, 6, 7*f*
  greater sciatic, 142*f*
  infraorbital, 112*f*
  intervertebral, 125
  lesser sciatic, 142*f*
  magnum, 112, 113*f*
  mental, 112*f*
  obturator, 139*f*, 141
  vertebral, 125, 125*f*, 126, 126*f*
forearm, 101*t*, 105*t*-106*t*
  arteries, 95*f*, 96
  bones, 61-64, 62*f*
  joints and ligaments, 64-65, 65*f*
  movements and muscles, 65, 65*f*, 66*f*, 69-70, 70*f*
  nerves, 91-94, 92*f*
  veins, 98, 98*f*, 99*f*
forefoot, 233
fossa, 6, 6*f*
  acetabular, 139*f*
  coronoid, 61, 62*f*
  glenoid, 6*f*, 40*f*, 42
  infraspinous, 40*f*, 42
  mandibular, 113*f*
  olecranon, 62, 62*f*
  radial, 61, 62*f*
  subscapular, 6*f*, 40*f*, 42
  supraspinous, 40*f*, 42
  trochanteric, 203*f*
fovea, 126, 127*f*
frontal (lateral) plane, 31, 31*t*
frontal bone, 111, 112*f*
frontal horizontal axis, 30*f*, 31, 31*t*
frontal lobe, 162, 162*f*
frontal plane, 30*f*
frontal sinuses, 118, 118*f*
frontal vein, 170-171
frontalis muscle, 121, 121*f*
fundamental movements, 31-34, 32*f*, 33*f*, 34*f*. *See also specific body parts*
fundamental position, 29, 29*f*
fusiform muscles, 13, 13*f*

G

galea aponeurotica, 121*f*
gastrocnemius muscle, 229*f*, 230, 244*f*, 247, 247*f*, 248*f*
gemellus inferior, 208*f*, 211
gemellus superior, 208*f*, 211
gender differences
  ACL injuries, 221
  pelvic anatomy, 140*f*
  Q angle, 227, 227*f*
genicular arteries, 261, 261*f*, 263*f*
genioglossus muscle, 123, 124*f*
geniohyoid muscle, 123, 124*f*
genitofemoral nerve, 256*f*, 257, 257*f*
genu articularis, 226-227, 227*f*
genu recurvatum (hyperextended knee), 219
genu valgum (bow legs), 219
genu varum (knock knees), 219
Gerdy's tubercle, 217, 218*f*, 229
glenohumeral (GH) joint. *See* shoulder joint
glenohumeral ligaments, 46, 46*f*
glenoid fossa, 6*f*, 40*f*, 42
glenoid lip (glenoid labrum), 46*f*, 47, 47*f*
glossopalatinus muscle, 123
glossopharyngeal nerve, 164*f*, 166, 166*t*
gluteal nerves, 256*f*, 257*f*, 258
gluteus maximus, 208*f*, 210
gluteus medius, 208*f*, 213-214, 213*f*
gluteus minimus, 208*f*, 214
gracilis
  at hip joint, 211*f*, 212-213
  at knee joint, 228, 228*f*
  locating, 213*f*
gray matter, 161
great (or greater) auricular nerve, 90*f*, 91*f*
great cardiac vein, 23*f*, 181
great saphenous vein, 23*f*, 265, 265*f*, 266*f*
greater multangular (trapezium) bone, 73*f*, 74
greater sciatic foramen, 142*f*, 242*f*
greater sciatic notch, 139*f*, 140
greater trochanter, 202*f*, 203*f*, 204*f*
greater tubercle, 6*f*, 7*f*, 40*f*, 42, 42*f*
groin muscles, 205, 211*f*, 212
groin strain, 214
growth (epiphyseal) plates, 4
gyri, 162, 162*f*

H

H-zone, 15, 16*f*
hamate, 73*f*, 74, 74*f*
hamatometacarpal ligament, 75*f*
hamstrings, 207-208, 209*f*, 211
hand. *See* wrist and hand
hangman's ligament, 129*f*
hard palate (palatine bones), 111, 112, 113*f*
head, 183*t*-192*t*
  arteries, 166-170, 167*f*, 169*f*

head, 183*t*-192*t* *(continued)*
  bones, 111-114, 112*f*, 113*f*
  brain structure, 159-164, 160*f*, 161*f*
  joints, 115-117
  ligaments, 117, 118*f*
  movements and muscles, 120-123, 121*f*, 122*f*
  nerves, 166*t*
  sinuses, 118*f*, 120
  trauma to, 160-161, 160*f*, 170
  veins, 170-173, 171*f*
head of femur, 201, 202, 203*f*, 204*f*
head of fibula, 217, 218*f*
head of humerus, 40*f*, 42, 42*f*
head of rib, 147, 147*f*
hearing, 114, 162, 163*f*, 166
heart
  air exchange, 182*f*
  arteries, 178-181, 179*f*
  nerves, 177-178, 178*f*
  sounds of, 155, 156*f*
  structure of, 153-156, 154*f*, 155*f*
  veins, 180*f*, 181
heart attack, 181
hematoma, 160, 160*f*, 170
hemorrhage, 160
hepatic portal vein, 23*f*
hepatic vein, 23*f*, 178*f*
hernia, inguinal, 142
herniated disc, 132
hilum, 157
hindfoot, 233
hinge joint, 9, 9*f*, 215
hip joint, 204*f*, 268*t*, 271*t*-273*t*. *See also* lower extremity
  bones, 201-203, 202*f*
  injuries, 141, 201
  ligaments, 203-205, 204*f*
  movements, 205
  muscles, 205-214, 205*f*, 208*f*, 211*f*, 254*t*
hip pointer injury, 141
homolateral, defined, 164
horizontal (transverse) plane, 31, 31*t*
horizontal plane, 30*f*
horns of meniscus, 224
humerus, 39*f*
  anatomical landmarks, 7*f*, 40*f*, 62, 62*f*
  description, 42, 61
  head of, 40*f*, 42, 42*f*
  ligaments of, 47, 47*f*
  muscles acting on, 50*f*, 52-53, 56, 56*f*
hyoglossus muscle, 123, 124*f*
hyoid bone, 111, 114, 116*f*
hyperextended knee, 219
hyperextension, 32, 52, 132
hyperflexion, 132
hypogastric (internal iliac) artery, 22*f*, 176, 176*f*
hypogastric (internal iliac) vein, 23*f*, 177, 178*f*, 265*f*
hypoglossal nerve, 91*f*, 164*f*, 166, 166*t*
hypothalamus, 164
hypothenar eminence, 79*f*, 83

I

I-band (isotropic band), 15, 16*f*
iliac arteries, 22*f*, 176, 176*f*, 178*f*, 261*f*
iliac veins, 23*f*, 177, 178*f*, 265, 265*f*
iliacus, 205, 205*f*, 206*f*
iliocostalis cervicis muscle, 137, 137*f*
iliocostalis dorsi muscle, 137
iliocostalis lumborum muscle, 137, 137*f*
iliofemoral (Y) ligament, 203-205, 204*f*
iliohypogastric nerve, 90*f*, 256*f*, 257, 257*f*
ilioinguinal nerve, 90*f*, 256*f*, 257, 257*f*
iliolumbar, 242*f*
iliolumbar ligament, 129, 130*f*, 141, 142*f*
iliopsoas, 205
iliotibial band or tract, 205*f*, 207, 208*f*, 214, 218*f*, 228-229, 228*f*, 229*f*
ilium, 139, 139*f*, 141
incus (anvil), 114, 115*f*
inferior (arcuate) pubic ligament, 142*f*, 143
inferior angle, 40*f*, 41, 41*f*
inferior articulating facets, 125
inferior cardiac vein, 181
inferior, defined, 30
inferior glenohumeral ligament, 46, 46*f*
inferior gluteal nerve, 256*f*, 257*f*, 258
inferior mesenteric artery, 22*f*
inferior mesenteric vein, 23*f*
inferior nuchal line, 113, 113*f*
inferior oblique muscle, 122*f*
inferior petrosal sinus, 173
inferior pharyngeal constrictor, 124*f*
inferior pubic, 242*f*
inferior pubic ramus, 141
inferior rectus muscle, 122*f*, 123
inferior sagittal sinus, 173
inferior transverse scapular ligament, 47, 47*f*
inferior vena cava, 23*f*, 155*f*, 177, 178*f*, 180*f*, 181, 182*f*
infraclavicular nerves, 92-94, 92*f*
infraglenoid tubercle, 40*f*, 42
infrahyoid muscles, 123
infraorbital foramen, 112*f*, 113
infraspinatus muscle, 50*f*, 54-55, 55*f*, 56*f*
infraspinous fossa, 6*f*, 40*f*, 42
inguinal hernia, 142
inguinal ligament, 142, 142*f*, 242*f*
injuries and trauma
  Achilles tendon, 247
  ankle joint, 238
  bursitis, 24, 214
  carpal tunnel syndrome, 77
  cervical spine, 174
  costochondral sprain, 149
  CPR and xiphoid process, 147
  crest of the ilium contusion, 141
  groin strain, 214
  hamstring strain, 211
  head trauma, 160-161, 160*f*, 170
  hip injuries, 141, 201
  inguinal hernia, 142
  knee, 221, 221*f*, 223, 227
  lower-leg trauma, 259
  pinched nerve, 94
  pitcher's elbow, 69
  plantar fasciitis, 252
  rotator cuff, 56
  sciatic neuritis, 258
  shin splints, 251
  shoulder dislocation, 59, 60*f*
  shoulder separation, 45, 45*f*
  slipped or herniated disc, 132
  tennis elbow, 71
  turf toe, 241
  varicose veins, 267
inner ear, 114, 115*f*, 116*f*
innominate bone, 139-141, 139*f*
innominate vein, 170
inspiration, 152, 153*t*, 178
intelligence, 163
interatrial septum, 153
interbody ligaments, 128
intercarpal joints, 75*f*, 76
intercarpal ligaments, 75*f*, 76
intercavernous sinuses, 173
interchondral (interarticular) ligament, 147, 148*f*
interchondral joints, 147, 148*f*
interclavicular ligament, 43, 43*f*

intercondylar (femoral) notch, *7f*, 217
intercondylar eminence, 217, *218f*
intercondylar tubercles, 217, *218f*
intercostal muscles, *148f*, *149f*, 150, *151f*, *153t*
intercostal nerves, 174, 178, *179f*
intercranial joints, 115
intercuneiform ligaments, *239f*, 240
interfacial joints, 115
intermaxillary suture, *113f*
intermediate dorsal cutaneous nerve, *260f*
intermetatarsal ligaments, 240
internal carotid artery, *22f*, *167f*, 168, *169f*, 170, 175
internal iliac artery, *22f*, 176, *261f*
internal iliac vein, *23f*, 177, *178f*, *265f*
internal jugular vein, *23f*, 97, *98f*, 170, *171f*, 177, *178f*
internal maxillary artery, 168
internal oblique muscle, 134, *135f*, *153t*
internal obturator (obturator internus), *208f*, 211
internal rotation, 32, *33f*
internal thoracic artery, *167f*
interosseous carpometacarpal ligament, 76
interosseous ligament (membrane)
  ankle, 236, *236f*
  elbow, *8f*
  foot, 240
  forearm, *64f*, 65
  pelvis, 141
interphalangeal (IP) joints
  foot, *239f*, 240, *240f*, 242
  hand, *75f*, 76, 78
interphalangeal ligaments, 238, 240
interpubic disc, *242f*
interpubic fibrocartilage disc, *142f*, 143
interspinous ligament, 128, *130f*
intertarsal joints, *269t*
intertarsal ligaments, 238
intertrochanteric ridge, 203, *203f*
intertubercular (bicipital) groove, 42, *42f*
interventricular septum, 154, *155f*
intervertebral disc, *130f*, 132
intervertebral foramen, 125
intrinsic muscles. *See* muscles
inversion, 32, 241, *242f*
inversion sprain, 238
IP (interphalangeal) joints. *See* interphalangeal (IP) joints
iris, 121, *122f*
irregular (diarthrodial) joint, 9, *9f*
irregular bones, *4f*, 5
ischemia, 251
ischial notch, *139f*, 140
ischial tuberosity, *139f*, 140, *142f*, *242f*
ischiofemoral ligament, 203-205, *204f*
ischium, *139f*, 140
isometric contraction, 15
isthmus, 125

J

joints
  ankle and foot, 234, 236-237, *237f*, *268t-270t*
  elbow and forearm, 64-65, *100t-102t*
  head, 115-117, *183t*
  knee, *268t*
  movements, 8, 11
  pelvis, *184t*
  range of motion, 34
  shoulder girdle, 43-46, *100t*
  spinal column, *183t*
  strength of, 11
  thorax, 147, *148f*, *184t*
  types of, 8-12, *9f*, *10f*, *11f*
  wrist and hand, 75-76, *75f*, *101t-102t*
joints hip. *See* hip joint
jugular vein
  external, *23f*, 171, *171f*, 177, *178f*
  internal, *23f*, 97, *98f*, 170, *171f*, 177, *178f*

K

kinetic chain, 34-35
knee, *268t*, *273t-274t*. *See also* lower extremity
  anatomical landmarks, *7f*
  bones, 215-219, *218f*, *219f*
  conditions in, 219
  gender differences, 221, 227, *228f*
  ligaments, *8f*, 219-223, *220f*
  menisci, *222f*, 223, *223f*, 224
  movements, 224-225
  muscles, 225-230, *228f*, *254t*
knock knees (genu varum), 219
kyphosis, 123, *124f*

L

lacrimal bones, 111, *112f*
lambdoidal suture, *112f*
lamina, 125, *125f*, *127f*
lamina papyracea, 120
laminectomy, 130
laryngeal prominence (Adam's apple), 123
larynx, 156, *156f*, *190t-192t*
larynx muscles, 123, *124f*
lateral angle, *40f*
lateral anterior thoracic nerve, 92, *92f*
lateral border, scapula, *40f*, 41, *41f*
lateral calcaneal artery, *264f*
lateral collateral ligament (LCL), 219, *220f*, *221f*, *222f*
lateral compartment, *244f*, 246-247
lateral condyles. *See* condyles
lateral cord, *91f*, *92f*
lateral, defined, 31
lateral dorsal cutaneous nerve, *257f*
lateral epicondylar ridge, 215, *216f*
lateral epicondyle. *See* epicondyles
lateral femoral circumflex artery, *261f*, 262
lateral femoral cutaneous nerve, *256f*, 257, *257f*
lateral flexion, 130, *131f*
lateral intercondylar tubercle, 217, *218f*
lateral malleolus, 231, *232f*, *233f*
lateral pectoral nerve, *91f*
lateral plantar artery, *261f*, 262
lateral plantar nerve, *257f*, 259, *260f*
lateral styloid process, *7f*
lateral superficial femoral vein, 267
lateral supracondylar ridge, 62, *62f*
lateral sural nerve, 258, *260f*
lateral tarsal artery, *261f*, 263, *263f*
lateral tilt, 48, 143
lateralis rectus muscle, *122f*, 123
latissimus dorsi, 56-57, *58f*, 138, *139f*, *153t*
LCL (lateral collateral ligament), 219, *220f*, *221f*, *222f*
left atrium, 154, *155f*
left brachiocephalic vein, *23f*
left renal artery, *22f*
left subclavian artery, *22f*
left subclavian vein, *23f*
left ventricle, 154, *155f*
leg bones, *6f*. *See also* lower extremity; lower leg
lens, 121, *122f*
lesser multangular (trapezoid) bone, *73f*, 74
lesser occipital nerve, *90f*, *91f*

lesser saphenous vein, 263, 265*f*, 266*f*, 267
lesser sciatic (ischial) notch, 139*f*, 140
lesser sciatic foramen, 142*f*, 242*f*
lesser trochanter, 202, 203*f*, 204*f*
lesser tubercle (humerus), 6*f*, 7*f*, 40*f*, 42, 42*f*
levator glandulae thyroideae, 123
levator labii superior muscle, 121, 121*f*, 122*f*
levator palpebrae superioris muscle, 121, 122*f*
levator scapulae, 50, 50*f*, 133, 133*f*, 135*f*, 139*f*
levator veli palatini muscle, 123
levatores costarum, 137*f*, 151, 151*f*, 153*t*
levers, 17-19, 18*f*, 19*f*
ligament of Wrisberg, 222, 222*f*
ligamentous joints, 11, 11*f*
ligaments, 8, 8*f*
  ankle, 234, 236-237, 237*f*, 269*t*
  elbow and forearm, 8*f*, 64-65, 64*f*, 65*f*, 100*t*-102*t*
  foot, 237*f*, 238-241, 239*f*, 269*t*-270*t*
  head, 118*f*, 120, 183*t*
  hip joint, 203-205, 204*f*, 268*t*
  knee, 8*f*, 219-223, 220*f*, 268*t*
  lower leg, 236-237, 236*f*, 268*t*
  pelvis, 141-143, 142*f*, 184*t*
  shoulder, 43-47, 43*f*, 44*f*, 46*f*, 47*f*, 100*t*
  spinal column, 126-130, 130*f*, 183*t*
  thorax, 147, 148*f*, 184*t*
  wrist and hand, 75-76, 75*f*, 101*t*-102*t*
ligamentum capitis femoris, 204
ligamentum flavum, 128-129, 130*f*
ligamentum nuchae, 129, 129*f*
linea alba, 134, 135*f*
linea aspera, 203, 203*f*
linea musculi solei, 217, 218*f*
lingual artery, 167, 167*f*
Little League elbow, 69
lobes, lung, 158
long bones, 4, 4*f*, 5*f*
long head of biceps femoris, 208, 208*f*
long plantar ligament, 237*f*, 239*f*, 240
long posterior sacroiliac, 242*f*
long sacroiliac ligament, 141, 142*f*
long thoracic nerve, 91*f*, 92, 92*f*
longissimus capitis muscle, 133*f*, 137, 137*f*
longissimus cervicis muscle, 137, 137*f*
longissimus dorsi (thoracis) muscle, 137, 137*f*
longitudinal arch, 240, 241*f*
longitudinal fissure, 162
longitudinal muscle fibers, 13, 13*f*
longitudinalis muscles, 123, 124*f*
longus capitis muscle, 132, 133*f*
longus colli muscle, 132, 133*f*
lordosis, 123, 124*f*
low-back pain, 258
lower extremity. *See also* ankle; foot; hip joint; knee; lower leg
  arteries, 261-264, 261*f*, 263*f*
  muscles, 254*t*
  nerves, 255-260, 257*f*
  summary tables, 268*t*-276*t*
  veins, 265-267, 265*f*
lower leg, 268*t*. *See also* ankle; foot
  bones, 231-233, 232*f*
  compartments of, 243-250, 244*f*
  ligaments, 236-237, 236*f*
  movements, 224-225, 241, 242*f*
  muscles, 225-230, 229*f*, 247-250, 247*f*
  trauma, 259
lower subscapular nerve, 94
lumbar plexus, 20, 175, 175*f*, 255-258, 256*f*
lumbar vertebrae, 123, 124*f*, 126, 127*f*
lumbosacral, 242*f*
lumbosacral plexus, 90*f*, 175, 255-260
lumbricales
  foot, 251*f*, 252
  hand, 79*f*, 84
lunate, 73*f*, 74
lungs
  air exchange, 182*f*
  arteries and veins, 181, 182*f*
  nerves, 177-178, 179*f*
  respiration, 145, 149-152, 153*t*
  structure of, 156-158, 156*f*

M

malleoli, 231, 232*f*, 233*f*
malleus (hammer), 114, 115*f*
mandible, 111, 112*f*, 114, 116*f*, 120
mandibular condyle, 116*f*
mandibular fossa, 113*f*
mandibular symphysis, 114
manubrium, 145, 146*f*
margin vein, 181
marginal artery, 179*f*, 180*f*
masseter muscle, 121, 121*f*, 122*f*
mastoid process, 112*f*, 113, 113*f*, 116*f*
maxilla bones, 111, 112*f*, 113*f*
maxillary artery, 167, 167*f*, 168
maxillary sinuses, 118*f*, 120
MCL (medial collateral ligament), 219, 220*f*, 221*f*, 222*f*
medial (ulnar) styloid process, 7*f*, 62*f*, 63*f*, 64
medial antebrachial cutaneous nerve, 92*f*, 93
medial anterior thoracic nerve, 92*f*, 93
medial border, scapula, 40*f*, 41, 41*f*
medial brachial cutaneous nerve, 91*f*, 92*f*, 94
medial calcaneal artery, 264, 264*f*
medial circumflex artery, 262
medial collateral ligament (MCL), 219, 220*f*, 221*f*, 222*f*
medial condyles. *See* condyles
medial cord, 91*f*, 92*f*, 93
medial cutaneous antebrachial nerve, 91*f*
medial cutaneous nerve of forearm, 93
medial, defined, 30
medial dorsal cutaneous nerve, 260*f*
medial epicondylar ridge, 215, 216*f*
medial epicondyle. *See* epicondyles
medial femoral circumflex artery, 261*f*
medial intercondylar tubercle, 217, 218*f*
medial malleolus, 231, 232*f*, 233*f*
medial pectoral nerve, 91*f*
medial plantar artery, 261*f*, 263*f*, 264
medial plantar nerve, 260*f*, 264
medial superficial femoral vein, 267
medial supracondylar ridge, 62, 62*f*
medial sural branch, 258, 260*f*
medial sural nerve, 90*f*
medial tarsal artery, 261*f*, 263, 263*f*
medial tilt of the scapula, 48-49
medialis rectus muscle, 122*f*, 123
median cubital vein, 23*f*, 97, 97*f*, 98*f*
median nerve, 90*f*, 91*f*, 92*f*, 93, 93*f*
median palatine suture, 113*f*
median vein, 97, 97*f*, 99*f*
mediastinal septum, 157
mediastinal surfaces, 157

mediastinum, 153, 157*f*
medulla oblongata (myelencephalon), 161*f*, 164, 172
membrana tectoria, 128, 129*f*
memory, 161
meninges, 160, 172
menisci (singular, *meniscus*), 222*f*, 223, 223*f*, 224
mental foramen, 112*f*, 114
mentalis muscle, 121, 121*f*, 122*f*
mesencephalon (midbrain), 161*f*, 164
mesenteric artery, 22*f*
mesenteric vein, 23*f*
metacarpal arteries, 97*f*
metacarpal bones
- foot, 234*f*
- hand, 73*f*, 74

metacarpophalangeal (MP) joints, 75*f*, 76, 78
metatarsal arch, 240, 241*f*
metatarsal bones, 233, 234*f*
metatarsophalangeal (MTP) joints, 239*f*, 240, 240*f*, 241
metatarsophalangeal ligaments, 238, 240
metencephalon (pons), 161*f*, 164
midbrain, 161*f*, 164
midcarpal joints, 75, 75*f*
middle cardiac vein, 180*f*, 181
middle cerebral artery, 168-169, 169*f*
middle ear, 114, 115*f*
middle facet, 7*f*, 42, 42*f*
middle glenohumeral ligament, 46, 46*f*
middle phalanges, 75, 234*f*
midfoot, 233
mitral valve, 155, 155*f*
mortise-and-tenon joint, 236, 237*f*
motor (efferent) nerves, 20*f*, 21, 25*f*, 166, 166*t*
motor cortex, 163*f*, 164
motor end plate, 20, 25*f*
motor neurons, 20, 89
motor units, 20, 24-27, 25*f*
movements
- anatomical location terms, 30
- ankle and foot, 241, 242*f*, 243*f*, 268*t*-270*t*
- elbow and forearm, 65, 65*f*, 100*t*-102*t*
- fundamental, 31-34, 32*f*, 33*f*, 34*f*
- head, 120-123, 183*t*
- hip joint, 204-205, 268*t*
- joints, 8, 11
- knee, 224-225, 229*f*, 268*t*
- lower leg, 224-225, 241, 242*f*, 268*t*
- muscle contractions and, 14
- pelvis, 143-144, 144*f*, 184*t*
- planes and axes, 30-31, 30*f*, 31*t*
- shoulder, 47-51, 48*f*, 49*f*, 59, 100*t*
- spinal column, 130, 130-138, 131*f*, 132*f*, 183*t*
- thorax, 149-152, 184*t*
- thumb, 87, 87*f*, 88*f*
- wrist and hand, 77-78, 77*f*, 85*f*, 101*t*-102*t*

MP (metacarpophalangeal) joints, 75*f*, 76, 78
multangular major (trapezium) bone, 74
multangular minor (trapezoid) bone, 74
multangulus major (trapezium) bone, 73*f*
multangulus minor (trapezoid) bone, 73*f*
multifidus muscle, 137*f*, 138
multipennate muscle fibers, 13, 13*f*
muscles
- ankle and foot, 242-254, 244*f*, 247*f*, 254*t*, 274*t*-276*t*
- contraction, 14, 15, 26-27, 26*f*
- elbow and forearm, 65-70, 66*f*, 70*f*
- fiber types, 16-17, 16*f*
- force of, 26, 26*f*
- functions of, 15
- head, 120-123, 121*f*, 122*f*, 185*t*-192*t*
- hip joint and upper leg, 205-214, 205*f*, 208*f*, 212*f*, 254*t*, 271*t*-273*t*
- knee and lower leg, 225-230, 228*f*, 229*f*, 254*t*, 274*t*
- lower leg, 247-250
- motor units, 20, 24-27, 25*f*
- naming criteria, 12
- number of, 12
- pelvis, 143-144
- of respiration, 148*f*, 153*t*
- shoulder girdle, 49-50, 50*f*, 103*t*
- shoulder joint, 51-59, 53*f*, 54*f*, 55*f*, 56*f*, 103*t*-105*t*
- spinal column, 130-138, 133*f*, 135*f*, 137*f*, 194*t*-197*t*
- strength of, 13
- structure of, 15, 16*f*
- tendons and, 13-14, 14*f*
- thorax, 149-152, 193*t*
- thumb, 79*f*, 81*f*, 84-88
- tissue types, 3, 12, 13*f*
- viscosity of, 18
- wrist and hand, 78-88, 79*f*, 81*f*, 106*t*-108*t*

musculocutaneous nerve, 90*f*, 91*f*, 92, 92*f*
musculospiral nerve, 94
myelencephalon, 164
myelin sheath, 20, 20*f*
mylohyoid muscle, 123, 124*f*
myocardium, 155*f*, 156
myofibrils, 15, 16*f*
myology, 12
myoneural junction, 20, 25, 25*f*
myosin, 15, 16*f*

N

nasal bones, 111, 112*f*
nasal concha, 111, 112*f*
nasalis muscle, 121, 121*f*, 122*f*
navicular (scaphoid)
- foot, 233, 234*f*, 235*f*
- wrist, 73*f*, 74, 74*f*

neck of femur, 202, 203*f*, 204*f*
neck of fibula, 217, 218*f*
neck of rib, 147, 147*f*
nerve tissue, 3
nerve to the piriformis, 259
nerves
- cranial, 164-166, 164*f*, 166*t*, 185*t*
- damage to, 258
- diagram of, 90*f*
- facial, 164*f*, 166, 166*t*, 185*t*-187*t*
- heart and lungs, 177-178, 178*f*, 179*f*
- lower extremity, 255-260, 257*f*, 271*t*-276*t*
- motor units, 20, 22-27, 25*f*
- pinched, 94
- spinal column, 173, 174*f*, 175*f*, 194*t*-197*t*
- structure and function, 20-21, 20*f*
- thorax and trunk, 173-175, 174*f*, 175*f*, 193*t*
- upper extremity, 89-94, 91*f*, 92*f*, 103*t*-108*t*

nervous system, 20, 90*f*, 164-166, 165*f*
neuroglia, 3
neuromuscular system, 25
neurons, 3, 20, 20*f*
nodes of Ranvier, 20, 20*f*, 25*f*
nonaxial joints, 10, 10*f*
nose, 172, 186*t*
notch, 6, 7*f*
- acetabular, 139*f*

notch, 6, 7*f* *(continued)*
- cardiac, 157*f*, 158
- clavicular, 43, 44, 145, 146*f*
- femoral (intercondylar), 7*f*, 217
- fibular, 231
- first costal, 145
- greater sciatic, 139*f*, 140
- lesser sciatic (ischial), 139*f*, 140
- radial, 63
- scapular, 40*f*, 41, 46
- subscapular, 6*f*
- suprasternal (jugular), 145, 146*f*
- trochlear, 62*f*, 64
- ulnar, 63*f*, 64
- vertebral, 125*f*, 127*f*

nucleus pulposus, 130, 130*f*

O

oblique cord, 8*f*, 65
oblique muscles, 134, 135*f*, 136*f*, 152*f*
oblique popliteal ligament, 220*f*, 222, 222*f*
oblique vein, 181
obliquus capitis muscle, 133, 133*f*, 137*f*
obliquus muscles (eyeball), 122*f*, 123
obturator externus, 208*f*, 211
obturator foramen, 139*f*, 141
obturator internus, 208*f*, 211
obturator nerve, 90*f*, 256*f*, 257, 257*f*
occipital artery, 22*f*, 167, 167*f*
occipital bone, 111, 112*f*, 113*f*
occipital condyle, 113*f*, 114
occipital lobe, 162, 162*f*
occipital sinus, 172, 173
occipital vein, 171
occipitalis muscle, 121
occipitoaxial ligament, 128, 129*f*
oculomotor nerve, 164*f*, 165, 166*t*
odontoid process (dens), 126, 126*f*, 128
OKC (open kinetic chain), 34-35
olecranon fossa, 62, 62*f*
olecranon process, 7*f*, 62, 62*f*, 64
olfactory center, 162, 163*f*
olfactory nerve, 164*f*, 165, 166*t*
omohyoid muscle, 123
open kinetic chain (OKC), 34-35
opponens digiti minimi, 79*f*, 83
opponens pollicis, 79*f*, 86, 87*f*
opposition, 83, 87, 88*f*
optic nerve, 122*f*, 164*f*, 165, 166*t*
orbicularis oculi muscle, 121, 121*f*, 122*f*
orbicularis oris muscle, 121, 121*f*, 122*f*, 124*f*
orbit, 172
orbitalis muscle, 123
orbits, 112*f*
osteoperiostitis, 251
osteoporosis, 5

P

paired sinuses, 172-173
palatine bones (hard palate), 111, 112, 113*f*
palmar (volar) interossei, 84, 84*f*
palmar accessory ligaments, 75*f*
palmar aponeurosis, 79*f*
palmar aspect, 30
palmar carpometacarpal ligaments, 75*f*
palmar metacarpal ligament, 75*f*
palmar ulnocarpal ligament, 75*f*
palmaris longus, 79-81, 79*f*, 80*f*
palpation, 21
papillary muscles, 155, 155*f*
parasympathetic (craniosacral) system, 19, 165, 165*f*
parietal bones, 111, 112*f*, 113*f*
parietal lobe, 162, 162*f*
parietal pleura, 157, 157*f*
parrot-beak tears of menisci, 223, 223*f*
pars interarticularis, 125
patella, 217-218, 219*f*
patellar facet, 216*f*, 217
patellar ligament, 220*f*, 222*f*, 223
patellar tendon, 205*f*
patellofemoral joint, 217
patellofemoral pain (PFP), 227, 228*f*
PCL (posterior cruciate ligament), 220-222, 220*f*, 222*f*
pectineal line, 203*f*
pectineus, 205*f*, 207
pectoralis major, 50*f*, 52, 52*f*, 153*t*
pectoralis minor, 49, 50*f*, 135*f*, 153*t*, 154*f*
pedicles, 125, 125*f*, 127*f*
pelvis, 184*t*
- anatomical landmarks, 142*f*
- bones, 138-141, 139*f*, 202*f*
- ligaments, 141-143, 142*f*
- movements and muscles, 143-144, 143*f*-144*f*

pennate muscles, 13
perforating arteries, 259, 263*f*, 264
perforating cutaneous nerve, 256*f*, 259
pericardium, 153, 154*f*
perimysium, 15, 16*f*
periosteum, 5, 5*f*, 16*f*
peripheral nervous system, 19, 164-166
peristalsis, 12
fibular artery, 261*f*, 263*f*
fibular communicating nerve, 260*f*
fibular retinaculum, 244*f*
fibular vein, 23*f*, 265*f*
fibularis brevis, 244*f*, 246*f*
fibularis longus, 244*f*, 246*f*
fibularis tertius, 244*f*
pes anserinus, 228, 228*f*
PFP (patellofemoral pain), 227, 228*f*
phalanges, 73*f*, 76, 233, 234*f*
pharyngeal artery, 167
pharyngopalatinus muscle, 123
pharynx, 156, 156*f*, 190*t*
pharynx muscles, 123, 124*f*
phrenic nerve, 91*f*, 174, 174*f*, 177, 179*f*
pia mater, 160, 160*f*
pinched nerve, 94
pinna (auricle), 114, 115*f*
PIP (proximal interphalangeal) joints. *See* proximal interphalangeal (PIP) joints
piriformis, 208*f*, 210-211, 259
pisiform, 73*f*, 74, 74*f*
pisohamate ligament, 75*f*
pisometacarpal ligament, 75*f*
pitcher's elbow, 69
pivot joint, 9*f*, 10
planes, 30-31, 30*f*, 31*t*
plantar aspect, 30
plantar fascia, 247-248, 250*f*
plantar fasciitis, 252
plantar flexion, 31, 241, 242*f*
plantar interossei, 252*f*, 253
plantaris, 229*f*, 230, 247-248, 247*f*
pleural cavities, 157
pleural membrane, 157
plexuses (nerves), 20, 91-94, 173, 174*f*
pneumogastric (vagus) nerve, 166
pons, 161*f*, 164
popliteal artery, 22*f*, 261*f*, 262, 263*f*
popliteal groove, 216*f*, 217
popliteal line, 217, 218*f*
popliteal surface, 216*f*, 217
popliteal vein, 23*f*, 265, 265*f*
popliteus muscle, 228, 229*f*
popliteus tendon, 220*f*
posterior auricular artery, 167*f*, 168
posterior auricular vein, 172
posterior band, 64, 64*f*

posterior cerebral arteries, 167, 169*f*, 170, 175
posterior chamber (eyeball), 123
posterior communicating artery, 168, 169*f*, 170
posterior cord, 91*f*, 92*f*, 95
posterior cruciate ligament (PCL), 220-222, 220*f*, 222*f*
posterior, defined, 30
posterior facial vein, 172
posterior femoral cutaneous nerve, 256*f*, 257*f*, 259
posterior inferior iliac spine, 139*f*, 140, 242*f*
posterior interventricular artery, 179*f*, 180*f*
posterior lateral malleolus artery, 264*f*
posterior longitudinal ligament, 128, 129*f*
posterior medial malleolus artery, 264*f*
posterior pubic ligament, 143
posterior sacroiliac ligaments, 141, 142*f*
posterior scapular nerve, 91*f*
posterior superficial sacrococcygeal, 142*f*, 242*f*
posterior superior iliac spine (PSIS), 139*f*, 140, 141*f*, 242*f*
posterior talofibular ligament, 237, 237*f*
posterior talotibial ligament, 236, 237*f*
posterior tibial artery, 261*f*, 264, 264*f*
posterior tibial nerve, 257*f*
posterior tibial recurrent artery, 261*f*, 263
posterior tibial vein, 23*f*, 264, 265*f*
posterior tibiofibular ligament, 236, 237*f*
prehensile hands, 73
prevertebral muscles, 132
prime mover muscles, 14
procerus muscle, 121
process, 5, 6*f*
profunda artery, 262
pronation
  defined, 32
  foot, 241
  forearm, 65, 66*f*
pronator quadratus, 70, 70*f*
pronator teres, 66*f*, 69, 70*f*
protraction, 48
protruded disc, 132
proximal, defined, 30
proximal facet, 7*f*, 42, 42*f*
proximal interphalangeal (PIP) joints, 76, 78, 84, 240, 242
proximal phalanges, 76, 234*f*
proximal radioulnar joint, 61, 64
PSIS (posterior superior iliac spine), 139*f*, 140, 141*f*
psoas major, 205*f*, 206, 206*f*
psoas minor, 205*f*, 206, 206*f*
pterygoid masseter muscles, 121
pubic bone, 139*f*, 141
pubic symphysis, 141, 143
pubic tubercle, 139*f*, 141, 142*f*, 242*f*
pubofemoral ligament, 203-205, 204*f*
pudendal (coccygeal) plexus, 20, 175, 175*f*, 259
pudendal nerve, 256*f*, 259
pulmonary arteries, 22*f*, 178, 179*f*, 180*f*, 182*f*
pulmonary semilunar valve, 181
pulmonary veins, 23*f*, 180*f*, 181, 182*f*
puncta maxima, 155, 155*f*
pupil, 122*f*, 123
Purkinje fibers, 177, 178*f*
pyramidal decussation, 164
pyramidal tracts, 164

Q

Q angle, 227, 227*f*
quadrate muscle fibers, 13, 13*f*
quadratus femoris, 206, 208*f*, 211, 226*f*, 273*t*
quadratus labii muscles, 121
quadratus lumborum muscle, 137*f*, 138, 151
quadratus plantae, 251, 251*f*
quadriceps femoris, 206, 225, 226*f*
quadriceps femoris tendon, 220*f*, 223

R

radial artery, 22*f*, 95*f*, 96, 96*f*
radial collateral ligament, 8*f*, 64, 64*f*, 75*f*, 76
radial deviation, 78, 78*f*
radial fossa, 61, 62*f*
radial head, 62*f*, 63, 63*f*
radial neck, 62*f*, 63
radial nerve, 90*f*, 91*f*, 92*f*, 94
radial notch, 63
radial styloid process, 62*f*, 63*f*
radial tuberosity, 7*f*, 62*f*, 63
radial vein, 97, 97*f*
radiate carpal ligaments, 75*f*
radiocarpal collateral ligament, 75*f*
radiocarpal joints, 75*f*, 76
radiohumeral joint, 61, 64
radius, 61, 62-63, 62*f*, 63*f*
ramus of mandible, 116*f*
range of motion (ROM), 34
recti muscles (eyeball), 123
rectus abdominis muscle, 134, 135*f*, 136*f*, 153*t*
rectus capitis muscles, 132, 133, 133*f*, 137*f*
rectus femoris, 205*f*, 206, 207*f*, 225-226, 225*f*, 226*f*
recurrent tibial nerve, 258, 260*f*
reflex center, 164
regurgitation, 155
renal artery, 22*f*
renal vein, 178*f*
repetitive motion injuries, 77
respiration, 145, 149-152, 153*t*
respiratory bronchioles, 157
respiratory exchange, 181, 182*f*
respiratory system, 164, 177, 179*f*, 182*f*
respiratory tree, 156, 156*f*
retina, 122*f*, 123
retinacula, 76, 76*f*, 244*f*, 247*f*
retraction, 49
rhomboids, 50-51, 50*f*, 133-134, 133*f*, 135*f*, 137*f*, 138, 153*t*, 154*f*
rib-out, 149
ribs, 128, 146-147, 146*f*, 147*f*, 148*f*, 150
right atrium, 154, 155*f*
right brachiocephalic vein, 23*f*
right common carotid artery, 22*f*, 167*f*, 176*f*
right subclavian vein, 23*f*
right ventricle, 154, 155*f*
risorius muscle, 121, 121*f*, 122*f*
ROM (range of motion), 34
rotation, 31*t*, 32, 33*f*
  hip, 205
  knee, 223
  lower leg, 224
  pelvis, 143, 144*f*
  shoulder, 48-49, 52
  spine, 131*f*, 132, 133
  thorax, 147, 148*f*
rotator cuff, 56, 56*f*

S

SA (sinoatrial) node, 177, 178*f*
sacculus, 114, 116*f*
sacral horn, 127*f*
sacral plexus, 20, 175, 175*f*, 256*f*, 258-259
sacral tuberosity, 128*f*, 140*f*

sacral vertebrae, 123, 124*f*, 126
sacroiliac ligaments, 141, 142*f*
sacrospinalis (erector spinae), 137*f*, 138, 153*t*
sacrospinous, 242*f*
sacrospinous ligament, 142*f*, 143
sacrotuberous, 242*f*
sacrotuberous ligament, 142*f*, 143
sacrum, 123, 124*f*, 126, 127*f*, 140*f*
saddle joint, 9-10, 9*f*, 76
sagittal (anteroposterior) plane, 30-31, 30*f*, 31*t*
sagittal horizontal axis, 30*f*, 31, 31*t*
sagittal suture, 112*f*
SAID principle, 15
salpingopharyngeus muscle, 123
saphenous nerve, 90*f*, 257, 257*f*
sarcolemma, 15, 16*f*, 25, 26*f*
sarcomeres, 15, 16*f*
sarcoplasm, 15, 16*f*
sarcoplasmic reticulum, 25, 26*f*
sartorius
  hip, 205*f*, 206, 207*f*
  knee, 225, 225*f*, 228*f*
SC (sternoclavicular) joint, 43, 43*f*, 44*f*
scalene muscles, 132, 134*f*, 135*f*, 151, 153*t*
scaphoid (navicular)
  foot, 233, 234*f*, 235*f*
  wrist, 73*f*, 74, 74*f*
scaption, 60
scapula, 5, 39*f*, 40*f*, 41-42
  ligaments, 46, 47*f*
  locating, 41*f*, 42*f*
  muscles acting on, 49-54, 50*f*, 135*f*
scapular notch, 40*f*, 41, 46
scapulohumeral rhythm, 59
sciatic nerve, 90*f*, 256*f*, 257*f*, 258, 260*f*
sciatic neuritis, 258
sclera, 122*f*, 123
scoliosis, 123, 125*f*
screw home movement, 224, 224*f*
second brain (diencephalon), 161*f*, 164
second-class lever, 18-19, 19*f*
semicircular canals, 114, 116*f*
semimembranosus, 208-209, 208*f*, 210*f*, 228, 228*f*
semispinalis capitis muscle, 133*f*, 137, 137*f*
semispinalis cervicis muscle, 137, 137*f*
semispinalis dorsi (thoracis) muscle, 137
semispinalis dorsi muscle, 137, 137*f*
semitendinosus, 208, 208*f*, 209*f*, 228, 228*f*
senses, as brain function, 162, 163*f*
sensory (afferent) nerves, 20*f*, 21, 165, 166*t*
septum, 112*f*
serratus anterior muscle, 49-50, 50*f*, 135*f*, 153*t*
serratus posterior muscle, 151, 152*f*, 153*t*
sesamoid bones, 4*f*, 5, 234, 234*f*
shin splints, 251
short bones, 4*f*, 5
short head of biceps femoris, 208, 208*f*
short posterior sacroiliac, 242*f*
short sacroiliac ligament, 141, 142*f*
shoulder. *See also* scapula; shoulder girdle; shoulder joint
shoulder (glenohumeral) joint, 6*f*, 39, 40*f*, 51*f*, 100*t*, 102*t*-105*t*
  bones, 39*f*, 40*f*, 42
  combined actions with shoulder girdle, 59-60
  dislocation of, 59, 60*f*
  ligaments, 46, 46*f*
  movements, 51-52
  muscles, 52-59, 53*f*, 54*f*, 55*f*, 56*f*
  rotator cuff, 56, 56*f*
shoulder girdle, 39, 100*t*, 103*t*
  bones, 39-42, 39*f*, 40*f*
  combined actions with shoulder joint, 59-60
  joints and ligaments, 43-46, 43*f*, 44*f*
  movements, 47-49, 48*f*, 49*f*
  muscles, 49-50, 50*f*, 51*f*
shoulder separation, 45, 45*f*
sinoatrial (SA) node, 177, 178*f*
sinuses, 118*f*, 120. *See also* coronary sinus; cranial sinuses
skeletal muscles, 13, 13*f*. *See also* muscles
skull. *See* head
sliding filament mechanism, 26
slipped disc, 132
slow-twitch fibers, 16, 16*f*
small cardiac vein, 23*f*
smell, 162, 163*f*
smooth muscles, 12-13, 12*f*
soft palate muscles, 123, 189*t*-190*t*
soleus, 247, 247*f*
somatosensory cortex, 163*f*
spear tackling injuries, 94
speech, 162
sphenoid bones, 111, 112*f*, 113*f*
sphenoid sinus, 118*f*, 120
sphenomandibular ligament, 118*f*, 120
sphenoparietal sinus, 172
sphincter pupillae, 123
spinal accessory nerve, 166, 166*t*, 173, 174*f*
spinal column, 124*f*, 183*t*, 194*t*-197*t*
  bones, 125-126
  curvatures of, 123, 124*f*, 125*f*
  injuries, 174
  ligaments, 126-130, 129*f*
  movements and muscles, 130-138, 131*f*, 133*f*, 135*f*, 137*f*
  nerves, 20, 173-175, 174*f*, 175*f*
  vertebra, 20
  vertebrae, 7*f*, 123, 125*f*
spinal nerves, 164, 174*f*, 175*f*
spinalis capitis muscle, 136
spinalis cervicis muscle, 136
spinalis dorsi (thoracis) muscle, 136, 137*f*
spine of ischium, 139*f*, 140
spine of scapula, 40*f*, 42, 42*f*
spinous process, 5, 7*f*, 125*f*, 126, 127*f*, 128
splenic artery, 22*f*
splenic vein, 23*f*
splenius capitis muscle, 122*f*, 133, 133*f*, 137*f*
splenius cervicis muscle, 133, 133*f*
sprains, 149, 238, 241
stapedius muscle, 121
stapes (stirrup), 114, 115*f*
stenosis, 155
sternal end of clavicle, 40, 40*f*
sternoclavicular (SC) joint, 43, 43*f*, 44*f*
sternoclavicular ligament, 43, 43*f*
sternocleidomastoid muscle, 122*f*, 132, 132*f*, 133*f*, 134*f*, 137*f*, 153*t*
sternohyoid muscle, 123
sternothyroid muscle, 123
sternum, 39, 145, 146*f*
styloglossus muscle, 123, 124*f*
stylohyoid muscle, 123
styloid process
  head, 112*f*, 113*f*, 114
  ulnar, 7*f*, 62*f*, 63*f*, 64
stylomandibular ligament, 118*f*, 120
stylopharyngeus muscle, 123, 124*f*
subarachnoid hemorrhage, 160, 160*f*
subarachnoid space, 160, 160*f*
subclavian artery, 22*f*, 95, 95*f*, 167*f*, 176*f*

subclavian groove, 40, 40*f*
subclavian nerve, 91*f*, 92, 92*f*
subclavian vein, 23*f*, 97, 97*f*, 98*f*, 170, 171*f*, 178*f*
subclavius muscle, 50, 50*f*
subcostal muscles, 151, 151*f*, 153*t*
subcostal nerve, 174, 256*f*
subdural hematoma, 160, 160*f*
subdural space, 160, 160*f*
subscapular fossa, 6*f*, 40*f*, 42
subscapular nerve, 91*f*, 92*f*, 94
subscapularis, 53, 54*f*, 56, 56*f*
subtalar joint, 235*f*
sulci, 162, 162*f*
superficial circumflex iliac artery, 261, 261*f*
superficial circumflex iliac vein, 267
superficial epigastric artery, 261*f*
superficial external pudendal artery, 261*f*
superficial fibular nerve, 90*f*, 257*f*, 260*f*
superficial posterior compartment, 244*f*, 247-248
superficial temporal artery, 167*f*, 168
superficial temporal vein, 171
superficial veins, 97, 171, 265, 265*f*
superficial volar arch, 95*f*, 97, 97*f*
superior angle, 40*f*, 41
superior articulating facets, 125
superior border, 40*f*, 41
superior, defined, 30
superior glenohumeral ligament, 46, 46*f*
superior gluteal nerve, 256*f*, 258
superior mesenteric artery, 22*f*
superior mesenteric vein, 23*f*
superior nuchal line, 113*f*
superior oblique muscle, 122*f*
superior ophthalmic vein, 171*f*
superior pharyngeal constrictor, 124*f*
superior pubic, 242*f*
superior pubic ligament, 142*f*, 143
superior pubic ramus, 141
superior rectus muscle, 122*f*, 123
superior sagittal sinus, 23*f*, 173
superior thyroid artery, 167*f*, 168
superior transverse scapular ligament, 46, 46*f*
superior vena cava, 23*f*, 171*f*, 177, 178*f*, 179*f*, 180*f*, 181, 182*f*
supination
  defined, 32
  foot, 241
  forearm, 65, 66*f*
supinator, 70, 70*f*
supraclavicular nerves, 90*f*, 91*f*, 92, 92*f*
supraglenoid tubercle, 40*f*, 42
suprahyoid muscles, 123
supraorbital ridge, 112*f*
supraorbital vein, 171
suprascapular nerve, 92, 92*f*
Suprascapular notch, 6*f*
supraspinatus muscle, 50*f*, 54, 55*f*, 56, 56*f*
supraspinous, 242*f*
supraspinous fossa, 6*f*, 40*f*, 42
supraspinous ligament, 129, 130*f*, 139*f*
suprasternal (jugular) notch, 145, 146*f*
sural nerve, 90*f*, 257*f*, 258, 260*f*
surgical neck, 40*f*, 42
suspensory ligament, 123
sustentaculum tali, 234, 234*f*, 235*f*
suture joints, 11, 11*f*, 115-117
sweat center, 164
sympathetic (thoracolumbar) nervous system, 19, 165, 165*f*
synapses, 25, 25*f*
synaptic cleft, 25, 25*f*
synarthrodial joints, 11, 11*f*
syndesmosis sprain, 238
synergist muscles, 14
synovial joints, 10, 10*f*
systole, 12

T

talocalcaneal joint, 234, 236
talocrural joint, 234, 236, 237*f*
talus, 233-234, 235*f*
tarsal bones, 233-234, 234*f*
tarsometatarsal joint, 239*f*, 240, 240*f*, 242
tarsometatarsal ligaments, 239*f*, 240
tarsometatarsophalangeal joints, 270*t*
taste, 162, 163*f*
tear ducts, 122*f*
temporal bones, 111, 112*f*, 113*f*
temporal lobe, 162, 162*f*
temporalis muscle, 121, 121*f*, 122*f*
temporomandibular joint, 117
temporomandibular ligament, 118*f*, 120
temporoparietalis muscle, 122*f*
tendon sheaths, 14, 14*f*
tendons, 13-14, 14*f*
tennis elbow, 71
tennis leg, 249
tensor fasciae latae, 205*f*, 207, 208*f*
tensor tympani muscle, 121
tensor veli palatini muscle, 123
teres major, 50*f*, 55*f*, 57, 58*f*
teres minor, 50*f*, 55-56, 55*f*, 56*f*
tetanus, 27, 27*f*
thalamus, 164
thenar eminence, 79*f*, 85
thigh. *See also* lower extremity
  bones, 6*f*
third-class lever, 19, 19*f*
thoracic aorta, 22*f*, 176
thoracic cavity, 157
thoracic nerves, 174
thoracic parietal branch, 176, 176*f*
thoracic vertebrae, 123, 124*f*, 126, 127*f*
thoracic visceral branch, 176, 176*f*
thoracodorsal nerve, 91*f*
thoracodorsalis, 92*f*, 94
thorax and trunk, 184*t*, 193*t*
  arteries, 175-176, 176*f*
  bones, 145-147, 146*f*
  joints and ligaments, 147, 148*f*
  movements and muscles, 149-152
  nerves, 173-175, 175*f*
  structures within, 153-158
  veins, 177, 178*f*
throwing action, 69
thumb, 102*t*, 108*t*
  joints, 76
  movements, 87, 87*f*, 88*f*
  muscles, 79*f*, 81*f*, 84-88
thyroarytenoid muscle, 123
thyroepiglotticus muscle, 123
thyrohyoid muscle, 123
thyroid cartilage, 123, 124*f*
tibia, 217, 218*f*, 219*f*, 231, 232*f*, 233
tibial collateral ligament, 220*f*, 222*f*
tibial condyle, 7*f*
tibial nerve, 90*f*, 257*f*, 258-259, 259*f*, 260*f*
tibial recurrent arteries, 261*f*, 263
tibial tuberosity, 6*f*, 7*f*, 217, 218*f*
tibialis anterior muscle, 243, 244*f*, 245*f*
tibialis posterior muscle, 244*f*, 247*f*, 249, 249*f*
tibiofibular joint, 231, 236*f*
tibionavicular ligament, 236, 237*f*
tilt, of pelvis, 143, 143*f*
toes. *See* foot
Tommy John surgery, 69
tongue muscles, 123, 124*f*, 189*t*
trachea, 156*f*, 157
transverse acetabular ligament, 204, 204*f*

transverse arch, 241, 241*f*
transverse band, 64, 64*f*
transverse cervical nerve, 91*f*
transverse ligament
  cervical spine, 128, 128*f*
  humeral, 47, 47*f*
  knee, 220*f*, 222*f*, 223
  metatarsal, 240
transverse processes, 124*f*, 125, 125*f*, 126*f*, 127*f*
transverse sinus, 172
transverse tubules, 25, 26*f*
transversus abdominis muscle, 134, 135*f*, 153*t*
transversus linguae muscle, 123, 124*f*
transversus thoracis, 152, 152*f*, 153*t*
trapezium, 73*f*, 74
trapezius, 50*f*, 51, 51*f*, 135*f*, 138, 139*f*, 153*t*
trapezoid bone, 73*f*, 74
trapezoid ligament, 44, 44*f*
trapezoid line, 40, 40*f*
trauma. *See* injuries and trauma
triangular muscle fibers, 13, 13*f*
triangularis muscle, 121
triaxial joints, 10*f*, 11, 205
tributaries, 170
triceps brachii, 54*f*, 58, 59*f*, 66*f*, 67-68, 68*f*
triceps surae, 247, 247*f*
tricuspid valve, 154, 155*f*
trigeminal nerve, 164*f*, 166*t*
triquetrum, 73*f*, 74
trochanteric fossa, 203*f*
trochanteric ridge, 203*f*
trochlea, 62, 62*f*
trochlear nerve, 164*f*, 165, 166*t*
trochlear notch, 7*f*, 62*f*, 64
trunk. *See* thorax and trunk
tubercle, 5, 6*f*
tuberosity, 5, 233, 234*f*
turbinates, 120
turf toe, 241
12th intercostal nerve, 90*f*
tympanic membrane, 114, 115*f*

U

ulna, 62, 62*f*, 63, 63*f*
ulnar (medial) styloid process, 7*f*, 62*f*, 63*f*, 64
ulnar artery, 22*f*, 95*f*, 97, 97*f*
ulnar collateral ligament, 8*f*, 64, 64*f*, 69, 75, 75*f*
ulnar deviation, 78, 78*f*
ulnar nerve, 90*f*, 91*f*, 92*f*, 93, 93*f*
ulnar notch, 63, 63*f*
ulnar tuberosity, 7*f*, 62*f*
ulnar vein, 97, 97*f*
ulnohumeral joint, 61, 64
uniaxial joints, 10, 10*f*
unipennate muscle fibers, 13, 13*f*
unpaired sinuses, 173
upper extremity. *See also* elbow; forearm; shoulder; wrist and hand
  arteries, 95-97, 95*f*
  nerves, 90*f*, 91-94, 91*f*, 92*f*
  summary tables, 100*t*-108*t*
  veins, 97, 97*f*
upper leg. *See* thigh
upper subscapular nerve, 94
upward rotation of scapula, 48
utriculus, 114, 116*f*
uvulae muscle, 123

V

vagus nerve, 164*f*, 166, 166*t*
varicose veins, 267
vasomotor center, 164
vastus intermedius, 205*f*, 206, 225*f*, 226
vastus lateralis, 205*f*, 206, 224, 225*f*, 226
vastus medialis, 205*f*, 206, 224, 225*f*, 226
veins
  diagram of, 23*f*
  function of, 21, 24*f*
  head and brain, 170-173, 171*f*, 185*t*-192*t*
  heart, 180*f*, 181
  lower extremity, 265-267, 265*f*, 271*t*-276*t*
  respiratory, 181, 182*f*
  thorax and trunk, 177, 178*f*, 193*t*
  upper extremity, 97, 97*f*, 98*f*, 99*f*, 103*t*-108*t*
  valve action, 24*f*
  varicose, 267
vena cava, 23*f*, 155*f*, 177, 178*f*, 179*f*, 180*f*, 181, 182*f*
venae cordis minimae, 181
venous arch, 97, 97*f*
ventricularis muscle, 123
venules, 21
vertebrae, 7*f*, 123, 125*f*, 127*f*
vertebral arteries, 167, 167*f*, 177, 179*f*
vertebral foramen, 125, 125*f*, 126, 126*f*
vertebral notch, 125*f*, 127*f*
vertebral trauma, 94
vertebral vein, 170, 171*f*, 177
vertical (longitudinal) axis, 30*f*, 31, 31*t*
verticalis linguae muscle, 123, 124*f*
vestibular nerve, 166*t*
vestibule, 114, 116*f*
vestibulocochlear nerve, 114, 164*f*
visceral membrane, 157, 157*f*
viscosity, muscle, 18
vision, 162, 163*f*
vitreous humor, 123
vocal muscle, 123
volar (palmar) interossei, 84, 85*f*
volar (palmar) radiocarpal ligament, 75*f*, 76
volar arches, 95*f*, 97, 97*f*
volar carpal arch, 97*f*
volar carpal branch, 97*f*
volar carpometacarpal ligament, 76
volar, defined, 30
volar radioulnar ligament, 65, 65*f*, 75*f*
vomer bone, 111, 112*f*, 113*f*

W

white matter, 161
Wolff's Law, 5
wrist and hand, 101*t*-102*t*, 106*t*-108*t*
  arteries, 97, 97*f*
  bones, 73*f*, 74
  carpal tunnel syndrome, 77
  grasping capability, 73
  joints and ligaments, 75-76, 75*f*, 76*f*
  movements, 78, 78*f*, 85*f*
  muscles, 78-88, 79*f*, 81*f*
  nerves, 91, 92*f*
  veins, 97, 98*f*

X

xiphoid process, 145, 146*f*, 147

Y

Y ligament, 204, 204*f*

Z

Z-lines, 15, 16*f*
zygomatic arch, 112*f*, 113*f*, 114
zygomatic bones, 111, 112*f*, 113*f*
zygomaticus muscle, 121, 121*f*, 122*f*

ABOUT THE AUTHORS

Courtesy of Rachel M. Koldenhoven.

Rachel M. Koldenhoven, PhD, ATC, is an associate professor of athletic training and prerehabilitation sciences in the department of health and human performance at Texas State University. She has taught a variety of anatomy courses since 2015. Her research focuses on biomechanics, rehabilitation, and wearable technology, with an emphasis on understanding movement in individuals with chronic ankle instability and developing strategies to improve gait biomechanics. In 2025, she received the Presidential Distinction Award for Excellence in Service from Texas State University. She was also the recipient of the College Achievement Award for Excellence in Scholarly Activity and the NATA Research and Education Foundation Free Communications Award for Early Career.

Courtesy of Robert Behnke.

Robert S. Behnke, HSD, is retired after 39 years of teaching anatomy, kinesiology, physical education, and athletic training courses at the high school and university levels. Behnke has been honored on several occasions for excellence in teaching, including receiving the Caleb Mills Outstanding Teaching Award at Indiana State University and the Educator of the Year Award from the National Athletic Trainers' Association (NATA). He served 11 years as chair of the NATA Professional Education Committee. In 1990, Behnke was inducted into the NATA Hall of Fame. He is also in the Indiana State University Athletic Hall of Fame, the Illinois Athletic Trainers Association Hall of Fame, and the Indiana Athletic Trainers' Association Hall of Fame.

Behnke spent most of his career at Indiana State University (1975-1998), where he was a full professor of physical education and athletic training and director of undergraduate and graduate athletic training programs; in retirement, he holds the rank of professor emeritus. He served as head athletic trainer and an instructor at Illinois State University from 1966 to 1969. He was the head athletic trainer and an assistant professor at the University of Illinois from 1969 to 1975. He was an athletic trainer for boxing, men's field hockey, team handball, ice skating, roller hockey, gymnastics, judo, and cycling at the U.S. Olympic Training Center in Colorado Springs in 1984 and 1991.